Droughts in Agricultural Production
Monitoring and Management

Droughts in Agricultural Production
Monitoring and Management

G.G.S.N. Rao
Former Project Coordinator (Agri. Met)
Central Research Institute for Dryland Agriculture
Santoshnagar; Hyderabad -500059, India

NEW INDIA PUBLISHING AGENCY
New Delhi – 110 034

NEW INDIA PUBLISHING AGENCY
101, Vikas Surya Plaza, CU Block, LSC Market
Pitam Pura, New Delhi 110 034, India
Phone: + 91 (11)27 34 17 17 Fax: + 91(11) 27 34 16 16
Email: info@nipabooks.com
Web: www.nipabooks.com

Feedback at feedbacks@nipabooks.com

Prepared with the financial assistance from DST, GOI under USERS Scheme during the period Dec 2010 to Nov 2012

ISBN: 978-93-85516-00-9

1st Declaration: "The author's current official title and current official address included in the book in good faith and intention. However, the views expressed in the book are of the authors only and does not reflect the views of the organization/countries etc".

2nd Declaration: "All figures/diagrams/tables included in the book were acknowledged in the reference section of each chapter as well as at the bottom of each figure/diagram/table and were used in good faith and intention of promoting science in the world in particular in the third world countries such as Asia-Pacific regions where information is not readily available".

Composed, Designed and Printed in India

Dedicated to
Goddess of Wisdom
"Gyan Saraswathi" of Basara in the state of Telengana

Preface

Drought is considered to be a normal feature of climate especially in tropical countries where occurrence of periods of dry weather during rainy season is common. Virtually droughts can occur in all climates and they can also be witnessed when the air temperatures are higher than the normal for a longer period that increases both evaporation and water deficit. It can be defined as a condition of abnormal dry weather that has serious impacts on hydrological balance, which in turn affect the agricultural productivity as well the water needs of both human and livestock. India being located in tropical region, it often experience droughts of different intensities in one corner or in other part of the country. Over the last 140 years (1870-2010) the country was affected by seven phenomenal droughts that seriously affected the lives of human as well the livestock population.

It is estimated that in India, a large chunk of arable land of around 90 m ha would continue to be dependent on rainfall even after achieving the fullest irrigation potential. Therefore, any deviation in rainfall over its mean during rainy season is bound to affect the nation's food security. Many researchers over the years have contributed significantly in understanding various mechanisms viz., occurrence, impacts, monitoring, and management associated with droughts. However, no publication is available as of now by integrating various topics on droughts in relation to agriculture including livestock and fisheries production. Therefore, attempts have been made through this publication to review the state of the art of information on droughts available with different countries in minimizing its impact on agricultural food production and on natural resources. The material collected for this publication has been incorporated in to eight chapters.

The first chapter deals with the introduction to the topic, historical evidences of drought and famines in India. It also discussed about the distribution of drought prone regions across globe and as well in the country and the associated weather parameters during drought years. Effect of droughts on water resources, food production and the economic impacts are also discussed.

The definition of droughts, criteria for classification of droughts, drought indices based on rainfall, climate derivatives and remote sensing are covered under chapter Two. Also, criteria for declaration of droughts, drought code adopted by local governments, frequencies and probabilities of droughts, drought vulnerability assessment, drought monitoring indicators, and their computation techniques are discussed in this chapter. Information on climate change with respects to climate trends, projections on temperature and rainfall at global as well at national level, consequences of climate change on droughts are presented in chapter Three.

Approaches on drought monitoring, adopted by various countries, international cooperation in drought monitoring and drought monitoring mechanism in India are presented in chapter Four. Decision Support Systems and applications of GIS for drought monitoring are also covered in this chapter. World food production and its variability, the impact of phenomenal drought during 2009 on food grain production in different countries, impacts of droughts on livestock, milk and fisheries production are covered under chapter Five. Advanced estimates of food grain production in India using rainfall indices are covered under this chapter. Global surface waters, availability of surface waters in India, influence of drought on fauna and flora of surface waters and global warming on water resources at global and national level is discussed in chapter Six.

Drought management strategies, drought planning and preparedness action plans at global as well at national level, indigenous knowledge on drought management, astrological techniques in weather forecasting and traditional wisdom on management of droughts in different nomadic communities in Asia and Africa are covered in chapter Seven. Drought mitigation through watershed developments, Farm ponds, rain fed technologies and contingency crop plans for different agro-climatic regions and farming systems approaches as drought proofing mechanism are discussed. Climate Risk management through Agro Advisory Services in India and in other Asian and African nations, role of ICT as climate risk management tool, drought mitigation strategies adopted by Indian farming community and traditional water harvesting methods are discussed in this chapter. Based on the information collected on various aspects of management of droughts followed at global level, improvements in the existing mechanism on management of droughts and improving the income from rural areas in the country has been suggested.

It is hoped that the information contained in this book shall be useful to the students and research scholars of Agrometeorology, Meteorology and Geography and all Research Scientists of ICAR Institutes, State Agricultural Universities dealing with Natural Resource Management.

Author

Acknowledgements

Droughts are not still fully understood with respect to the global and regional weather phenomenon. Efforts have been made worldwide by many researchers to forewarn the droughts across the region by developing some teleconnectivity relationships with the local weather phenomenon. The author reviewed the latest information on droughts, its monitoring methods both traditional and advanced techniques followed by various nations and current technologies to adapt to the drought situation and the long term plans proposed to mitigate droughts. In this endeavor, the sanction and support received in the form of grants from Department of Science and Technology, Government of India under the "Utilization of Scientific Expertise of Retired Scientists" scheme for a period of 2 years is duly acknowledged with thanks. I also take this opportunity to profusely thank Dr. B.Venkateswarlu, Director CRIDA, Hyderabad for permitting me to work at CRIDA after my retirement and providing me with all necessary help and facilities. I am deeply indebted to Dr. V.U.M. Rao, Project Coordinator (Ag.Met) who has provided me all facilities at AICRPAM Coordinating Unit and rendered necessary help as and when required in completing the task as per time schedule. I am also equally thankful to the Scientists of the unit viz., Shri. AVM. Subba Rao, Dr. P. Vijaya Kumar, Dr. B.Bapuji Rao and the entire staff, who have lent their valuable support and discussions, for improving the content throughout my post- period retirement stay at CRIDA.

My special thanks to Sri. Manikandan, Asst. Professor (Agri. met) IGKVV, Raipur and other Research staff at AICRPAM for their help in compiling the information.

I wish to place on record my special thanks to Dr. P.Sanjeeva Rao, Scientist,-G of DST New Delhi for his encouragement since the beginning of this work. I also wish to place on record my sincere gratitude to Dr. Y.S. Ramakrishna, Former Director, CRIDA, and coordinator of AICRP on Agrometeorology for his constant encouragement throughout my scientific career. I am also thankful to Prof BV Ramana Rao former cordinator of AICRPAM, CRIDA, Hyderabad and Prof. P.S.N. Sastry Retd. Principal Scientist (Agri. met) IARI, New Delhi for going through the manuscript and improving the content of the Book by their critical observations. My sincere thanks to Mrs. Padmini and to Miss D. Harini for their Secretarial help and typing the manuscript.

Finally I wish to thank sincerely my wife Smt. G. Nagalakshmi and my family members who understood and supported me fully during the study period. Without their support, I would not have completed the work.

Author

Contents

List of Figures

□□□□

List of Tables

□□□□

CHAPTER 1

Introduction

Weather played a dominant role in settlement of human population in different parts of globe and the human evolution has faced many extreme weather conditions due to which some civilizations vanished over time and some migrated to better climates. The variability of weather over a place is dependent on its geographical location over the globe. For example, stable weather is observed over higher latitudes while it is highly variable in tropical regions due to differential energy levels received from the sun because of changes in earth's axis over its journey around the sun. Therefore, weather aberrations like droughts, floods, violent tropical storms, etc., are mostly confined to tropical and to some extent sub-tropical and extra tropical regions. Among these, drought which is broadly associated with water deficit conditions resulted from the receipt of abnormally low rainfall during the rainy season is insufficient to meet the human and other environmental demands. Many of the countries located in the latitudinal belt from 0 to 30° in both the hemispheres have fixed rainy period and the duration of the period could vary from 12 months to less than a month in the extremely arid parts. The General Circulation of the atmosphere indicates subsidence of air around 25 to 30° latitude resulting in formation of arid and desert regions over these latitudinal belts, which are more vulnerable to droughts. However, the presence of monsoon system for a period of 4-6 months in a year in certain regions over globe has prevented the formation of extremely dry regions around tropics in both the hemispheres. Especially it can be seen in the South Asian region covering the parts of Bangladesh, India and to some extent in Pakistan. The influence of South-West monsoonal winds covering India weakens after reaching Pakistan and the adjacent countries to west of Pakistan remain arid throughout the year. Therefore, the importance of understanding the variability of rainfall over a region, its impacts on human settlements and other eco-system is essential to develop large scale planning. Advancement in research and technological innovations, in various disciplines viz: atmospheric sciences, agricultural and physical sciences and exchange of information among the countries improved the status of human life to a great extent in spite of increased frequencies of climate extremes in recent times. However, we are yet to understand a lot the secrets of the mother earth for a comfortable living.

As per world meteorological organization (WMO) drought is a temporary aberration in rainfall over a region and should be considered a 'relative' rather an 'absolute' condition. Drought is not a disaster by itself; it becomes so depending upon the extent of damages it causes to environment and economy of people. It occurs in all the climates of the world, both in low and high rainfall areas. Drought severity varies on seasonal and annual time scale (WMO, 2006). Drought occurs in each year over a portion of big countries like USA, Australia, China and India. Drought is a regional phenomenon, the characteristics and their impacts differ from one climate regime to another. Significant departure of rainfall over a period from its long-term average would affect the water related sectors, *viz*., agriculture, drinking water, discharges into river, fisheries, etc. The key to understand the nature of drought is to have comprehensive knowledge on its natural and social dimensions. Therefore, drought impacts, assessment, vulnerability, monitoring and development of early warning systems form major components of Drought Research Program. Before understanding the droughts in totality, an understanding of the general climate of different regions of the world, the spatial and temporal distribution of weather parameters are essential and is covered in the following section.

1.1 General Circulation of the Atmosphere

The General Circulation of the atmosphere represents the atmosphere's response to its global setting such as (1) Large latitudinal variation of atmospheric heating, (2) North and southward oscillation of maximum heating zone, (3) Strengthening of north-south gradient of heating in winter hemisphere and its weakening in the summer hemisphere (4) The distribution of continents and oceans and (5) The rotation of the earth.

Hadley in 1735 was the first to explain the existence of persistent northeast winds known as "Trade winds" in the tropics, north of the equator. He concluded that strong solar heating in the equatorial latitudes causes general rising motion of the air and flows towards the pole at higher elevations. The air moving towards the pole cools, becomes heavier and sinks. At low elevation the air then flows equator ward to replace the rising air at the equator, thus, forming a continuous circulation cell oriented north-south along the meridians. This cell is known as Hadley cell. It accomplishes certain atmospheric processes such as (1) Transportation of energy from equator by upper level branch to about 30° latitude in both the hemispheres. (2) Lower branches transport moisture from subtropics to equatorial region. (3) Rising of warm air in subtropics is a process that transforms total potential energy in to Kinetic Energy; thereby making Hadley cells a major source of atmosphere's Kinetic energy in low latitudes.

Several theories of General Circulation were put forwarded by other researchers and they often postulate a three cell circulation between equator and pole due to earth's rotation. Hadley type of cells was thought to exist in low latitudes and again in high latitudes. All the low rainfall regions are concentrated in latitudes between 20 to 30^0 in both the hemispheres due to sinking of air, which prevents the formation of clouds. In the mid latitudes, an indirect meridian cell (Sinking warm air and rising cold air) exists. The cell theory of General Circulation of the atmosphere is shown in the Figure 1.1.

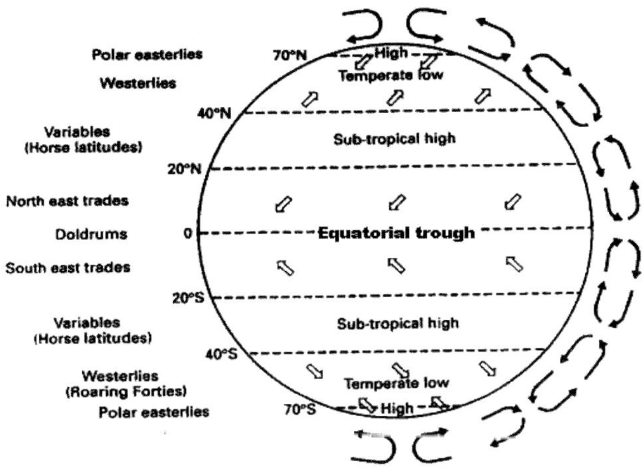

Fig. 1.1: General Circulation of the atmosphere

http://weather.mailasail.com/franks-weather/transatlantic-crossing-weather

1.2 World Climates

Due to General Circulation pattern of atmosphere over the globe, different climate types are found, such as one area of the world is a desert, another a grassland, and another a rainforest etc. Even within broad climates, there is a wide variability, with respect to its flora and fauna. Therefore there is a need to delineate climates over globe based on weather parameters. The most important parameters that determine an area's climate are air temperature, precipitation and solar radiation. The climate of a region determines the distribution of plants and what animals would inhabit it. The distribution of annual rainfall and average temperature over the globe are shown in following Figures 1.2 and 1.3.

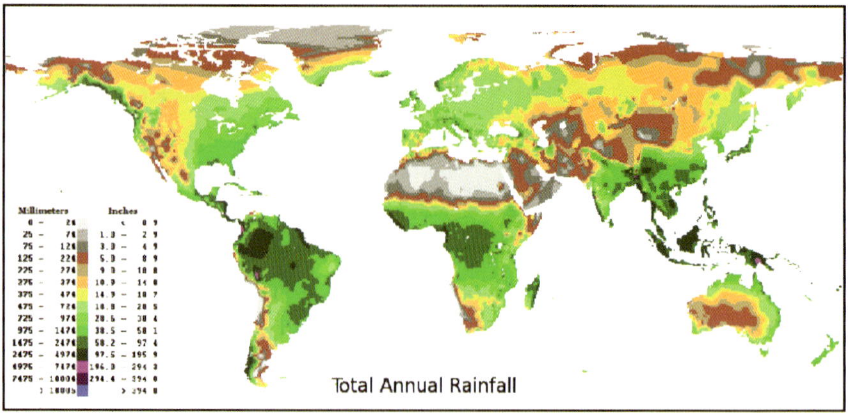

(*Source*: en.wikipedia.org/wikil)

Fig. 1.2: Annual Rainfall distributions over globe

High rainfall zones in the world are located along the equatorial tropical belt in which the annual totals vary from 2500 to 7500 mm/year. Similarly, the driest low rainfall regions are concentrated in the Southwest Asia, North Africa, South Africa, South America, and west USA and in some parts of Australia. The polar and sub polar regions receive practically nil rainfall and receive precipitation only in the form of snow.

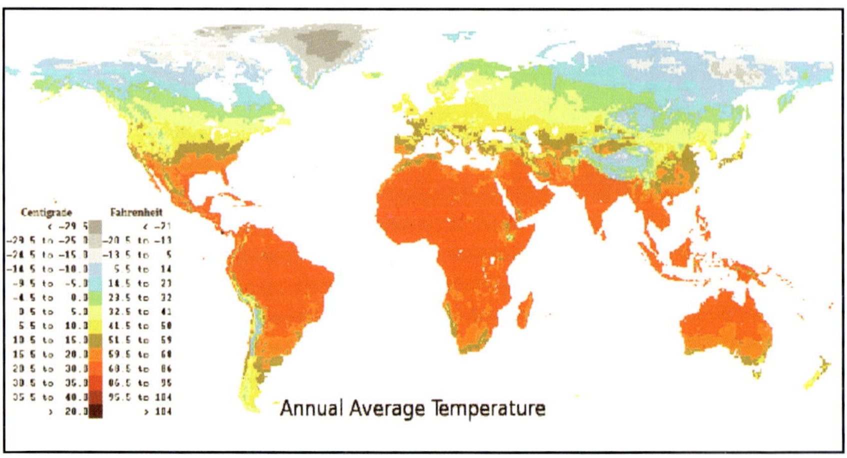

(*Source*: en.wikipedia.org/wikil)

Fig. 1.3: Annual average Temperature distributions over globe

From the average annual temperature over the globe (Figure 1.3) it is observed that the annual average temperatures around the tropics are in range from 15 to 35°C. As expected that highest air temperatures are observed in the driest areas viz., North African region where world's largest desert is located. The temperature over mid latitudes vary between -5 to 15°C while in the polar region it is around <-20°C. As the African continent spread in both north and southern hemispheres, the variability in temperature across the continent is high.

1.3 World Climate Classification by Köppen's method

Spatial distribution of weather parameters over the globe shows large variation due to general circulation of atmosphere and the location of continent and oceans. Grouping such parameters shall help in identifying the common constraints and help to formulate plans to improve the living conditions. Many scientists have tried to identify areas having similar characteristic of climate, vegetation etc. Among the all, the Köppen's Climate Classification System is the most widely used for classifying the world's climates. Most classification systems used today are based on the one introduced in 1900 by the Russian-German climatologist Wladimir Köppen. He divided the Earth's surface into climatic regions that generally coincided with world patterns of vegetation and soils. The Köppen's system recognizes five major climate types based on the annual and monthly averages of temperature and precipitation and each type is designated by a capital letter. The distribution of different types of climates as per Köppen's classification is given in the Figure 1.4. The following nomenclature in the figure identifies different climate types such as

a. ***Moist Tropical Climates*** - They are known for their high temperatures year round and for their large amount of year round rain.

b. ***Dry Climates*** - They are characterized by little rain and a huge daily temperature range. There are two subgroups,

   S - Semiarid or steppe, and

   W - Arid or desert, are used with the B climates.

c. ***Humid Middle Latitude Climates***: These climates have warm, dry summers and cool, wet winters.

d. ***Continental Climates***: These can be found in the interior regions of large land masses. Total precipitation is not very high and seasonal temperatures vary widely.

e. ***Cold Climates***: These climates are part of areas where permanent ice and tundra are always present. Only about four months of the year have above freezing temperatures. Further subgroups are designated by a second, lower case letter which distinguishes specific seasonal characteristics of temperature and precipitation.

f. Moist with adequate precipitation in all months and no dry season.

This letter usually accompanies the A, C, and D climates.

m - Rainforest climate in spite of short dry season in monsoon type climate. This letter only applies to A type climates.

s - There is a dry season in the summer of the respective hemisphere.

w - There is a dry season in the winter of the respective hemisphere.

To further denote variations in climate, a third letter was added to the code.

a. Hot summers where the warmest month is over 22°C (72°F). These can be found in C and D climates.

b. Warm summer with the warmest month below 22°C (72°F). These can also be found in C and D climates.

c. Cool, short summers with less than four months over 10°C (50°F) in the C and D climates.

d. Very cold winters with the coldest month below -38°C (-36°F) in the D climate only.

e. Dry-hot with a mean annual temperature over 18°C (64°F) in B climates only.

f. Dry-cold with a mean annual temperature less than 18°C (64°F) in B climates only.

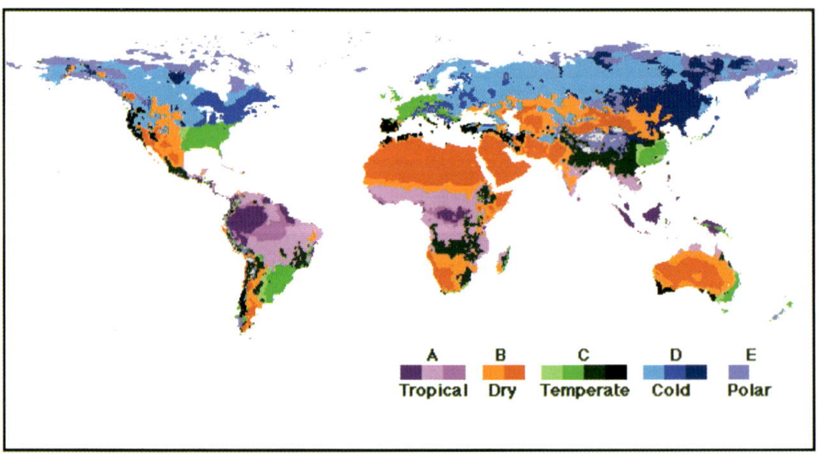

(*Source:*en.wikipedia.org/wiki/climate)

Fig. 1.4: Köppen's climate classification

It can be seen that the area under dry and tropical climates are higher compared to other climate types.

1.4 Climatic Classification over India

Many researchers over the past century have used rainfall and temperature values for classification of climates in the earlier studies. However, Thornthwaite (1948) was the first to propose the concepts of Potential evapotranspiration (PET) and developed methodology for rational classification of climates using climatic Water Balance computational procedure. The moisture regime classification suggested by Thornthwaite has six climatic types viz., per-humid, humid, moist sub-humid, dry sub-humid, semi-arid and arid. Subrahmanyam (1956) has given the first rational climatic classification of India, using the Thornthwaite method. Krishnan and Singh (1968) classified the climates following the same method but used PET computed by Penman's (1948) method. Rao, K.N. *et al* (1972) classified the climates of India following the revised method of Thornthwaite and Mather (1955) for both moisture and thermal regimes utilizing PET computed by Penman's (1948) method and 'available water capacity of the soil on the basis of soil type and crop. Their classification of climates of India as per the moisture regime is depicted in Figure 1.5.

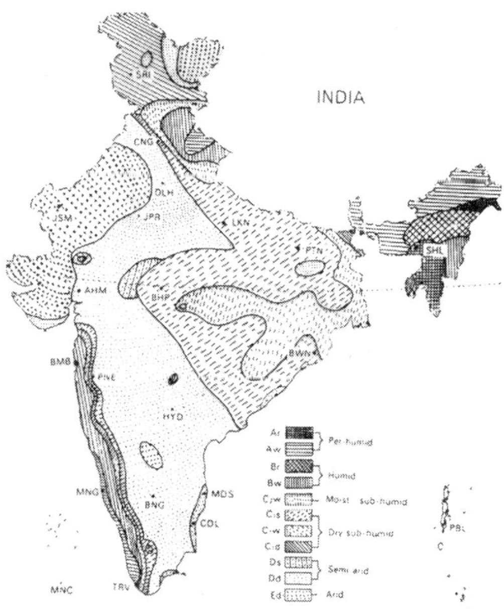

(*Source*: Rao. *et.al*, 1972)

Fig. 1.5 : Climatic classification of India by Thornthwaite & Mather (1955) method

It is seen that all six types of climates viz: per-humid, humid, moist sub-humid, dry sub-humid, semi-arid and arid are present in India. The NE region, west coast and adjoining hills fall under per-humid and humid zones, while, Saurashtra, Kutch, western Rajasthan, Bellary-Anantapur area in the states of Karnataka and A.P., and Tirunelveli region of Tamil Nadu fall under the arid climate. The area extending from Kanyakumari in the south to the Punjab in the north and covering practically the whole of the Peninsula east of Western Ghats and Gaya-Jumai area in Bihar fall under the semi-arid climate. Northern portions of Punjab, Haryana, U.P., Bihar, West Bengal, Orissa, M.P., Vidarbha and northern parts of A.P. – all these areas come under sub-humid (moist or dry) climate. The coastal areas of Tamil Nadu extending from Chennai to Nagapattanum also fall under this zone. Utilizing the Water Balance approach, the regions of water surplus and water deficit during *kharif* season were demarcated by Rao and Rajput (1986). Their analysis on rice productivity across the country with reference to these regions brought out that productivity of irrigated rice was higher in the water deficit region, due to better radiation regime associated with clear skies that enhanced photosynthetic activity. Also, lower average RH values during the season compared to other parts of the country reduced the pest damage considerably during monsoon season.

Other methods of classification, *viz*., Troll (1965), Cocheme and Franquin (1967), Hargreaves (1971), Papadakis (1975), etc., based on duration of moisture availability periods were utilized for agro climatic characterization. Overlaying maps of climatic parameters, soils and physiographic characteristics, 15 agro climatic zones (Figure 1.6) have been demarcated by the Planning Commission (Government of India, 1987).

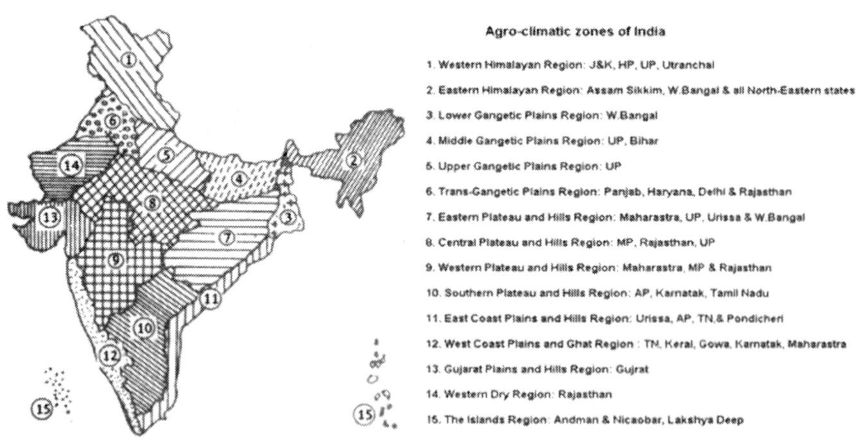

(*Source*: GOI, 1987)

Fig. 1.6: Agro-climatic Zones as per Planning Commission

The National Agricultural Research Program (NARP), based on Physiography, rainfall, soils, cropping patterns and administrative boundaries, divided the country into 127 agro-ecological zones (Ghosh, 1990). Sehgal *et al.* (1990) and Sehgal & Mandal (1996) have identified 20 homogeneous agro-ecological regions (Figure 1.7) and 60 sub-regions (AESR) based on topography, climate, soil and effective growing days.

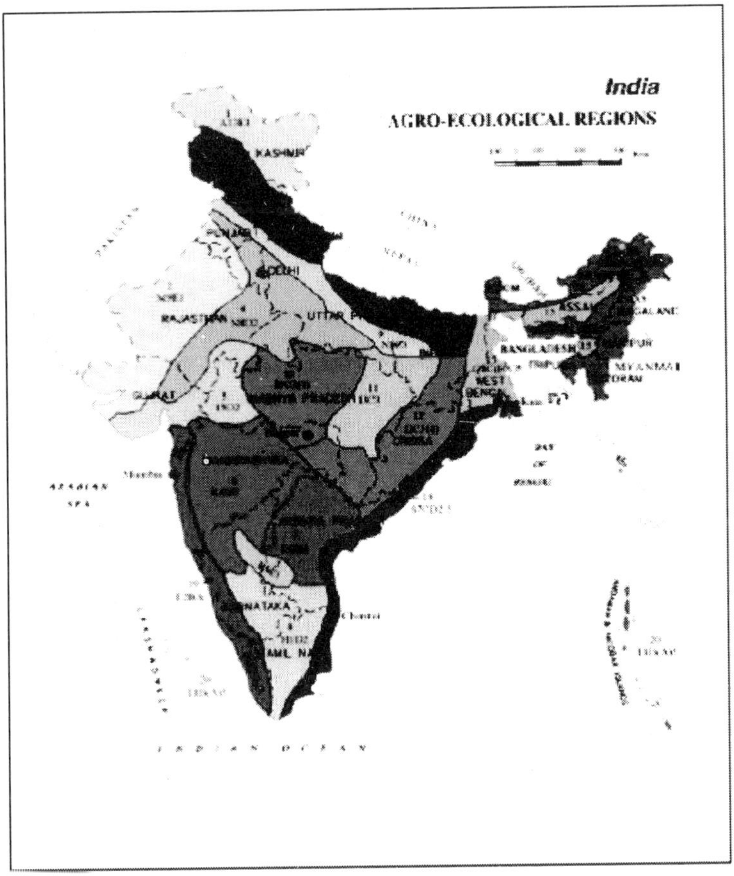

(*Source*: Sehagal *et al.*, 1990)

Fig. 1.7: Agro Ecological Regions of India

In view of the variability in the area of rainfed crops and their growing conditions under semi-arid and sub-humid climates, these climatic regions have been regrouped in relation to Rainfall, Length of Growing Period (LGP) (Table 1.1) by All India Coordinated Research Project on Agrometeorology.

Table 1.1: Regrouping of Agro-climatic regions in the country

Type of climate	Annual rainfall (mm)	Moisture index Im (%)	Growing period (days)	Agro-eco /Sub-region	Physiography	Area (m ha)
Arid						
Cold arid	<500	> -66.7	60-90	1.1,1.2	Western Himalayas, parts of U.P. and J & K	15.2
Hot arid	<500	> -66.7	< 90	2.1,2.2, 2.3, 2.4, 3.0	Western Plains and Kutch, Peninsula, Deccan Plateau	36.8
Semi arid						
Dry	500-700	-66-7 to –55.8	90-120	4.1,4.2,5.1,6.1, 7.1,8.1 14.1,18.1	Deccan Plateau, Northern Plains and central highlands including parts of Gujarat Plains. Central (Malwa) highlands, Gujarat Plains and Kathiawar Peninsula, eastern ghats and Tamil Nadu uplands	116.4
Moist	700-1000	-55.7 to –33.3	90-150	4.3,4.4,5.2 5.3,6.2,6.3 7.2,8.2,8.3,9.1,18.2		
Sub-humid						
Dry	1000-1200	-33.3 to 0.0	150-180	6.4,7.3,9.2 10.1,10.2 10.3,11.0 12.3,14.2, 18.3	Northern Plains, Central highlands (Malwa and Bundelkhand), eastern Plateau (Chattisgarh region and Chota Nagpur), parts of western Himalayas, eastern ghats and plains and parts of western Himalayas	94.9
Moist	1200-1600	0.0 to 20.0	150-210	10.4,12.1 12.2,13.1 13.2,18.4		
Humid and Per-humid						
Humid	1600-2000	20.0 to 100.0	180-270	15.2,18.5 19.1,19.2 19.3,20.2	Parts of western and eastern Himalayas, Bengal basin and Assam Plains	33.3
Per-humid	> 2000	> 100	> 270	14.3,14.4 14.5,15.3 15.4,16.1 16.2,16.3 17.1,17.2 20.1	Parts of eastern Himalayas, northeastern hills, western ghats and coastal plains.	31.3

(*Source*: Ramakrishna *et.al.* 2000)

1.5 Asian Summer Monsoon and Rainfall Distribution

The term "monsoon" is derived from Arabic words "*Mausin*" which is called a "seasonal wind" that reverse its direction between winters and summer seasons. In the year1686, Edmund Hillary, an Englishman explained the Asiatic monsoon to be due to thermal contrasts between the continents and the adjoining oceans. The eastern and southern Asia has the earth's largest and best monsoon circulation (Trewartha & Horn, 1980). The huge land mass of Asian continent, intensifies the contrasts arising due to heating and cooling differences between land and sea. Further the large extent of huge land mass in east-west direction in the northern hemisphere and the oceans to its south in the equatorial and southern hemisphere latitudes has made this atmospheric event a huge phenomenon in the global atmospheric circulation. The broad area of the summer monsoon activity extends between 30° N to 30° S and from 30° W to 165° E. Nowhere else in the other parts of the world such an extensive land area is affected by the seasonal wind reversal, as in the eastern and southern mainland of Asia. There are two branches of Asian monsoon systems viz., East Asia monsoon and South Asian monsoon system. These two Asiatic monsoon systems differ in the locations of their centers of action. China, Taiwan, Japan and Korea are influenced by East Asian monsoon system whereas monsoons of South Asia influence the Southeast Asia and south Asian countries.

The seasonal distribution of rainfall in south Asian region covering India, Bangladesh, Pakistan and Sri Lanka is determined by the 'advance' and 'retreat' of southwest monsoon. Much of bulk of annual precipitation is received during Southwest monsoon period (June-September). However, the regions in south east India and Sri Lanka receive substantial rainfall during November and December in association with the retreat of monsoon. Southeast Asia receives abundant rainfall of 1000-4000 mm/year. These regions are one of the wettest regions in the world.

1.6 The Indian Monsoon

India receives its rainfall mainly from the SW monsoon and partly from the NE monsoon. In a broad sense, the climate of India is controlled by monsoon circulation. The Indian sub-continent comes under the influence of the SW monsoon during June to September and NE monsoon during October to December. The SW monsoon is also called as summer monsoon.

Features of Monsoon

A number of interesting synoptic features occur in the circulation of the atmosphere when the summer monsoon sets in over India (Ramakrishna *et.al.*, 2000). These are

Heat low: With the beginning of summer season the continental area around the Arabian Sea begins to receive large amounts of heat. Due to this intense heat, a trough of low pressure (heat low) forms over the region extending from Somalia northwards across Arabia into Pakistan and northwest India. By the end of May the low is well established and southwest wind spreads northwards over Arabian Sea, Bay of Bengal and the Indian sub-continent.

Near equatorial troughs: Before the onset of the monsoon, a low pressure zone forms on either side of equator roughly along 5°N and 5°S. This is cften referred as equatorial double trough. Prior to monsoon, the trough north of 5°N weakens but the trough at 5°S is fairly active. It is believed by some that the trough at 5°N moves northwards with the progress of monsoon and after merging with the heat low; it forms a quasi-stationary monsoon trough. However, others believe that the monsoon trough is different from that of the near equatorial trough (5°N). Some also feel that the onset of monsoon is related to a sudden acceleration of air from southern hemisphere (SH) towards India across the equator. In the SH, the monsoon circulation is dominated by anticyclone (high pressure) circulation near Madagascar known as the 'Mascarenes High'. When this high pressure system becomes more pronounced it leads to delay in the onset of monsoon.

Jet streams: Equally interesting changes takes place in the upper atmosphere with the advent of the summer monsoon. Towards the end of May, a narrow belt of strong winds in the upper air around 300 hPa, blowing from west to east known as westerly jet stream suddenly weakens and moves to a new location far to the north of Himalayas. This shifting has a strong association with the onset of monsoon. As this jet moves northwards a new jet stream sets in over India, but at a higher altitude (100-250hPa) flowing from east to west known as the easterly jet stream. It prevails only during the 100 days of monsoon regime around 15-16°N and shifts periodically north and south of its mean position, influencing the monsoon activity.

The Tibetan high: A dominant feature associated with the onset of monsoon is the emergence of a zone of high pressure over the plateau of Tibet. This plateau is at a height of 4 km (corresponding to 600 h Pa pressure level). This high elevated land acts as a direct source of heat (heat island) creating strong thermal convection and the ascending air rapidly spreads outwards to both the north and south of the plateau leading to formation of anticyclone over Tibet around 300-200 h Pa (9-12 km).

Monsoon Circulation

The southwest monsoon thus is an important meteorological phenomenon of great physical and economic importance. The heat liberated in the atmosphere by the monsoon rains over the Indian region is about 8.4 x 10^{21} J in the monsoon season, equivalent to the total short wave radiation received by the same area. The average transport of moist air by the monsoon westerlies across the Indian region is 7 x 10^{15} kg per day while the transport of air by the upper easterlies over India is 12 x 10^{15} kg per day (Subbaramayya and Ramanatham, 1981). This mass circulation by the monsoon is comparable to that of Hadley Cell on the global scale, which is 19 x 10^{15} kg per day (Kidson*et al*., 1969). Four significant features of the monsoon are of importance to both Agro meteorologists and agriculturists. These are (i) date of onset, (ii) periods of active monsoon conditions, (iii) break monsoon conditions, (iv) date of withdrawal.

Onset of Monsoon

The onset of monsoon is meteorologically associated with certain changes in the upper air circulation features (Yin, 1949). Formation of monsoon trough at 700 h Pa (2.1 km) level is associated with steady westerly winds over Kerala (Pant, 1964) and upper troposphericwesterlies over north India (Anantha Krishnan, 1968), besides movement of synoptic and sub-synoptic rain bearing systems progressing northwards and westwards (Subbaramayya and Ramantham, 1981). The event is in general linked to the first showers of significant amount of rainfall with persistent cloudy conditions at a given region. Based on the movement of the low pressure (monsoon trough) across the country and thc associated rainfall, the northern limit of monsoon (NLM) onset is defined. The IMD has carried out detailed studies on the long-term pattern of the onset and withdrawal phases of monsoon.

The monsoon rainy season begins earliest in south China towards the beginning of May. Over the Indian main continent the monsoon sets in either by the end of May or the beginning of June along the Malabar Coast (it sets in earlier by mid-May over Andaman and Nicobar islands). The normal date of onset of monsoon over India is 1st June (Figure 1.8). The onset, often called as the 'burst of the monsoon' is often accompanied by a cyclone (depression or storm) either in the Bay of Bengal or in the SE Arabian Sea, associated with heavy rainfall over the Kerala and coastal Karnataka region. It then spreads in the peninsular region and slowly in the southwestern and southeastern parts of India (Arabian branch up to Bombay and the Bay of Bengal branch up to West Bengal) come under the influence of monsoon by 10th June. The filling up of monsoon current over the rest of the central and NW India is gradual and is associated with the merging of these two branches and the monsoon enters the western part of Rajasthan by 1st of July and the extreme western parts and into southern parts of Kashmir by 15th July. Thus, barring the Kashmir and in the extreme north and parts of Tamil Nadu and the southern peninsular regions, most of the annual rainfall over the country occurs during the SW monsoon season.

Though the date of onset of monsoon is considered regular occurring around 1st June, there are many occasions when not only the onset over Kerala coast is delayed but also the movement of the monsoon across the country is sluggish. Under such weak monsoon conditions, the delay in onset may be of few days (< week) over Kerala while it could be delayed by 30 to 35 days over the arid parts of western Rajasthan.

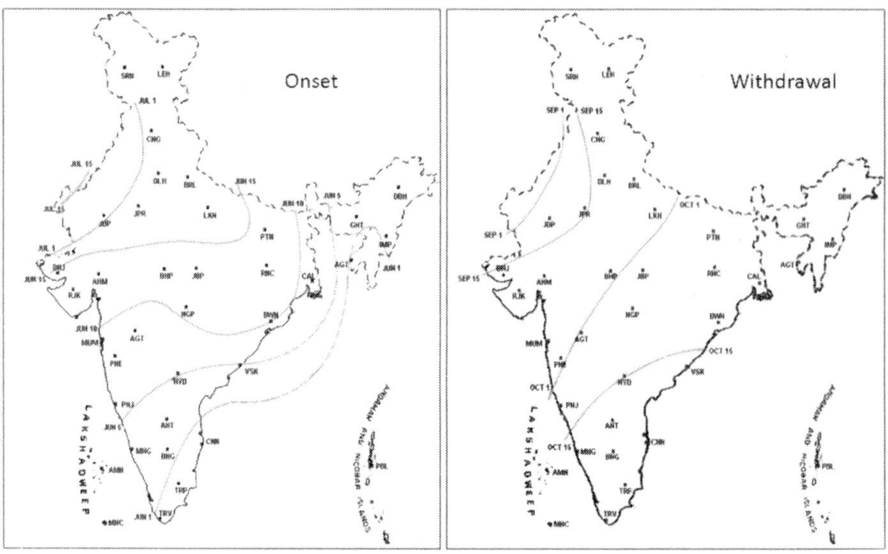

Onset Withdrawal

(*Source*: IMD)

Fig. 1.8: Onset and Withdrawal of monsoon

Withdrawal of Monsoon

The change from summer to winter type of monsoon circulation is gradual compared to the sudden onset of monsoon in the beginning of June over Malabar Coast. The monsoon begins to withdraw from extreme arid areas of western Rajasthan by 1st September and in general from northern India around mid-September (Figure 1.8). The withdrawal phase extends rapidly over western India up to west Uttar Pradesh, west Madhya Pradesh and Gujarat by 1st October.

The withdrawal phase is slow to the east of 80° E over the Gangetic valley, NE India, eastern Madhya Pradesh, Vidarbha and other adjoining areas of Maharashtra and interior peninsula. By the end of October it withdraws to south of 15°N. About the time of withdrawal of monsoon, the westerly jet again reappears over the Indian sub-continent and moves southward to its normal position by October end, and the easterly jet disappears by early October. The NE monsoon also becomes active over the Peninsular India by October and the NE monsoon rains merge with the withdrawing SW monsoon along the coastal Andhra Pradesh, Tamil Nadu and Kerala. Thus, it often becomes difficult to distinguish between the onset of NE monsoon and withdrawal of SW monsoon. During the period of October and November, occasional low pressure systems form in the Bay of Bengal and at times intensify into cyclonic storms and lash the coastal regions of Andhra Pradesh and Tamil Nadu. Sometimes, these storms move northwards and cross inland over Orissa as was witnessed during the late October period of 1999, resulting in extensive damage.

The monsoon finally withdraws from the Indian Peninsula by 15th December. Thus, the total duration of the monsoon activity across the country varies from 122 days over the Kerala coast (onset to withdrawal) while over the extreme parts of western Rajasthan, the duration is as low as 45 days as in this region it reaches last and withdraws first. Variability in the duration of monsoon, however, is higher in the region, east of 80° E in the Gangetic valley, which is one of the major crop production zones. In this region, the onset in some years (with weak monsoon activity) can be delayed by 15 days and also the withdrawal (cessation of rainfall) can be early by 15 days thereby reducing the total duration of the rainy season by about a month and adversely affecting the agricultural productivity.

Southwest Monsoon Rainfall

Nearly 70 per cent of the annual rainfall in India is recorded during the SW monsoon. The regions that receive the largest rainfall are along the west coast of India and the states of Assam and West Bengal in NE India. In these regions, orographic features play an important role. Interestingly, Bombay on the western side of Western Ghats receives 187.5 cm rain while Pune at a distance of 160 km on the eastern side (leeward of mountains) receives only 50 cm rain. The largest amount of rainfall is at Mawsynram on the southern slopes of Khasi hills which records 11871 mm rainfall annually. Over other regions of the country the rainfall varies from above 1000 mm and decreasing to as low as less than 100 mm in the NW part of the extreme western Rajasthan (Thar desert).

Active phase of Monsoon

Once the monsoon sets in over the Kerala coast, it extends to different parts of the country providing the much needed rains. The activity of the monsoon depends on the strength of the monsoon current. The monsoon current is pulsatory in nature and advances into different areas in two or three stages. In the first stage, the rains cover the west coast up to Konkan and the Peninsular India and the northeast region. This stage is accomplished in about 10 days after the onset of monsoon over the Kerala coast. However, if monsoon current is weak, this movement is slowed down or initiates late as a result of which the onset over the peninsular region can be delayed by a few days. The second stage usually follows after a gap of about 7 days, when a cyclonic disturbance forms in the North Bay of Bengal or in North Arabian Sea and then the rains advance into the Gangetic valley up to Uttar Pradesh and into Gujarat region (Sikka, 1996). Again after a gap of about 7 days, the rains further advance westward and northward often associated with the formation of another cyclonic disturbance in the Bay of Bengal. On some occasions the western disturbances moving across Kashmir region also help in accelerating the movement of the monsoon towards eastern Rajasthan. The movement of monsoon across the western part of Rajasthan is slow as it takes about a fortnight (1st to 15th July) to cover this region. By the time the whole country is covered, the monsoon trough takes its normal position with the trough extending from the Ganganagar region in Rajasthan, to the northern portion of the Bay of Bengal. The western end of the trough corresponds to a heat wave circulation with shallow convergence limited to half a kilometer or so and clear skies. On the other hand, the eastern end that extends over the warm water of Bay of Bengal is a dynamic trough and acts as a source of heat and moisture to the monsoon flow. It is thus characterized by moisture convergence and low tropospheric cyclonic vorticity. The monsoon trough during the active monsoon period (July to September) is a quasi-stationary large feature of synoptic scale (Figure 1.9) and its position oscillates around this normal location over the Gangetic valley, either to the south across 20-24°N or to the north close to the foothills of Himalayas. Active rain spells occur when the monsoon trough is close to its normal position.

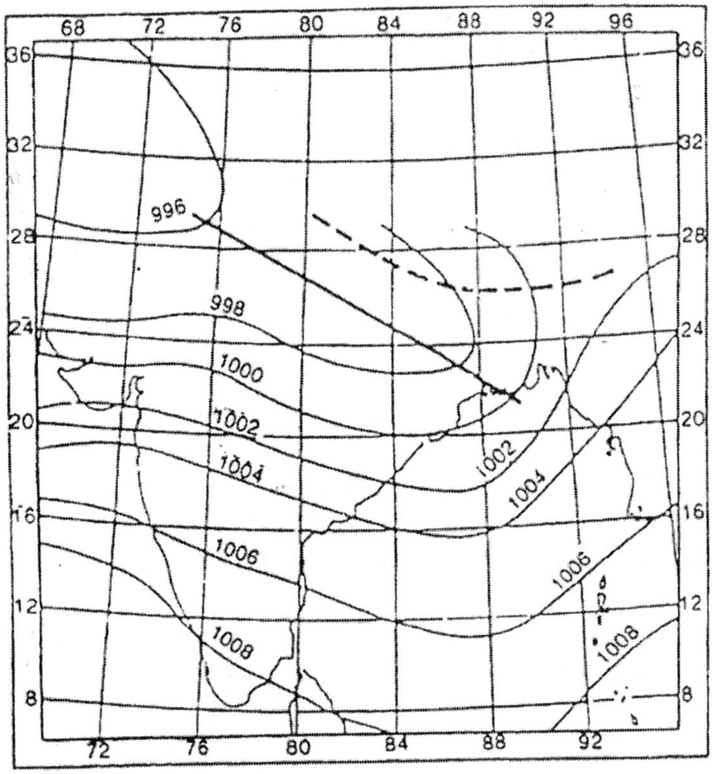

Source: Sikka (1996)

Fig. 1.9: The monsoon trough over India

Thick line indicates the normal position of the trough and the dotted line is the through position under break monsoon conditions

Often rainfall during the active monsoon season is associated with depressions passing over the country. Most of these depressions form in the Bay of Bengal and then intensify as cyclones and enter into the country along the east coast and provide rainfall as they move over the land region and weaken. Occasionally, one or two storms, which form over the Arabian Sea close to the west coast, provide substantial rainfall in Gujarat and south Rajasthan regions. Intense rainfall is associated with such low pressure systems. Lower the pressure at the centre of the storm, the more intense is the wind regime, occasionally reaching even up to 300 km/h and associated with high intensity rainfall. In general, 5 to 6 storms of different intensities cross the coast during good monsoon year providing adequate rainfall. Also the movement of the monsoon trough provides rainfall in its wake. Often low pressure systems are seen embedded in the monsoon trough. These are not as intense as the normal cyclonic systems but at times they also provide good rainfall.

Break Monsoon Conditions

'Break' conditions prevail in monsoon rainfall when the monsoon trough moves northwards to the foothills of Himalayas. During such periods the rainfall is limited mainly to the NE regions of Assam and adjoining regions while the rest of the country goes dry. Even in such occasions some rainfall is recorded along the west coast region under the influence of waves in the easterlies at mid level.

Winter Monsoon

In comparison to summer (SW) monsoon, the winter (NE) monsoon is not one of uninterrupted rains. In this, rainfall occurs in spells, often accompanied by surges of cold air that emanate from a large anticyclone over Siberia and adjoining parts of China.

The winter monsoon often known as 'NE monsoon' because of the reversal of wind direction to northeast provides rainfall mainly to the Peninsular India in the months of October to December. The Tamil Nadu region, which is shielded by the Western Ghats from the rains in SW monsoon season, depends mainly on NE monsoon rainfall for its agriculture. There is no clear-cut distinction between the withdrawal of the SW monsoon and the onset of NE monsoon. Most of the rainfall in NE monsoon is associated with westwards passing disturbances in the 5 to 15°N belt.

1.7 Causes for Failure of Monsoon

The seasonal rainfall (monsoon rains) over the Indian sub-continent is a global phenomena associated with large-scale hemispherical movement of air masses. Therefore, identification of the major atmospheric phenomenon that influences the monsoons over Indian sub-continent is essential in understanding and management aspects of drought research. Two such relationships, viz., (i) sea-surface temperature anomaly around the Indian sub-continent in relation to atmospheric circulation, and (ii) large-scale pressure oscillation in atmosphere over southern Pacific Ocean are found to be useful in this context. The El Nino event is one such phenomenon, which has profound influence on the monsoon activity over Indian Sub-continent. The Southern Oscillation Index (SOI) is one important parameter in the predictive sixteen parameters model used by IMD for long-range forecasting purposes. The study of the Indian summer monsoon over the country by India Meteorological Department showed that all the drought years are El Nino years whereas all the El Nino years are not drought years indicating thereby that various other factors also equally influence the monsoon over the sub-continent. In this context, the winter circulation over the sub-continent, extended period of occurrence of western disturbances (late in the season), strengthening of heat low over NW India in summer and shifts in zonal cells over India are some of the important parameters that influence monsoon system over the country. Some of the researchers are of the opinion that the sea-surface temperature anomaly in the monsoon path is more important in predicting the monsoon rather than the pressure difference at far off places on the globe. Such studies have been initiated, but definite conclusions are yet to be arrived at.

Prediction of Onset of Monsoon Rainfall

The onset of Indian summer monsoon over Kerala is the most anxiously awaited weather phenomena in the Indian subcontinent as it heralds the rainy season and marks the end of the hot summer. The onset of Indian summer monsoon represents significant transitions in the large scale atmospheric and oceanic circulation in the Indo-Pacific region. There is no widely accepted definition of this monsoon transition. However, at the surface, monsoon onset is recognized as a rapid substantial and sustained increase in rainfall. The first rains of monsoon occur over Burma and Thailand and in the Andaman by third week of May and subsequently extend to the northwest. In association with the monsoon onset around 1 June, heavy rains lash south peninsula after the cross-equatorial low-level jet is established across the Somali coast into the near-equatorial Arabian Sea. The northward progression of monsoon is symptomatic of a large scale transition of a deep convection from the equatorial to continental regions (Rao 1976; Sikka and Gadgil 1980; Webster *et.al* 1998). By middle of July, monsoon covers the whole country.

The India Meteorological Department (IMD) has been issuing Long Range Forecasting (LRF) of the southwest monsoon rainfall since 1986. It was, however the extensive and pioneering work of Gilbert Walker (1923, 1924), that led to the development of the first objective models based on statistical correlations between monsoon rainfall and antecedent global atmosphere, land and ocean parameters. Since then, IMD's operational LRF system has undergone changes in its approach and scope from time to time. There are many reviews on the LRF of Indian southwest monsoon rainfall (ISMR). In a very recent study, Gadgil *et al.* (2005) addressed the major problems of the statistical and dynamical methods for LRF of monsoon rainfall in view of the recent forecast failures in 2002 and 2004. Their analysis revealed that IMD's operational forecast skill based on statistical methods has not improved over seven decades despite continued changes in the operational models. For the LRF of the ISMR, three main approaches are used. The first is the statistical method, which uses the historical relationship between the ISMR and global atmosphere–ocean parameters. The second approach is the empirical method based on a time series analysis. This method uses only the time series of past rainfall data and does not use any predictors. The third approach is based on the dynamical method, which uses general circulation models of the atmosphere and oceans to simulate the summer monsoon circulation and associated rainfall. In spite of its inherent problems, at present, statistical models perform better than the dynamical models in the seasonal forecasting of ISMR. The dynamical models have not shown the required skill to accurately simulate the salient features of the mean monsoon and its interannual variability (Krishnamurti *et al.* 2000; Kung and Sharif, 1982;Gadgil*et al.* 2005; Wang *et al.* 2005).

During the period of 1988–2002, IMD's operational forecasts were based on the 16-parameter power regression and parametric models (Gowariker *et al.* 1989, 1991). The forecasts issued during this period were qualitatively correct. However, the mean forecast error during this period was more than the mean error of the forecasts based on climatology alone. This model failed to predict the severe drought of 2002. Following the failure of forecast in 2002, a critical evaluation of the 16-parameter power regression and parametric models was made and in 2003, and two new models (8 and 10 parameter models) were introduced for operational work. Further a two-stage forecasting strategy was also adopted with the provision for a forecast update by end of June/first week of July (Rajeevan *et al.* 2004). Accordingly, IMD's operational forecasts for the seasonal ISMR for the country as whole are issued in two stages. The first forecast is issued in mid April and an update or second forecast is issued by the end of June. While the 2003 and 2005 operational forecasts for the southwest monsoon rainfall based on these new models were accurate, the forecast for the 2004 monsoon came up false.

1.8 Spatial Distribution of Weather Parameters over India

Rainfall

The rainfall received during the southwest monsoon season is the most important of all the weather elements as the country's agricultural production is totally dependent on its quantum and distribution. The average annual rainfall of the country is 1200 mm amounting to 400 M ha m of rainwater over the country's geographical area which is about four per cent of the global precipitation. However, the spatial distribution across the country varies from less than 100 mm in extreme arid areas of western Rajasthan to greater than 3600 mm in the NE states and 1000 mm from east coast to 2500-3000 mm in the west coast. The average rainfall received during different seasons in four regions of the country is shown in the Table 1.2. Similarly the spatial distribution of mean annual rain fall, seasonal rainfall during South-West monsoon season are shown in Figure 1.10.

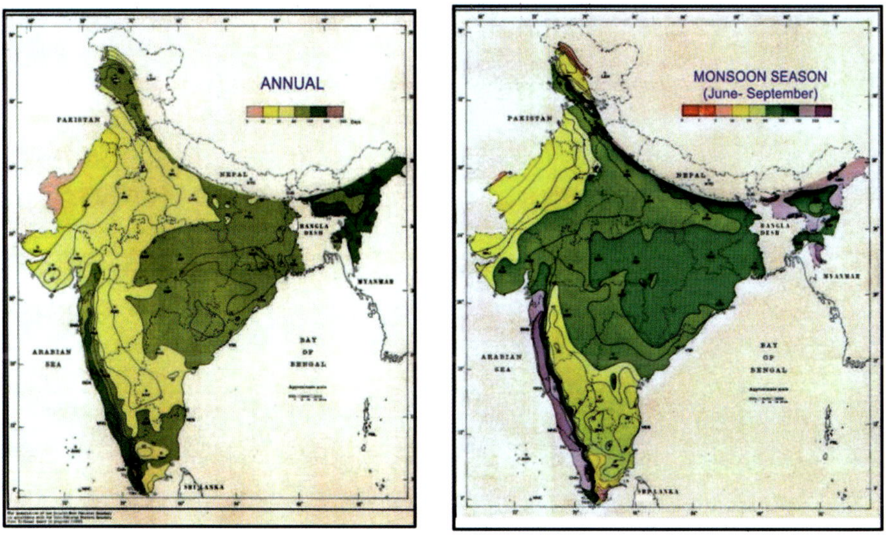

(*Source*: IMD)

Fig. 1.10: Mean annual and seasonal rainfall over India

Table 1.2: Average rainfall (mm) in different regions of the country

S.No	Region	Annual	Pre monsoon	Monsoon	Post monsoon	Winter
1.	NW India	865.8	110.9	615.0	62.7	772
2.	Central India	1106.6	40.1	974.2	79.2	15.8
3.	South Peninsula	1129.2	123.9	716.3	273.7	15.3
4.	NE India	2036.3	371.9	1437.8	171.2	55.5
5.	Country Average	1186.2	131.2	887.0	127.1	40.9

(*Source*: IMD)

Temperature

Due to the latitudinal extent from near equatorial to sub-tropical, the country experiences wide range of temperature conditions. The mean annual temperature across the country varies from less than 10°C in the extreme north to more than 28°C in the southern parts. As such, temperature is not a limiting factor for agriculture over a major part of the cultivated area. The spatial distribution of monthly temperature of January, April, July and October representing Winter, Summer, monsoon and post monsoon are given in Figure.1.11. The temperatures start to increase over the country from March onwards and reach a peak in

May and June. By May, many parts of India including Rajasthan, Uttar Pradesh, Madhya Pradesh, Vidarbha in Maharashtra, Telangana and Rayalaseema regions of Andhra Pradesh record mean daily maximum temperatures above 40°C. On individual days temperatures can be over 46 to 49°C. During May and June, heat low forms over Rajasthan and the adjoining regions of Pakistan which helps usher in the monsoon over the country. Once the monsoon sets in over the country, temperatures sharply decrease in the peninsular region to less than 30 to 35°C while higher temperature conditions prevail over northern parts of India till the end of June or early July when monsoon current reaches this region bringing in rain and extensive clouding.

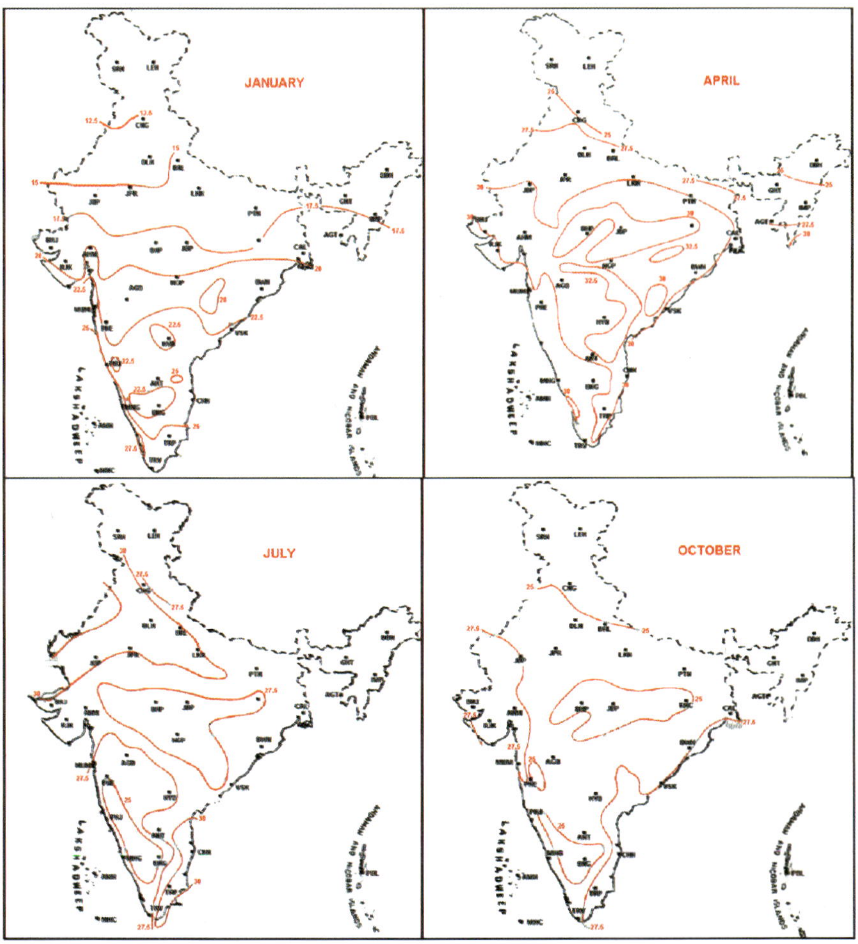

(*Source*: Ramakrishna *et.al.*, 2000)

Fig. 1.11: Temperature distribution over India during different months

July and August are the active monsoon months during which period the mean temperatures over various regions of the country remain around 28 to 29°C. Tamil Nadu in the south, however, is characterized by higher temperatures during this period also due to low monsoon activity over the region. With the withdrawal of monsoon over northern India from mid-September, a second peak of temperatures is observed in October with day temperatures touching up to 40°C. However, nights remain cooler. From November onwards a gradual decrease in both day and night temperatures is observed. In winter season, low temperatures are observed in January with mean temperature around 14°C in the northern most parts of India and around 27°C in the southern most parts. During winter, the northern states like Bihar, Uttar Pradesh, Punjab, Himachal Pradesh, Rajasthan, northern parts of Madhya Pradesh experience minimum temperatures below 10°C. These regions are prone to occasional frost during the *rabi* season. The lowest temperatures in winter are observed in Kashmir region which record below freezing temperature while the lowest are recorded in Ladakh region of Dras.

Humidity

Atmospheric humidity plays an important role in plant growth as it not only influences the crop water needs but also decides the congenial atmospheric conditions for the buildup of pests and diseases. Except for the per-humid north eastern region, the island and coastal regions, relative humidity (RH) over the country exhibits seasonal variation and is influenced on spatial basis by the monsoon air mass movements to a great extent. RH values are lowest during the summer period and they range from 20 to 40 per cent in the central and western parts of the country. The western parts of arid Rajasthan record the lowest, with the afternoon RH values at times going as low as 2 to 4 per cent. With the approaching monsoon, sharp rise in RH values are observed and the RH in June remains in the range of 30 to 60 per cent in most parts of the country. As the monsoon advance over the country almost all locations record mean RH values ranging from 50 to 80 per cent. However, in southern Peninsula, where monsoon activity is low during this period, the RH remains low during this period, i.e., between 40 to 60 per cent. The RH decreases with the withdrawal of monsoon current and in October it remains around 30-60 per cent in West Uttar Pradesh, West Madhya Pradesh and Punjab and 60-70 per cent over the rest of the country. The RH increases in the peninsular region of Tamil Nadu during this period and in November this region records higher values of RH over the country.

Sunshine and Radiation

Solar radiation is the primary source of energy reaching earth's surface. Out of the total spectrum, the visible part of the spectrum (0.4 to 0.7 m) constitutes about 50 per cent of total global radiation, which affects a number of plant functions and regulates the plant growth and development. In India, measurements on global solar radiation are scanty compared to duration of bright sunshine hours. The mean daily duration of bright sunshine hours during different seasons recorded at various regions in the country indicated that sunshine is not a limiting factor during greater part of the year at majority of these places in spite of good cloud cover during monsoon season. On an average, greater part of the country has about 7-8 hrs of duration of bright sunshine, whereas it varies between 6.7 hrs per day in the NE part of the country. Highest number of sunshine hours are recorded in April, while lowest was found in monsoon months (July). The duration is about 3 hrs per day in the west coast of the country and less than 4 hrs per day in the NE region during peak monsoon month.

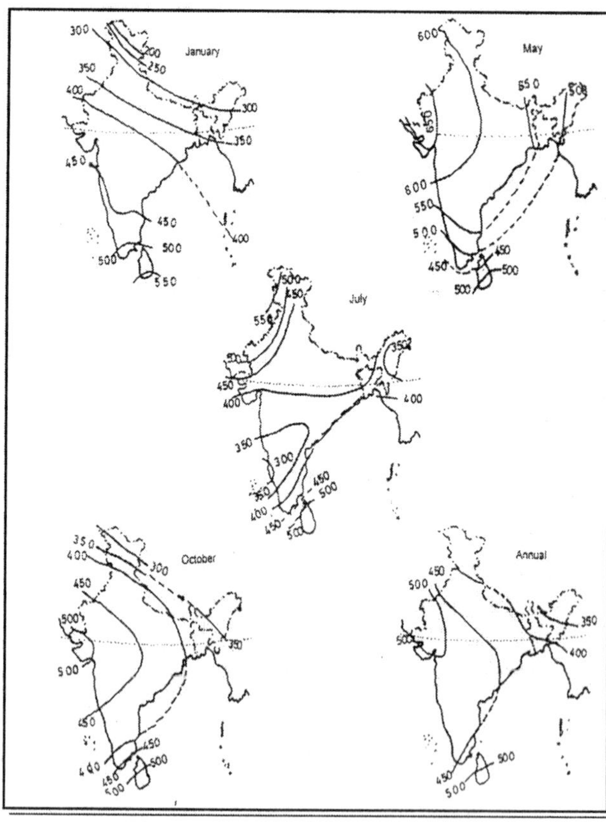

(*Source*: Ramakrishna et al., 2000)

Fig. 1.12: Mean daily global solar radiation over India in different months

The solar radiation received in different months is shown in Figure 1.12. On an average, the radiation received on a horizontal surface varies between 400-500 Cal. cm^2 day^{-1}, highest being recorded in the NW part of the country. Higher values of radiation recorded in May throughout the country with the peak values of 600-650 Cal cm^{-2} day^{-1} in NW parts of India. With the advancement of monsoon, the radiation values drastically reduce to 300-350 Cal. cm^{-2} day^{-1} in the west coast and in the NE regions of the country.

Wind Speed

Wind regime influences the evaporative demand of crop plants to a great extent. Stronger the wind, water requirements of crops shoot up considerably. Besides, stronger wind regime leads to plant injury, and depletion of surface moisture and removal of fertile soil through wind erosion. The mean wind speeds are lowest during the winter months (3-6 kmph) and the wind regime builds up with the initiation of summer season and is at peak during the pre-monsoon periods of April to June (6-10 kmph). Stronger wind regimes are observed in western Rajasthan, Gujarat and parts of Tamil Nadu. Wind speeds, however, can be extremely high for shorter periods during intense cyclonic activity, resulting from the atmospheric pressure gradients causing damage to standing crops and vegetation.

1.9 Famines in India

Famines are resultant of occurrence and successive drought years over a region where deaths due to starvation and migration of both human and livestock population are common. It was reported that about 90 famines occurred over the past 2800 years in India and about 60 million deaths occurred over the 18, 19 and early 20th centuries (***en.wikipedia.org***). A better record keeping of famines were taken up during British Rule. The major famines are Bengal famine of 1770, the Chalisa Famine, the Doji Bara Famine, the great famine of 1876-78 and the Bengal famine of 1943.

Historically, the rulers of the country took several famine relief measures during famine period such as free distribution of food grains, throwing open the grain stores and kitchen to the people, remission of revenue and taxes, construction of public works, canals and embankments and digging of wells etc. The late 18 and 19 the centuries experienced the worst famines which had a remarkable impact on the long term population growth in the country. The Bengal famine in 1770 was estimated to have taken the lives of one third population of the region (about 10 million). The revenue decline to East India Company was put at £174300 during 1770-71. The stock price of the East India Company fell sharply and it obtained a loan of one million pounds from Bank of England.

The famines viz., Orissa famine of 1866-67, and the famines recorded in western Ganges region, Rajasthan and Central India during 1868-1870, Bengal and Eastern India famines during (1873-74), Deccan famine (1876-78) were attributed to uneven rainfall and the economic and administrative policies of the British rules. The Great Famine of 1876-78 caused a large death toll of about 10.3 million population and large migration of agricultural laborers and artisans to British tropical colonies where they worked as plantation laborers. Florence Nightingale pointed out that famines in British India were not caused by shortage of food in particular geographical region, but were caused by inadequate transportation. (Wikipedia). She observed two types of famines viz., "grain famine" and "money famine". Money was drained from the farmer to the landlord, making it impossible for the peasants to procure food.

There were some serious efforts by the British Government in providing mitigation measures to safe guard the public as well their army from famines. After the devastating famine in 1770 in eastern parts, the Governor-General of India Warren Hastings, ordered the construction of big food grain storage bin (capacity 140,000 tons) at Patna for the people and to the army which was completed in 1786 by Capt. John Garstin, an Engineer in East India Company (Figure 1.13).

(*Source*: Wikipedia)

Fig. 1.13: Golaghar for storing food grains in Patna

Similarly the construction of various irrigation systems like, upper Ganga canal, upper Bari Doab canal, Krishna and Godavari Delta system between1836-1866 were helpful in managing the shortage of food and reducing human causalities. The gross irrigated area in British India by the end of 19th century was about

7.5 Mha (Ministry of Water Resources Govt. India). Amartyasen (1981) through his classical publication on poverty and Famines observed that the famines in British era were due to lack of serious effort on the part of British Government to prevent famines; he further linked it to the absence of democracy in the country. The Bengal Famine of 1943 was the last catastrophic famine in India and it holds a special place in history of famines. Roy, Tirthanker (1999) attribute the famines were due to environmental factors, which are inherent characteristics of India's monsoon system and argues that massive investment was required to reduce the impacts. Agriculture and irrigation sector were given top priority after the Independence. It was not until 1970, the investment on the above sector did not yielded the results and the country became famine free.

Mc Alpin and Michelle Burge (1979) had argued that economic changes in India during 19th century significantly contributed in reducing the famines. Diversified economy in 20th century lead to reduced dependency on agriculture, which results in less mortality during time of scarcity. Also, construction of Indian Railways between 1860 to 1920 provided opportunities to export the produce for better economic returns and rapid mobility of the food grains from surplus to deficit regions. Though railways were intended to safe guard against famines, they were used by merchants to ship grain from drought stricken districts to central depot for hoarding and the telegraph systems served to provide the latest price information of essential commodities which resulted in increased prices of food, that was out of reach to the outcaste labourers and poor farmers (Davis, 2001). Also the large grain market created by Railways encouraged the poor farmers to sell off their reserve stocks of grain. The North Eastern state like Mizoram has a peculiar problem with flowering of Bamboo plantations. A common belief among the tribal population that bamboo flowering is followed by an increase in rat population follows famine and unrest amongst the people. Though it cannot be scientifically proved, a possible evaluation for this phenomenon was provided by John and Nadgauda (2002). According to them, flowering is followed by a large quantity of bamboo seeds on the forest floor which causes a spurt in the rat population, which feeds on these seeds. With the rains, the seeds germinate and force the mice to migrate to nearby agricultural fields in search of food and feed on crops and grains stored in granaries, which lead to shortage of food. To solve such problems; scientists suggest switching over to cultivation of other crops such as ginger and turmeric as they are not consumed by rats.

There has been no famine of catastrophic nature after 1943 except for severe droughts. The famine like conditions of 1967, 1973, 1979, 1987 in Bihar, Maharashtra, West Bengal and Gujarat were successfully controlled by the

Government of India. Distribution of food through Public Distribution Systems (PDS) at cheaper rates controlled the escalation of prices of essential commodities. Also strengthening of Agricultural Research programs with establishments of new Research Institutes under ICAR and establishment of Agricultural Universities and collaboration with International Agencies helped in developing high yielding varieties, and drought resistant improved agronomic and soil management practices and increased area under cultivation and irrigation; easy credit policies of GOI for farmers made the country self sufficient in agricultural production. (Green Revolution). The country did not import any food grains during the two major droughts of the current century in 2002 and 2009. However, threats from changing climate due to global warming is looming over for the past several years and is likely to bring in conditions similar to pre-independence era unless measures to contain the global warming with strict policies and programs are implemented at national and at International level. People participation over the globe across boundaries in conserving resources shall only minimize the impact of global warming on human population.

1.10 Historical Evidences of Droughts

Historical review of drought has brought out that it existed in the form of famine and famines occurred in Northern India, Deccan region, parts of Eastern India and parts of Western India and also in Central India. There have been instances when Indian rulers provided certain relief measures, adjustments in the treasury and allowed the farmers certain exemptions from paying land and crop taxes during famine. The occurrences of droughts during the period 1276 to 1870 A.D. as compiled by Sikka (1999) are given in (Table 1.3).

Table 1.3: Historical Infamous Monsoon Droughts followed by Famines over different parts of India (Period 1276-1870)

S.No.	Year of drought /Famine	Area of Drought/Met. Sub Division/ Homogenous Rainfall Region of India	Regional /All India Drought
1.	1290-1291	Punjab, Haryana and Delhi	Regional
2.	1326-1327	Haryana and West Uttar Pradesh	Regional
3.	1334-1335	Haryana, West Uttar Pradesh	Regional
4.	1362-1363	Punjab, Haryana and West Rajasthan	Regional
5.	1471-1472	Marathwada, Madhya Maharashtra and North Interior Karnataka	Regional
6.	1508-1509	North Interior Karnataka and South Interior Karnataka	Regional
7.	1555-1556	Delhi , Haryana and West Uttar Pradesh	Regional
8.	1630-1631	Gujarat, Saurashtra Kutch, West Madhya Pradesh, Madhya Maharashtra and Marathwada	All India
9.	1648-1649	West Rajasthan and East Rajasthan	Regional
10.	1649-1650	West Uttar Pradesh, Gujarat, Saurashtra and Kutch	Regional
11.	1659-1660	Most of India , Severe in Maharashtra and Rajasthan	All India
12.	1685-1686	Madhya Maharashtra, Marathwada, Rayalaseema and Telangana	Regional
13.	1718-1719	Gujarat, Saurashtra Kutch, West Madhya Pradesh	Regional
14.	1729-1731	South Peninsular India Homogenous Region	Regional
15.	1747-1748	Guajarat, Saurashtra Kutch, West Madhya Pradesh, Marathwada, West &East Rajasthan	All India
16.	1769-1770	Bengal, Bihar, Jharkhand, Orissa, Punjab, Haryana, West & East Uttar Pradesh	All India
17.	1782-1783	Punjab, Haryana, West Uttar Pradesh, East Uttar Pradesh, West MP, Vidarbha, Bihar Rayalaseema and North Interior Karnataka	All India
18.	1791-1792	West Central India and South Peninsula India Homogenous Regions	All India
19.	1801-1802	South Peninsular India and West Central India Homogenous Regions and West Uttar Pradesh, East Uttar Pradesh, and Bihar	All India
20.	1803-1804	West and East Uttar Pradesh and adjoining areas	Regional
21.	1805-1806	Tamil Nadu, South Interior Karnataka and Coastal Andhra Pradesh	Regional
22.	1812-1813	North west India and West Central India Homogenous Regions	All India
23.	1819-1820	Uttar Pradesh and Haryana	Regional
24.	1823-1824	Tamil Nadu, Telangana, Coastal Andhra Pradesh and Rayalaseema	Regional

(*Source*: Sikka -1999)

Past All India Droughts during period 1870-2001based on the All India Monsoon Rainfall series were computed and the area affected given by Parthasarthy *et al.* (1995), Bhalme and Mooley (1981) and Choudhary *et al.* (1989) has been updated by AICRPAM, CRIDA (2010) and given in (Table 1.4).

Table 1.4: Drought years during 1870-2010

Year	Severity of drought	Area extent of Indiaaffected	Homogeneous Regions affected simultaneously
1873	Mild	33%	Central NE India, NE India and S. Peninsular India.
1877	Phenomenal	67%	Northwest India , West Central India and Central NE India
1899	Phenomenal	62%	North west India, West Central India and S. Peninsular India.
1901	Severe	66%	North West India and Central NE India
1904	Mild	35%	North West India and West Central India.
1905	Severe	41%	North West India, West Central India and S. Peninsular India
1911	Moderate	42%	North West India, West Central India, South Peninsular India
1918	Phenomenal	69%	North West India and West Central India, Central NE India
1920	Severe	37%	West Central India
1925	Mild	31%	North West India and NE India
1941	Moderate	38%	West Central India
1951	Moderate	37%	North West India, West Central India and Central NE India.
1965	Severe	42%	North West India, West Central India and NE India.
1966	Moderate	35%	West Central India and Central NE India
1968	Mild	25%	North West India, West Central India and Central NE India
1972	Severe, close to Phenomenal		46% North West India, West Central India, S. Peninsular India, NE and Central NE India.
1974	Close to moderate	32%	North West India, West Central India and Central NE India.
1979	Severe	43%	West Central India and Central NE India
1982	Moderate	41%	North West India and NE India
1985	Mild	30%	North West India and West Central India.
1986	Moderate	35%	North West India and NE India
1987	Severe	53%	North West India , West Central India, Central NE India and S. Peninsular India
2002	Severe	29%	Northwest India, central India, Peninsular India
2009	Phenomenal	59%	Northwest, Indo-Gangetic plains, Orissa, and Peninsular India

(*Source*: Choudhary *et al.*, 1989 and AICRPAM 2010)

Mostly north and north western parts of the country witnessed more number of drought years compared to their counter parts elsewhere in the country. Severe to phenomenal drought affecting more than 50% of area was observed in 1877, 1899, 1901, 1918, 1972, 1987 and 2009. The droughts had serious impacts on countri's economy.

The departures of per cent average monsoon rainfall and area affected under the phenomenal droughts from years 1872 to 1990 reported by Kulshreshtra (1997) and for the period 1991-2010 given by AICRP on Agrometeorology is given in the following (Table 1.5).

Table 1.5: Years of phenomenal All-India Droughts (modified by AICRPAM, (CRIDA)

Year	Departure of Average monsoon < –20% season's rainfall (%)	Area(%) under deficient monsoon seasonal rainfall > 47.7%
1877	-29.1	66.8
1899	-26.2	83.0
1918	-23.9	68.2
1972	-23.3	49.5
1987	-19.3**	64.3
2002	-19.0**	29*
2009	-23.0	59.0

(*Source*: Kulshreshtra, 1997)

* 52% of the total districts were drought affected ** Marginal Case

There are six phenomenal droughts that affected more than 60 per cent area of the country with the departure of average seasonal rainfall beyond 20% of LPA. Though the drought in 1899 affected large area of the country, the intensity of drought with greater departures of rainfall in 1877 was the worst hit which led to serious famine condition.

Drought intensity of the all India drought years for the year (1877-2009) computed using Standardized Precipitation Index (SPI) by IMD, (2010) are given in the table 1.6. The intensities computed by rainfall departures from the normal as given in Table 1.4 are matching with the values computed using SPI.

Table 1.6: Drought Intensity over India based on Standardized Precipitation Index (SPI)

Year	Seasonal (June –Sept.) Rainfall (cm)	SPI value Drought intensity	Category
1877	58.7	-3.47	ED
1899	62.1	-3.01	ED
1901	76.5	-1.21	MD
1904	77.6	-1.08	MD
1905	72.7	-1.66	SD
1911	75.1	-1.38	MD
1918	66.1	-2.48	ED
1920	73.3	-1.59	SD
1941	76.3	-1.23	MD
1951	71.5	-1.81	SD
1965	72.0	-1.75	SD
1966	76.4	-1.22	MD
1972	67.0	-2.37	ED

(Contd.)

(*Contd.*)

Year	Seasonal (June –Sept.) Rainfall (cm)	SPI value Drought intensity	Category
1974	77.4	-1.11	MD
1979	71.4	-1.82	SD
1982	75.2	-1.36	MD
1986	76.8	1.18	MD
1987	70.9	-1.88	SD
2002	71.3	-1.83	SD
2004	76.6	-1.20	MD
2009	69.8	-2.02	ED

(*Source*: IMD, 2010)

ED = Extreme Drought (SPI: More than -2.0)SD = Severe Drought (SPI: -1.50 to -1.99)
MD = Moderate Drought (SPI: -1.0 to -1.49)

1.11 Distribution of Drought Prone Regions Over the Globe

All the developing countries, being primarily agrarian, are very much dependent on the vagaries of seasonal rainfall and climatic conditions and hence more vulnerable to droughts. On an average, severe drought occurs once in every five years in most of the tropical countries, though often they occur on successive years causing severe losses to agriculture and allied sectors (Murthy and Seshasai, 2010).

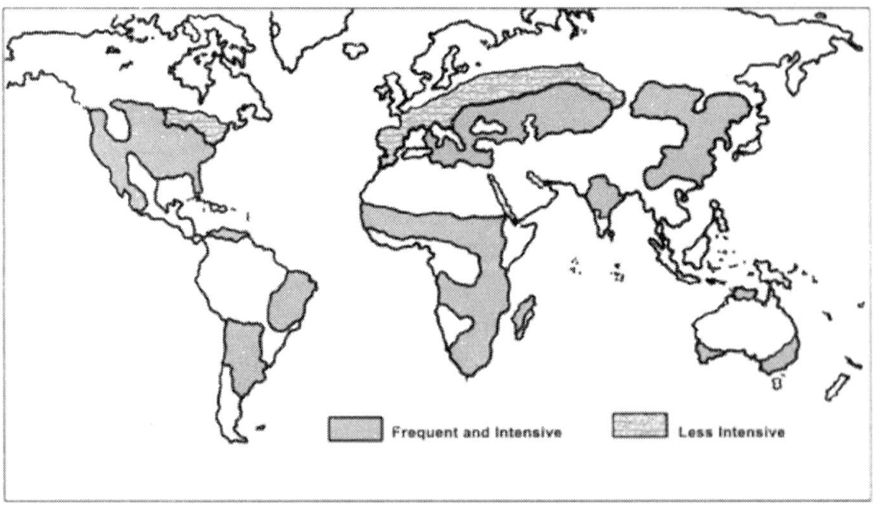

Fig. 1.14: Drought prone areas of the World (Goldsberg, 1972)

The distribution of drought prone countries over the globe (Fig1.14) indicated that nearly 50 per cent of the world's most populated areas are highly vulnerable to drought and more importantly, almost all of the major agricultural lands are located in these regions and more than 500 million people live in the drought prone areas of the world (Murthy and Seshasai, 2010). Since this delineation was done using sparse weather data, it may not represent the true conditions. It would be more realistic to use current weather data sets for identification of drought prone regions over the globe, that would help the international organizations/local governments in prioritizing the allocation of funds for drought relief measures.

1.11.1 Occurance of drought in different countries

Number of drought events occurred over the period 1974-2004 as compiled by the IRI , Columbia University, USA during 2009 using International disaster data base as presented by Daniel Tsegai (2014) at Regional workshop on Capacity development to Support National Drought Management Policies during 6-9 May 2014 at Hanoi, Vietnam is presented below.

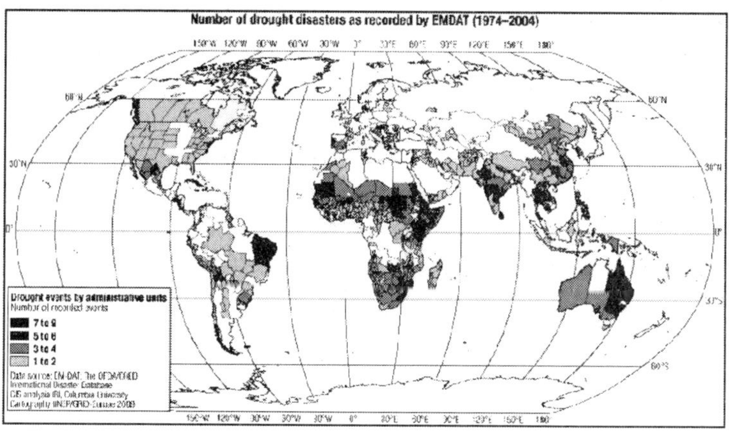

(*Source*: Daniel Tsegai, 2014)

Fig. 1.15: Number of drought events occurred for the years 1974-2004

It can be seen that drought affects almost all climatic regions over the globe during the period 1974-2004.The most affected parts in the countries are located in Africa, Australia, Brazil, Central Asia, China, India, and S.E. Europe and USA. No drought events were noticed in northern Europe, Canada, countries in and around equator in Africa and in south America

In 2004, wide spread drought in much of Asia resulted in loss of agricultural production of hundreds of million dollars. In Thailand, drought hit 70 of 76 provinces and affected more than 8 million people. Southern Chinese island of Hainan suffered its worst drought in 50 years during 2004. Vietnam's eight central highland provinces suffered due to severe drought in 2004.Drought is typical for North American climate and occurs almost every year. USA had experienced many droughts and dry spells in the last century which had significant impact on economic, social and environmental sectors. The 1988 drought ranks as one of the nation's greatest disasters of 20th century. In Europe, drought is a problem in the southern countries along the rim of the Mediterranean region. Over the past 1000 years of Russian history, catastrophic droughts occurred eight to twelve times in every century (Kogan, 2000). In Latin America and the Caribbean, the drought associated with the El Nino has severe impact. The Caribbean countries that are highly vulnerable to drought are Antigua, Cuba, Guyana, Haiti and Jamaica. West Asia is arid and vulnerable to drought and nearly 80% of the region is classified as semi-desert or desert.

1.12. Distribution of Drought Prone Regions in India

The drought prone areas in the country are classified on annual rainfall departures from its normal value which fall either in arid, semi-arid and dry sub-humid regions where droughts occur frequently. The probabilities are high in the arid zone (North-Western India) compared with other sub-divisions. However, droughts occur at random and no periodicity has been noticed. The drought prone districts based on classification of Irrigation Commission of India (1972) which identifies the drought prone area when annual rainfall is less than 75 per cent of the normal in 20 per cent of the years examined and has less than 30 per cent of the cultivated area under irrigation. Out of 325 districts in 1972, a total of 725 talukas in 99 districts have been identified (Table-1.7) as drought prone having an area of 51012 m ha which is approximately one sixth of the total geographic area (Ministry of Water Resources) of the country drought prone districts identified by above criteria are shown in Figure 1.16.

Fig.1.16: Drought prone districts in India (*AICRPAM, CRIDA*)

Table 1.7: Administrative districts frequentely affected by drought

State	Districts
Andhra Pradesh	Anantapur, Chittoor, Cuddapah, Rangareddy, Kurnool, Mehaboobnagar, Nalgonda, Prakasam
Bihar	Munger, Nawadah, Palamau, Rphtas, Bhojpur,Aurangabad, Gaya
Gujarat	Ahmedabad, Amrely, Banaskanta, Bhavanagar, Bharuch, Jamnagar, Kheda, Kutch, Mehsana, Panchmahal, Rajkot, Surendranagar
Haryana	Bhiwani, Gurgaon, Mahendragarh, Rohtak
Jammu &Kashmir	Doda, Udhampur
Karnataka	Bangalore, Belgaum, Bellary, Bijapur, Chitradurga,Chickmangalur, Dharwad, Gulbarga, Hassan, Kolar,Mandya, Mysore, Raichur, Tumkur
Madhya Pradesh	Betul, Datia, Dewas, Dhar, Jhabuva, Khandak, Khargaon, Shahdol, Shahjapur, Sidhi, Ujjain
Maharashtra	Ahmednagar, Aurangabad, Beed, Nanded, Nashik,Osmanabad, Pune, Parbhani, Sangli, Satara, Sholapur
Orissa	Phulbani, Kalahandi, Bolangir, Kendrapada,
Rajasthan	Ajmer, Banaswada, Barmer, Bikaner, Churu, Dungarpur, Hanumangarh, Jaisalmer, Jalore, Jhunjunu, Jodhpur, Nagaur, Pali, Sriganganagar, Sikar, Udaipur
Tamil Nadu	Coimbatore, Dharmapuri, Madurai, Ramanathapuram, Salem, Tiruchirapali, Tirunelveli, Kanyakumari
West Bengal	Bankura, Midnapore, Purulia
Jharkhand	Palamu
Chhattisgarh	Khargaon

1.13 Weather Phenomena Associated with Droughts in India

Droughts are characterized by deficient supply of water mostly due to failure of seasonal rains. The droughts can be due to

1. Late onset of monsoon over a region.
2. Slow movement of monsoon systems over the land.
3. A weak monsoon current that fails to penetrates and brings in sufficient moisture laden air masses.
4. Absence of formation of low pressure and cyclonic systems in the head bay during the rainy period.
5. Break monsoon conditions due to northward shift of monsoon trough over Himalayan region.
6. Early withdrawal of monsoon from the region affecting crop growth and water filling in reservoirs.

Monsoon being a global phenomenon, the teleconnection between the monsoon system and other global weather systems were analyzed by many researchers in the world. Sikka (1999) analyzed about 17 All India level drought years between 1877 to 1987 and worked out morphology of each drought year (Table 1.8) and identified a close relationship with the parameters viz., ENSO, SOI, winter Euro Asian Snow Cover, April 500 h Pa ridge along 75°E, Surface pressure anomaly which is given in the Table 1.8. It is observed that Indian droughts have close association with SOI, Euro Asian snow cover and the position of 500 h Pa ridge along 75°E.

Table 1.8: Atmospheric conditions favourable for droughts in India

Year	Rainfall departure from LPA	ENSO	Southern Oscillation Index(SOI)	Winter Eurasian snow cover	April 500 h Pa ridge along 75°E	Surface pressure anomaly*
1951	-13%	Warm	Negative	Excess	South	Positive
1965	-15%	Warm	Negative	Excess	South	Positive
1966	-12%	Normal	Normal	Normal	Normal	Positive
1972	-25%	Warm	Negative	Excess	South	Positive
1974	-11%	Normal	Normal	Excess	South	Normal
1979	-15%	Normal	Normal	Excess	South	Normal
1982	-13%	Warm	Negative	Excess	South	Normal
1986	-13%	Normal	Normal	Excess	South	Positive
1987	-19%	Warm	Negative	Excess	North	Positive
2002**	-19%	Warm	Negative	Excess	NA	NA

*NW and Central India (Apr-May) **2002 added by AICRPAM, CRIDA

(*Source:* Sikka et.al., 1999)

1.14 Effect of Droughts on Natural Resources

The most typical aspect of drought is that it does not appear all of a sudden. It is a slow moving disaster covering large area. Apart from the natural causes, the antecedent conditions viz., soil moisture, water availability, soil cover etc. have profound impact on the further development of drought. Kulshreshtra (1997) divided the impact of droughts into two categories *viz.*, 1) impacts on environment and 2) impacts on the society, which is diagrammatically given below

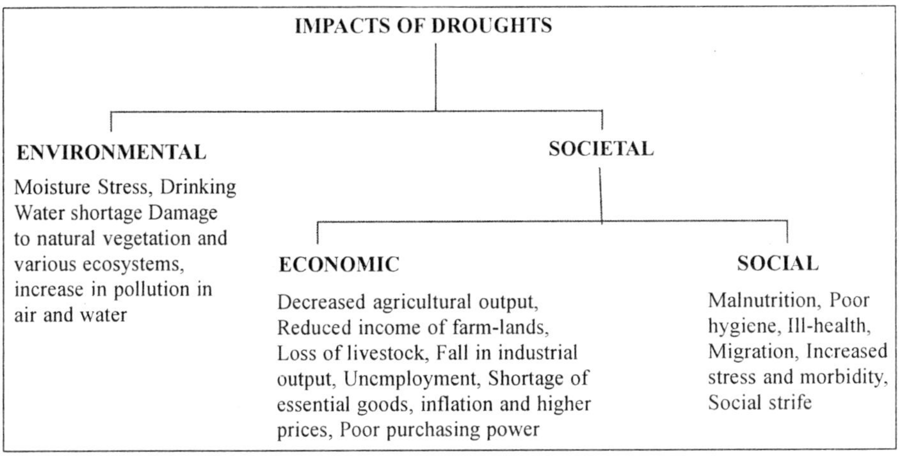

Some of the factors that aggravate the impact of drought are

1. Climate of the area
2. Size of farm holdings
3. Shortage of farm inputs and seeds
4. Shortage of water resources
5. Shortage of power
6. More number of small and marginal farmers
7. Lack of drought management plans
8. Lack of credit facilities
9. Malnutrition and diseases

1.14.1 Surface water resources

The Central Water Commission (CWC) of India has been given task of monitoring the levels of the 71 major reservoirs in the country since 1983. During the month of July 2002, when the monsoons failed, the total storage build-up was about 2% of the full reservoir level (FRL). It was 13% of FRL in the 1987 drought for the same month. (Drought in India, 2008).

As an example, to study the impact of recent droughts on water resources, the water levels in major reservoirs viz., Nagarjuna Sagar, Srisailam, Tungabhadra and Nizam Sagar, Singur, Sri Ramsagar, Kandaleru and Somsila in Andhra Pradesh under Krishna, Godavari and Penna River systems respectively during July, August and September months for the drought years 2002, 2004 and 2009 were collected (Table 1.9) from website "irrigation.cgg.gov.in". The per cent deviation of the gross capacity levels of these reservoirs at the end of rainy season (September) is also shown in the table.

It is observed that the impact of the drought on departures of the reservoir capacity level during the year 2002 was more than the other two drought years viz., 2004 and 2009, in respect of reservoirs constructed across Krishna River. On the reservoir capacities at the end of the season, about 31 to 79% of deficit was noticed in these dams under Krishna during 2002, while it was only 0 to 16 during 2004 and 0 to 34 per cent up to end of August during 2009 respectively. It is also seen that the Sri Ram Sagar (SRS) water reservoirs under Godavari and Penna River Basin Systems were more affected during the drought years under study period compared to the reservoirs of Krishna River Basin Systems. The deficits were 8, 86, and 71 for SRS reservoir, 91, 98, and 71 for Kandaleru reservoir and 88, 82, and 70 for Somsila reservoir for the years 2002, 2004, 2009 respectively.

Drought during 2009 ended abruptly with the receipt of torrential rains in the catchment areas of Krishna basin by end of September and early October 2009 resulted in flash floods and all the dams reached to their peak capacity by the end of September. Therefore the impact of drought on water resources by the end of the season could not be felt though 2009 happened to be a phenomenal drought year. However, the deficits noticed by the end of August 2009 for Nagarjuna Sagar, Srisailam and Tungabhadra reservoirs were 25, 34 and 0%. Similarly, deviations of water levels to its capacity in Jawai Dam in Rajasthan (Table 1.10) for the years 2002, 2003, 2004 and 2009 were obtained from **www.rajasthan.gove.in.**

Table 1.9 : Monthly Major Reservoir Water Levels and per cent deviation to the Gross storage Capacity at the end of Season during drought years (2002,2004 and 2009) in Andhra Pradesh

Basin wise Reservoir Name	Gross Capacity (TMC)	Gross Storage (TMC) 2002			Percent Capa-city (%)	Gross Storage (TMC) 2004			Percent Capa-city (%)	Gross Storage (TMC) 2009			Percent Capa-City (%)
		Jul	Aug	Sep		Jul	Aug	Sep		Jul	Aug	Sep	
Krishna Basin													
Nagarjuna-Sagar	312.04	195.69	198.40	198.01	63	176.27	262.77	262.54	84	190.65	235.00	262.09	84
Srisailam	263.63	31.17	120.06	55.63	21	42.97	254.70	240.55	91	120.98	174.54	263.63	100
Tungabhadra	100.85	25.89	74.19	69.86	69	50.31	111.50	103.68	103	103.75	103.79	104.06	103
Godavari Basin													
Nizam Sagar	17.80	0.09	0.5	0 83	5	0.08	0.52	0.15	1	1.25	1.8	1.66	9
Singur	29.91	14.88	15.94	16.73	56	8.18	8.43	8.14	27	9.3	10.10	12.76	43
Sri Ram Sagar	90.31	22.92	49.45	82.81	92	4.56	4.48	12.53	14	4.41	12.43	26.47	29
Pennar Basin													
Kandaleru	68.03	6.94	6.32	6.2	9	1.45	1.41	1.46	2	22.45	21.78	19.78	29
Somasila	73.89	9.05	9.27	3.97	12	1.59	1.11	13.14	18	14.87	13.14	22.4	30

(*Source*: http://cadarsms.cgg.gov.in/Login.do)

Table 1.10: Monthly (Jul - Sept) Water storage Levels and its per cent deviation during 2002-2004 and 2009 in Jawai Dam - Rajasthan

Designed Capacity(Mm^3)	Jul		Aug		Sep	
	Capacity (Mm^3)	Deviation (%)	Capacity (Mm^3)	Deviation (%)	Capacity (Mm^3)	Deviation (%)
			2002			
207.52	17.77	91	0.32	100	0.3	100
			2003			
	53.29	74	82.22	60	88.24	57
			2004			
	16.45	92	46.87	77	48.03	77
			2009			
	35.28	83	45.39	78	43.51	79

(*Source*: http://wris.rajasthan.gov.in/)

It shows that during the drought years the deviations in the reservoir storage levels with respect to its capacity by the end of season (sept) vary from 57 to all most 100 per cent in these years. The maximum deviation was observed during 2002, followed by 2009 drought years.

The monthly water storage levels (Jul-Nov) along with percent deviations at the end of November with respect to Full Resources Level (FRL) in respect of few reservoirs in Tamil Nadu for the years 2010 to November 2012 are shown in Table 1.11.

As the southern parts of peninsular India receive greater portion of its annual rainfall during October-November (North East monsoon season), the reservoir levels at the end of NE monsoon season have been considered to compute the per cent deviation of current reservoir level with FRL. It is observed that though the years 2010 and 2011 are not drought years, still the deficits in reservoir levels especially of Bhavanisagar, and Periyar dam were 38 to 39 and 41 to 53 per cent respectively during the above years. However, severe deficit rainfall condition existed in catchment area of these river basins up to October 2012 resulted in a deficit 53 to 82 per cent to its FRL. This alarming situation resulted in initiating the agitation of farmers in Tamil Nadu for release of water from the upstream reservoirs constructed in Karnataka state on Kaveri and other rivers.

Table 1.11: Water Storage Levels in Selected Reservoirs in Tamil Nadu during 2010-12 (Jul-Sep)

Name of Reservoir	Capacity at F.R.L (M.Cft)	Jul Water level (M.Cft)	Aug Water level (M.Cft)	Sep Water level (M.Cft)	Oct Water level (M.Cft)	Nov Water level (M.Cft)	% Deviation w.r.t FRL
				2010			
Mettur	93470	43326	49208	36694	28657	91114	3
Bhavanisagar	32800	13481	14358	11291	9327	19888	39
Periyar	10570	3064	2448	2141	2231	4954	53
Parambikulam	13408	5917	8685	9164	10425	12465	7
				2011			
Mettur	93470	45430	41777	49528	53998	72590	22
Bhavanisagar	32800	16936	17098	17345	16253	20461	38
Periyar	10570	2806	2945	2267	3321	6219	41
Parambikulam	13408	13155	13390	13353	12846	12826	4
				2012			
Mettur	93470	36648	38615	36749	29327	-	69 *
Bhavanisagar	32800	1484	1797	1792	5939	-	82 *
Periyar	10570	1627	2213	2303	3024	-	71 *
Parambikulam	13408	7198	6892	6212	6238	-	53 *

* For October Values Only.

(*Source* : http://164.100.167.12/pwd/public_view.php)

1.14.2 Groundwater resources

The Central Ground Water Board (CGWB), with over 15,000 hydrograph stations across the country, is monitoring the country's groundwater levels. The monitoring, usually done four times a year, which is basically a recording of the response of the groundwater regime to natural and artificial conditions, or recharge and discharge. The sustainability of the groundwater table has been a major concern during drought years as well after the drought period. Analysis of May 2002 values with the decadal mean data of May showed that in nine of the 10 drought-affected states, over 50% of wells experienced a drop in groundwater levels. The drought that followed in the subsequent months had considerably depleted storage. In November, the analysis showed that water in more than 55% of wells dropped in all 10 states. In Rajasthan and Gujarat, 22% of wells, showed a decline in ground water table over 4 m. A poor monsoon failed to replenish the aquifers.

1.15 Impact of Latest Droughts on Crop Production in India

Droughts have considerable impact on agricultural production. Continuous droughts during 1965-66 and 1966-67 forced the country to import food grains to avoid famine type conditions and starvation deaths. Similarly all the drought years have recorded decline in production over the previous years. However, no other drought year in the recent past led to such a drop in food production as that of in 2002. Food grain production dipped by 29 million tons (183 million tons) compared to212million tons in 2001(*DAC, Ministry of Agriculture – 2010)*. Over 18 million hectares of cropped area were left unsown during the *kharif* season. The percentage fall of *kharif* crop acreage, as compared to the normal, was highest in Kerala (-59.3%), followed by Rajasthan (-40.9%), Tamil Nadu (-27.3%) and Uttar Pradesh (-19.4%). During the *rabi* season also, the acreage was reduced by 52.1% in Rajasthan followed by Gujarat (-27.9%) and Tamil Nadu (-24.6%).

The fall in kharif food grain production over its mean due to SW monsoon rainfall deficit during drought years is given in the following Table 1.12. Also the departure of monthly rainfall during July for the whole country is also given. It can be observed that the fall in kharif food grain production varies from 5.9 to 19.1% over the drought years considered from 1972 to 2009. The departures of July rainfall over its average for the above years vary around -4 to -49%.

Table 1.12 : Seasonal Rainfall impact on Kharif output in drought years

Year	SWM Rainfall Departure (%)	July month Rainfall Departure (%)	Fall in Kharif Production (% Decline)
1972-73	-24	-31	-6.9
1974-75	-12	-4	-12.9
1979-80	-19	-16	-19
1982-83	-14	-23	-11.9
1986-87	-13	-14	-5.9
1987-88	-19	-29	-7.0
2002-03	-19	-49	-19.1
2009-10	-23	-4	-7.0

Source: DAC, Ministry of Agriculture – 2010

It was further reported by DAC, GoI, that production of rice fell drastically to 71.82 million tons (2002-03) as against 93.08 million tons during the previous year. Pulses fell to a level of 11.31 million tons. As for commercial crops, production of oilseeds declined by 13.7% during the 2002-03 *rabi* season. Cotton and sugarcane also recorded negative growths of 7.7% and 7.2% respectively. *Kharif* production during 2002 was badly affected, as July, normally the wettest month and crucial for a good kharif crop, received the lowest rainfall (-49%). Even the *rabi* production fell by 8%, which shows the vulnerability of irrigated areas also to drought. During 2009-10, the total food grain production fell by about 7 per cent (218.11MT) compared to its previous year 2008-09 (234.47MT). The fall in rice was roughly about 10.0 MT, coarse cereals by 7.5 MT, oilseeds by 3.16MT, cotton 1.95MT compared to its previous year . The Rabi production was good as rains received at the end of September 2009 and beginning of October compensated the fall in production in *Kharif.* Thus, the fall in agricultural production during 2009 was comparatively less when compared to the magnitude of 2009 drought.

1.16 Economic Impacts of Droughts in India

Drought has a direct and indirect impact on the economic, social and environmental conditions in the country. Depending on its spread and intensity it could bring about economic instability. The immediate impact of monsoon failure leading to drought is felt by the agricultural sector. The impact slowly passes on to other sectors, including industry, through one or more of the following routes as given in Drought in India (2008) is as follows.

- A shortage of raw material supplies to agro-based industries.
- Reduced rural demand for industrial/consumer products due to reduced agricultural incomes.

- Potential shift in public sector resource allocation from investment expenditure to financing of drought relief measures.

Most major droughts in India have been followed by recession in growth rate. Annual Gross Domestic Product (GDP) growth was negative in 1957-58, 1965-66, 1972-73 and 1979-80. However, the decade 1990s, reveal a different picture. Unlike in the '50s when a fall in agricultural production had a significant bearing on both industrial and overall GDP, the variability in monsoon did not affect the Indian economy as in the past. Drought-related shocks and their adverse effect on the economy were limited to a mere 1% of GDP in 2002-03 (Drought in India 2008). This can be attributed to the fact that the share of agriculture in national income has gradually declined over the decades – from 57% in 1961 and 35% in 1987-88, to 22% in 2002 and to 14.6% in 2009. Even within the manufacturing sector, the share of agro-based industries has come down from 44% in 1961 to 11.4% in 2002.Another reason why the economy was not affected due to drought was the fact that the country has developed its own financial resources and aquired the ability to self finance the drought relief operations without diverting resources from public investment expenditure in productive sectors.

Apart from the direct impact, droughts influences on a host of areas that can alter the dynamics of rural and urban economies. While losses in crop production are evident (the value of crop production losses during the 1987-88 drought was Rs 24,349 crore), its impacts on the lives of millions are well documented. The drought of 1987, affected 15 states and six UT damaging crops in an area of about 59 M ha spread over 267 districts and approximately 285 million persons and 168 million cattle were affected. The assistance from central government to state government was put as Rs.1471.3 crore (Indu Prakash 1994). Loss in rural employment, during 2002-03, drought year is a major issue for the Government. A rough estimate on unemployment caused by drought was computed from the reduction of coverage in area under various crops. During 2002-03 drought, a total of 18.4 million hectares were left unsown. The man-days required for raising crops range from 70 to 140 days/ha for variety of crops (Drought in India 2008). Assuming that on an average of 100 man-days per hectare, are necessary to grow crops and the total loss in number of man days due to a decline in sown area was estimated to be as 1,840 million man-days. Further the economic loss was estimated by considering the average wage rate for agricultural labour at Rs 60 put the total loss of rural employment to be Rs 35,000 crore. This was without considering the losses from crops. Using this criteria of the total loss estimates in agricultural production in major drought years, ranged from Rs 9,289 crore in 1972-73 to Rs 37,382 crore in 2002-03 (at 1994-95 prices) (Droughts in India 2008).

1.17 Economic Impacts of Droughts in Other Nations

In general, the economic the losses occurred during a drought year focuses mostly on agricultural production losses and do not consider the broad range of impacts resulting from drought. In this context few studies have been done in systematic manner by few countries. Combining of agricultural production losses, indemnity payments and relief costs are often quoted by the media as a loss incurred due to prevailing drought conditions. Drought impacts on agricultural sector can be easily visualized with dried crops, abandoned crop lands, soil drying, withering plants. To understand the full scope of drought on economic impacts and assessment methodologies, a review of literature on the above was done by Ding et.al. (2010). They argue that drought induced water deficiency affects agricultural production, sales and business operations in different industries as direct economic impacts while, indirect economic impacts are derived from the interactions among various sectors of industries. An overview of the drought economic impact as viewed by Ding *et al.* 2010 is given below (Figure 1.17).

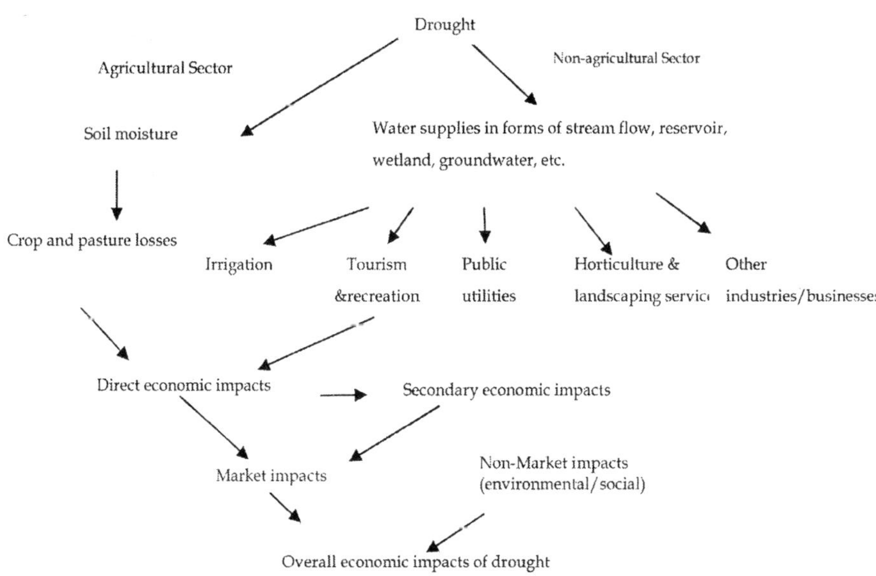

(*Source*: Ding *et al.* 2010)

Fig. 1.17: An overview of economic impacts of droughts

The economic impacts of drought can be studied either through modeling approach or by empirical approach. The commonly used models to estimate the secondary effects viz., interactions and transactions among industries and sectors is Input-Output model (I-O), which is based on interdependences, between industries and the sectors within an economic zone. Another model viz., Computable General Equilibrium (CGE) model, an advanced version of I-O model (Rose, 1995). The empirical approach compares the values of economic variables under drought with those under normal weather conditions. Diersen*et.al.* (2002) examined the economic impact of drought in South Dakota using I-O model as $1.8 billion and later revised to $1.4 billion due to $100 million federal aid and improved market conditions. The Food and Agricultural Policy Research Institute at the University of Missouri, USA estimated the direct losses due to 2002 drought as $251 million in the agricultural sector. Based on the water quality reduction on consumer welfare during drought period in 1990 in Seville (Spain) and the water demand functions, the average welfare losses in 2001 was estimated to be $124.5 per user (Garcia-Valinas, 2006). National level impact of 1988 US drought by Riebsame *et.al* (1991) put it as $39.2 billion which has taken into account the impacts on several sectors viz. agricultural, transportation, power generation, recreation, commerce and industry. Kulshreshta*et.al.* (2003) estimated the economic cost of droughts in Canada during 2001 and 2002 as lost gross domestic product at material level was $2.34 billion and loss of 23,777 jobs. Damages in the Andean community countries (Bolivia, Colombia, Ecuador, Peru and Venezuela) due to the 1997-98 EI Nino was estimated at more than US$7500 million (www.octi.gov.ve).

Horridge*et.al,* (2005) measured the impacts of 2002-03 droughts in Australia. The direct losses from droughts were introduced as output shocks to CGE model and impacts were projected at regional and at national level. They found a reduction of GDP by 1.6% of which 1% was directly related to agricultural sector and the remaining 0.6 was due to multiplier affects.

It was reported in World Conference on Disaster Reduction during 1994, that over the last three decades, droughts have reduced GNP at least by 1 per cent in countries in East Africa, Europe, North America, South America, South Eastern Asia/Australia, Southern Asia and West Africa (www.isse.uear.edu/sadc/chap3). In 1984, drought in Sub-Saharan Africa, the GDP for Mali, Niger and Ethiopia fell by 9, 18 and 17 per cent respectively. Similarly, Zimbabwe GDP fell by 3 per cent after 1983 drought and enormous agricultural losses of about $360 million and $120 million were observed due to droughts. In Niger, (a dualistic economy) the GDP fell by 17 per cent in 1984 with a 35 per cent decline in agricultural sector. The economic recovery may occur relatively quicker in these countries due to simplified sectoral needs (www.isse.ucar.edu/sade/chapter 3). The impact of droughts on the livestock sector in Baluchistan province of Pakistan as per the official sources were of Rs.15 billion. Out of this the Punjab suffered a loss of Rs 5.5 billion, NWFP Rs 4.4 billion,

Baluchistan Rs 3.5 billion and Sindh Rs.1.63 billion (Shafiq and Azamkakar, 2007). Using large scale Rural and Regional family Survey data, estimates of social and economic impacts of drought on families were reported in www.pc.gov.au/data/assests. Using the social definition of drought, drought was estimated to reduce farmer's equivalent household income by $20,483 per annum compared to those families who have not experienced drought in the last 3 years.

References

Amartya Sen. 1981. Poverty and famines: An Essay An entitlement and deprivation; Oxford University Press; Delhi.257pp.

Bhalme, H.N, and D.A. Mooley. 1981. Cyclic fluctuations in the flood/drought area and relationship with the double (Hale) sun spot cycle. J.Appl.Meteor., 20, 1041-1048

Choudhary, A., M.M. Bavedkar and P.S. Raut, 1989.Variability in drought incidence over India-a statistical approach. Mausam, 40, 207-214.

Cocheme, J. and Franquin, P. 1967. An agro-climatological survey of a semi-arid area in Africa South of Sahara.*Tech. Note* 86, *World Met. Org.*, Geneva.

Daniel Tsegai, 2014. Capacity Development to Support National Drought Management Policies, Power point presentation ***in*** Regional Workshop: Asia-Pacific UN-Water Initiative WMO, UNCCD, FAO, CBD and UNW-DPC, 06-09 May 2014, Hanoi, Viet Nam

Davis. Mike. 2001. Late Victorian Holocausts: Elnino Famine and the Making of the Third World, London; New York: Verso Books, ISBN 978-1-85984-739-8.

Diersen, M.A., Taylor, G. and Mary.A. 2002. Direct and indirect effects of drought on south Dakota economy. Economics commentator, 432, August 26, 2002.

Ding Ya., Hayes M.J. and Widhalm Melissa. 2010. Measuring economic impacts of Drought: A. Review and Discussion. Papers in Natural Resources: Paper 196. http://digitalcommons.unl.edu/natrespaeprs/196

Drought in India: Challenges and Initiatives. 2008. Poorest areas civil Society (PACS) Programme 2001-2008. New Delhi.48p.

Gadgil, Sulochana, M. Rajeevan and R. Nanjundiah. 2005. Monsoon predictions: Why yet another failure? Current Science. 88:1389-1400.

Garcia-Valinas, M. 2006 Analyzing Rationing policies Droughts and its effects on urban users welfare. Applied Economics, 38:955-965.

Ghosh, S.P. (ed.) 1990. Agro-climatic zone specific research. Indian perspective under NARP. Indian Council of Agricultural Research, New Delhi, India, 539p.

Goldberg I, Ed., 1972. Agro climatic atlas of the World, Hydrometizdat, 133p.

Government of India. 1987. Agro-climatic regions planning: An Overview. Planning Commission New Delhi, India.

Gowarikar, V., V. Thapliyal, R. P. Sarker, G. S. Mandel, and D. R. Sikka, 1989: Parametric and power regression models: new approach to long range forecasting of monsoon rain in India. Mausam, 40:115-122.

Gowariker, V., Thapliyal, V., Kulshreshtha, S.M., Mandal, G.S., Sen Roy, N. and Sikka, D.R. 1991. A power regression model for long range forecast of southwest monsoon season over India. Mausam. 42(2): 125-130.

Hargreaves, G.H. 1971. 'Precipitation dependability and potential for agricultural production in North-east Brazil'.EMBRAPA and Utah State University Publication No.74-D159, 123p.

Horridge, M. Madden, J. and Willwer, G. 2005. The impacts of 2002-2003 droughts in Australia,

Journal of Policy Modelling. 27:285-308.

InduPrakash. 1994. Disaster Management Societal Vulnerability to natural calamities and Manmade Disasters-preparedness and response. Chapter-3, Rashtraprahari Prakashan. Ghaziabad.

Johna, C.K. Naagauda, R.S. 2002. Bamboo flowering and famine.Current Science. http//www/ias.ac.in/currsci/Feb102002/261.pdf.

Kidson, J.W., Vincent, D.G. and Newell, R.E. 1969. Observational studies of the general circulation of the tropics: Long-term mean values. Quart. J. Roy. Meteor. Soc., 95: 258-87.

Kogan, F. N. 2000. Contribution of remote sensing to drought early warning. In Early warning systems for drought preparedness and drought management, (eds). D.A. Wilhite and D.A. Wood.75–87. Geneva: World Meteorological Organization.

Krishnamurthy V and Shukla J 2000 Intra-seasonal and inter- annual variability of rainfall over India; J. Climate 13:4366—4377.

Krishnan, A. and Mukhtar Singh. 1968. Soil climatic zones in relation to cropping pattern. Proceedings of Symposium on Cropping Pattern, Indian Council of Agricultural Research, New Delhi, 172-185.

Kulshrestha, S.M., 1997. Drought management in India and potential contribution of climate prediction, Joint COLA/CARE Technical Report No. 1, Centres of the Institute of Global Environment and Society, September 1997, 105p.

Kulsreshtha, S.M., Grant, W. Martean, R. and Guenther.E. 2003. Tech. Report. Canadian Drought of 2001 and 2002. Saskatchewan Res. Council.

Kung,E.C and T.A.Sharif.1982. Regression forecasting of Indian Summer monsoon with antecedent upper air conditions. J.Appl.Meteor.19:370-380.

McAlpin, Michelle Burgee. 1979. Death, Famine and Risk: The Changing impact of crop failures in western India 1870-1920, J.Econ. History 39:143-157.

Mckee, T.B., Doesken, N,J. and Kleist, J. 1993. The relation of drought frequency and duration to time scales. Preprints, 8th conference on Applied Climatology, 17-22 Jan, Anaheim, CA 179-184.

Murthy C.S. and Seshasai, M.V.R. 2010.Agricultural Drought Monitoring and Assessment in : Remote sensing Applications,Eds. P.S.Roy, R.S.Dwivedi, and D.Vijayan. National Remote Sensing Centre, Hyderabad.303-330.

Pant, P.S. 1964. Onset of monsoon over India. Ind. J. Meteor. Geophys., 15(3): 375-380.

Papadakis, J. 1975. Climates of the world and their potentialities. Buenos Aires.

Parthasarathy, B., A.A. Munot and D.R. Kothawale, 1995.Monthly and seasonal rainfall series for all-India, homogeneous regions and meteorological sub-divisions. IITM Research Rep. No,.RR 065.

Penman, H.L. 1948.Natural evaporation from open water, baresoil and grass.Proc. Royal Soc. of London Series A. 193: 120-145.

Rajeevan. M., D.S. Pai, S. K. Dikshit, and R. R. Kelkar, 2004: IMD's new operational models for long-range forecast of southwest monsoon rainfall over India and their verification for 2003. Curr. Sci., 86(3): 422-431.

Ramakrishna, Y.S. G.G.S.N. Rao, KesavaRao, AVR. And Vijaya Kumar, P. 2000. Weather Resources Management, Chapter -3 in Natural Resource Management for Agricultural Production in India, eds J.S. P. Yadav and G.B.Singh. Int. Conf. on managing Natural Resources for sustainable Agricultural production in 21st Century. Feb 14-18, 2000. New Delhi. 245-370.

Rao, K.N., C.J,George and Ramasastri, K.S. 1972. Agroclimatic classification of India. Agric. Met. Sci. Rep. No.1, IMD, Pune.

Rao. Y.P. 1976. Southwest monsoon: India Meteorological Department Meteorological Monograph, Synoptic Meteorology, No.1/1976. Delhi, 367p.

Rao., G.G.S.N. and Rajput, R.K. 1986. Rice yield during kharif season as influenced by weather. In: Proc. Int. Sym. On Water Management in Arid and Semi-arid Zones, CCS Haryana Agricultural University, Hisar.100-108.

Report of Irrigation Commission. 1972. Ministry of Irrigation and power, over Delhi. Vol-1-3

Riebsame, W.E., Changnon, S.A., Jr and Karl.T.R. 1991. Droughts and Natural Resources management in United States, Impacts and Implication of the 1987-89 drought. West View pren, Boulder, 10:174p.

Rose, A. 1995.Input-output Economics and computable General equilibrium models. Structural change and economic Dynamics, 6:295-304.

Roy, Tirthankar. 1999. Traditional Industry in the economy of colonial India, Cambridge University Press.

Sehgal, J.L. and Mandal, D.K. 1996.Agro-ecological regions of India and Climate Change. In: Climate variability and Agriculture. [Y.P. Abrol, S. Gadgil and G.B. Pant, eds.], Narosa Publishing House, New Delhi, India, 204-222.

Sehgal, J.L. Mandal, D.K, Mandel, C. and Vadivelu, S. 1990. Agroecological regions in India, Publicaiton 24: , NBSSLUP, Nagpur 130p.

Shafiq M. and M. Azam, Kakar. 2007. Effects of drought on livestock sector in Balochistan province of Pakistan. Int. J. Agri. Biol. 9(4): 657-665.

Sikka, D. R and Sulochana Gadgil. 1980. on the maximum cloud zone and the ITCZ over India longitude during the southwest monsoon; Mon.Wea.Rev.108:1840-1853.

Sikka, D.R. 1996. Monsoon variability and its forecasting for agriculture. In: Climate Variability and Agriculture. [Y.P. Abrol, S. Gadgil and G.B. Pant. eds.], Narosa Publishing House, New Delhi, India, 26-50.

Sikka. 1999. Monsoon drought in India, Report no.2, Joint Report of COLA/CARE Tech. Report, 270p.

Subramanyam, V.P. 1956. Climate types of India according to the national classification of Thronthwaite. Indian J. Met.Geophy.7:253.

Thornthwaite, C.W, 1948. An approach towards a rational classification of climate. Geogr.Rev. 38:85-94.

Thornthwaite, C.W, and Mather, J.R. 1955. The Water Balance.In Climatology, VIII (1) Drexel Institute of Technology, New Jersey, USA, 104.

Trewartha G.T and L.H. Horn. 1980. An Introduction to Climate. Int. Student Edition. (5 Edition) McGraw Hill International Book Company, 416p.

Troll, C. 1965. Seasonal climates of the earth.In World Maps of Climatology. (Eds) E, Rodenwaldt and H. Jusatz.Springer-Verlag, Berlin, 28p.

Walker, G.T, 1923. Correlations in seasonal Weather, VIII, Mem.Ind.Met.Department. 24: 275-282.

Walker, G.T. 1924. Correlation of seasonal variations of weather, the local distribution of Monsoon rainfall. Mem.Ind.Met.Dept.21, 30-40.

Wang B, Webster P J and Teng H. 2005. Antecedents and self-induction of active-break south Asian monsoon unraveled by satellites; Geophys.Res.Lett.32 doi:10.1029/2004. GL 020996.

Webster P J, Maga V, Palmer T N, Shukla J, Tomas R A, Yanai M and Yasunari T 1998 Monsoons: Processes, predictability, and the prospects for prediction; J. Geophys. Res. 103(14):451-510.

WMO, 2006. Drought monitoring and early warning: concepts, progress and future challenges. World Meteorological Organization.WMO.No.1006. 24p.

Yin, M.T. 1949. A synoptic aerological study of the onset of the summer monsoon over India and Burma J. Meteor: 6: 393-400.

□□□□

CHAPTER 2

Drought Assessment and Vulnerability

2.1 Definition of Droughts

Drought occurs when there is less rainfall than normal over an extended period of time, usually a season or more. It is a temporary condition even though it may extend to longer periods. Human activity increases the impact of drought due to higher water use when the supply through rainfall falls below its normal. It can occur in both high and low rainfall regions. The deficit rainfall conditions increase the air temperature and decrease the relative humidity. A strong wind regime associated with the above conditions increases the water demand considerably both for plant population and human beings. As drought affects many economic and social sectors, the approaches taken to define drought reflect the regional differences. Also the impacts due to drought that are governed by physical, economic and social resources differ from one location to the other. Hence, several definitions have been developed over the period of time by various of disciplines. Wilhite (1985) has reviewed the literature and identified about 150 definitions of droughts.

2.2 Types of Droughts

Drought definitions are categorized as conceptual or operational. Conceptual, referring to those definitions formulated in general terms as given in dictionaries such as "long period with no rain during crop season" which does not provide any guidance for real time assessment. The operational definitions identify various phases of drought like its onset, movement, severity and the termination (WMO. 2006). These definitions provide the rate of loss of soil moisture by considering the daily rainfall and evapotranspiration that are used in drought studies for identification of historical droughts and also for assessing current drought conditions.

Droughts were also defined disciplines-wise, *viz.*, meteorological, climatological, atmospheric, agricultural, hydrological and water management (Subrahmanyam, 1967). Economic and socio-economic sectors were also included for determination of drought occurrence (Gibbs, 1975). Though number of definitions for drought is available, it is now broadly categorized into four types, *viz.*, Meteorological, Agricultural, Hydrological and Socio-economic which have international acceptance (WMO, 2006).

2.2.1 Meteorological Droughts

The definition of meteorological droughts was formulated solely on the basis of intensity of dryness and its duration. Some of the definitions developed in various countries in the world (Wilhite, 1985) are as follows:

United States	→	Less than 2.5 mm of rain in forty eight hours (Blumenstock, 1942)
Britain	→	Fifteen days, none of which received as much as 0.25 mm of rainfall (British Rainfall Organization, 1936)
Libya	→	When annual rainfall is less than 180 mm (Hudson, 1964)
India	→	Actual seasonal rainfall is deficient by more than twice the mean deviation (Ramdas, 1960)
Bali	→	A period of six days without rain (Hudson, 1964)

Definition developed for one region is not applicable to another on meteorological conditions that results into drought since the perception conditions of these regions are variable.

Other definition of drought compares the degree of dryness to a long-term average or normal. Palmer (1957) defines drought as temporary departure of climate from the average climate towards drier conditions. The Palmer Drought Severity Index (PDSI) developed by Palmer (1965) relates the drought severity to the accumulated weighted differences between actual precipitation and the precipitation needs of evapotranspiration. It evaluates the periods of abnormal wet or dry weather. Monthly PDSI maps are published by the USDA in its Weekly Weather Crop Bulletins. It is also used for monitoring drought situations over US by the Drought Monitoring Network. In Australia, classification of droughts was studied using the concept of deciles of precipitation (Gibbs and Maher, 1967). The monthly and annual rainfall totals are ranked highest to lowest and the decile ranges are computed from the cumulative frequencies of the distributions. The first decile shows the rainfall values in the lowest 10 percent distribution while the tenth decile range would represent the highest 10 percent of the rainfall values in the distribution. Using the above criteria, Australian Drought Watch System (Lee, 1979) classified the intensity of droughts as severe when the rainfall over a 3-month period does not exceed the fifth decile range.

Some meteorological definitions of droughts were also developed using atmospheric parameters other than rainfall. Popov (1948) used wet bulb depression and Ivanov (1948) included humidity and temperature as indicators of drying power of atmosphere. Some others, *viz*., Levitt (1958) defined drought as proportional to vapor pressure deficit. However, these definitions have not become popular among drought researchers for wider applicability.

The meteorological droughts which are widely used are dependent on the departures of annual rainfall with respect to its long-term average. India Meteorological Department defines a drought year as moderate when the annual rainfall is deficit by 25-50 per cent of normal and severe when the deficit exceeds 50 percent of its normal annual value.

2.2.2 Agricultural Droughts

The definitions of Agricultural Droughts identify various meteorological drought conditions that affect the agricultural crops thereby the economic sector of the country. Those countries where agriculture contributes major share to its GDP are the worst affected and many of the developing nations including India are facing great threat in terms of food security to its large population. As per the IMD classification, Agricultural Drought occurs when weekly rainfall was less than half of its normal (>5mm) consecutively for a four week period during May to October. It signifies the periods of availability of soil water at different crop stages to support healthy crop growth resulting in better crop yields. The water deficit due to absence of rainfall affects the annual crops differently. Shallow rooted crops are the worst affected compared to deep rooted crops. However, prolonged period of water deficit affect the deep rooted crops also. It is also important to identify the occurrence periods of water deficit at different crop growth stages. For example, deficient sub-soil moisture at an early crop stage will have little impact on final yield, if sufficient soil moisture is available in the top soil (0-20 cm depth). If the deficiency in sub-soil continues, yield losses are very high. Similarly, if the crop had sufficient soil moisture up to vegetative phase and if water deficit conditions occur at crucial growth stages, *viz*., reproductive and physiological maturity, it is likely to affect the final crop yield. Based on distribution of rainfall over the crop season, the following diagram (Figure 2.1) identifies the early middle and terminal droughts for pearl millet crop in Jodhpur.

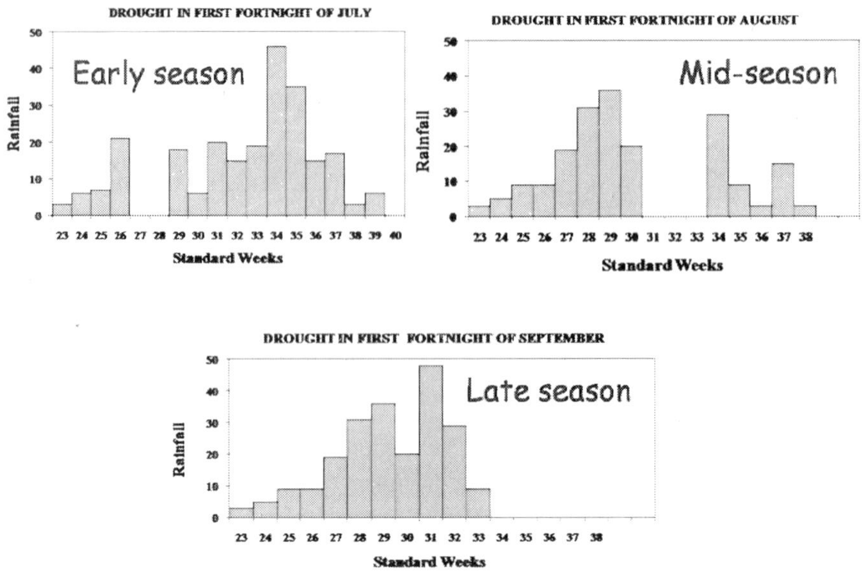

(*Source*: Ramakrishna, et al., 2000b)

Fig. 2.1: Weekly rainfall distribution in different Agricultural drought situations for pearl millet crop in Jodhpur

In 1968 Palmer modified the PDSI to represent Agricultural Drought situations. He has computed Crop Moisture Index (CMI), which defines the drought in terms of ET deficit. This deficit is the difference between the actual and expected weekly ET. The expected ET varies according to weekly temperature departures from the normal. Weekly CMI maps are being published by USDA. Heathcote (1974) defined Agricultural Drought as shortage of water harmful to man's agricultural activities.

2.2.3 Hydrological Droughts

It is normally defined by the departure of surface and sub-surface water supplies from some average conditions at various points in time (WMO, 2006). Unlike agricultural droughts, it has no direct relationship with precipitation amounts received locally and the status of water supplies in reservoirs, streams, etc. There is a time lag between departure of rainfall and the point at which these deficiencies become evident. Recovery of these deficiencies is time consuming due to long recharge periods. They are often out of phase with both meteorological and agricultural droughts (WMO, 2006). The definitions of Hydrological droughts are developed based on the effects of dry spells on surface and sub-surface hydrological systems (Wilhite, 1985). Linsely (1975) defined Hydrological Droughts as periods during which stream flows are inadequate to supply established users under a given water management system. Whipple (1966) defined a drought year as one in which the aggregate runoff is less than the long-term average runoff. A Surface Water Supply Index (SWSI) was developed during 1981 by the University of Colorado to provide information on drought condition in higher elevation river basins that are dependent on snow melt as major source of water supply (Denzman *et al*, 1982). It integrates the historical stream flow data with current figures on reservoir storage, stream flow and rainfall into a single index number. Drought assessment and action plan is implemented if SWSI and PDSI values exceed specified threshold values. For example, SWSI value of -1.0 alerts the Water Availability Taskforce of Colorado for assessing the current situation and prepare action plans.

2.2.4 Socio-economic Droughts

It differs markedly from other types of droughts as it reflects the relation between supply and demand for some commodity or economic goods such as water, livestock, forage, food grain availability, and hydroelectric power (WMO, 2006). Gibbs (1975) defined it as interaction between agricultural activities and natural events which are dependent upon the distribution of plant, animal and human population, their life style and their land use pattern. Supply is a function of rainfall which often fluctuates. Similarly, the demand also varies annually which often shows a positive association due to increasing population and their requirements. The inter-relationships as envisaged by WMO between these types of droughts is shown in the following Figure 2.2.

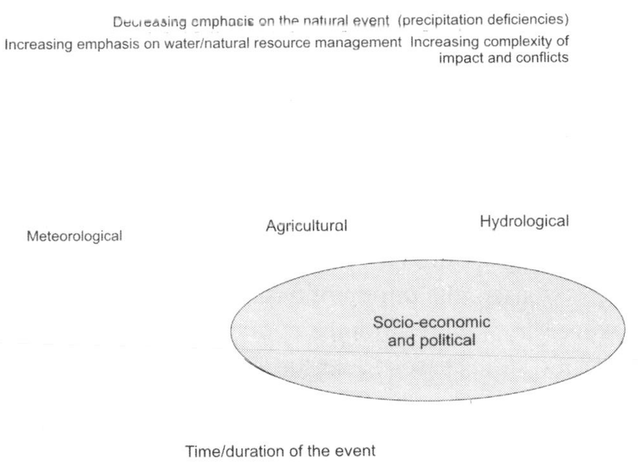

Fig. 2.2: Interrelationships between meteorological, agricultural, hydrological and socio-economic droughts. (WMO - 2006)

Hydrological and socio-economic droughts occur less frequently than meteorological and agricultural droughts as impacts in these sectors are related to the availability of surface and sub-surface water supplies (WMO, 2006). Changes in the management of water supplies can either decrease or increase the impacts of droughts. An appropriate soil and water conservation measures and planting of drought resistant varieties may reduce the impacts of droughts considerably.

In some cases, the land utilization pattern can create drought like situations or making the existing ones worse. The Dust Bowl years in 1930, the Sahelian droughts in West Africa in early 1970, Ethiopian droughts in later part of 1970 are some of the examples of symbiosis between drought and human activities (Wilhite, 1985). The worst Indian droughts during 1965 and 1966 have eroded the food resources and the country had to depend on imports that had put greater pressure on the economic front. Droughts should not only be linked to precipitation (supply) but also to the trends or fluctuations in demand as well as to other factors other than weather which influence food supply (Stanford, 1979).

2.3 Criteria for Classification of Droughts

Thornthwaite (1947) classified droughts into four kinds, *viz*., permanent droughts, seasonal drought, contingent droughts and invisible droughts.

- ***Permanent droughts*** are the characteristic of direct climate.
- ***Seasonal droughts*** are found in climates which have well defined rainy and dry season and most of arid and semi-arid regions fall in this category.
- ***Contingent droughts*** refer to irregularity of rainfall, which is not regular to occur in any definite season.
- ***Invisible droughts*** occur in humid areas where rains do not supply enough water to counteract the water loss by ET.

Three characteristics, of droughts are *viz*., intensity, duration and spatial coverage. The severity of drought impact is associated with the shortfall in rainfall within the season, the timing of onset and withdrawal of rainy season. The areas covered by severe drought evolve gradually and the maximum intensity centre shift from season to season and year to year. To evolve suitable computational methods for identification of seventy of droughts, several researchers and organizations over globe developed procedures using basic weather parameters and their deviations over its mean. Since soil moisture and temperature affect the agricultural crops, climate derivatives computed from water balance and cumulative temperature differential of air and canopy temperature over a period were used in deriving methods for computing agricultural droughts. Remote sensing derived parameters were also put into use for defining drought conditions over a region in estimating the water stress on crops.

Based on the rainfall departures alone, the Drought Research Wing of IMD categorized droughts as follows:

Drought week	→	When rainfall in a week is less than half of its normal
Agricultural droughts	→	When four drought weeks (normal > 5 mm/ week) occur consecutively during mid May-October
Seasonal droughts	→	When seasonal rainfall is deficient by more than 2 σ
Drought year	→	When annual rainfall is deficient by 25 percent or more of the normal value
Severe drought year	→	When annual rainfall is deficient by 50 percent or more of the normal

Meteorological Drought

The criteria followed in India for identification of different scales of meteorological droughts (Kulshreshta, 1997) based on rainfall departures are given below:

Scale of drought Spatial	Temporal	Definition of drought in terms of % departure of rainfall from long-term average
Meteorological sub-divisions	Weekly andMonthly	Less than 25%
Meteorological sub-divisions	Seasonal	Less than 25%
Group of homogeneous neighbouring met sub-divisions, *viz.*, northwest, Peninsular India, etc.	Weekly andMonthly	Less than 25%
Group of homogeneous neighbouring met sub-divisions, *viz.*, northwest, Peninsular India, etc.	Seasonal	Less than 10%
India as a whole	Seasonal	Less than 10%

India Meteorological Department declares a year as All India Drought Year when the average annual rainfall of the current year falls below 10 percent of long-term mean annual rainfall and the area affected by drought conditions (either moderate or severe or combined) must be above 20 percent of the total geographical area of the country. The Irrigation Commission of India identified a region as drought prone when 20 percent of data recorded years experienced meteorological droughts over the period. Similarly, chronically drought prone regions are identified when more than 40 percent of the recorded years experience meteorological drought. Subrahmanyam, V.P. (1964), defined the drought intensities based on the departures Aridity Index (Ia) which is defined as the ratio of Water Deficit (WD) to the Water need (WN) where WD = AE-PE and WN=PE. AE and PE are actual and potential evapo-transpirations computed from Water balance method of Thornthwaite. Using the standard deviation values (σ) of Ia as given below he had analyzed the occurrence of droughts over India.

Departures of Aridity Index (Ia) from normal		Drought intensity
Less than ½ σ	→	Moderate
½ to σ	→	Large
σ to 2σ	→	Severe
$> 2\sigma$	→	Disastrous

For arid regions of Rajasthan, Krishnan and Thanvi (1971) classified drought intensity using the *kharif* season Aridity Index (Ia) as 90-95 percent of annual rainfall over greater part of the country accounts in this season.

Departures of Aridity Index (Ia) from normal		Drought intensity
50-60	→	Slight
60-70	→	Moderate
70-80	→	Severe
>80	→	Disastrous

Krishnan and Thanvi (1972) computed the normal crop growing season for different regions of the country with nil or slight water stress under rain fed farming using the monthly water balance approach, Thornthwaite and Mather (1955). The period of nil or slight water stress under rainfed conditions was defined as when AE > PE/2 where AE and PE are actual and potential evapotranspirations.

Moderate stress or drought occurs

When AE lies between $\frac{1}{4}$ and $\frac{1}{2}$ of PE values $\left(\frac{PE}{2} > AE \geq \frac{1}{4}PE\right)$.

The period of severe drought occurs

When AE $< \frac{1}{4}$ PE values.

In Australia, the limits of the decile values derived from the cumulative frequency curve are used in drought monitoring studies. These are developed to alert the nation and monitor its progress. The criteria used by Australian Weather Watch System is as follows, (Michale I. Caughlan, 1981):

- A severe deficiency exists at a location if the rainfall for a period of three or more months does not exceed the fifth percentile value for that period.
- A serious deficiency exists if the rainfall total lies between the fifth percentile and the first decile.
- Rainfall is considered to be average if it exceeds the third decile and is less than seventh decile.

In USA, weekly drought maps are prepared and displayed in US Drought Monitoring Website which summarizes information from several systems on to a single and easy to read color maps. Drought intensity categories are based on five key indicators, *viz*., (1) Palmer Drought Index, (2) CPC Soil Moisture Model Output, (3) USGS Weekly Stream Flow, (4) Standardized Precipitation Index (SPI), and (5) Short and Long Term Drought Indicator Blends and numerous supplementary indicators. Short-term drought indicator blends focus on 1-3 month precipitation. Long-term blends focus on 6-60 month precipitation. Additional indices used, mainly during the growing season, include the USDA/ NASS Topsoil Moisture, Keetch-Byram Drought Index (KBDI), and NOAA/ NESDIS satellite Vegetation Health Indices. These indices used primarily during the snow season and in the West include snow water content, river basin precipitation and the Surface Water Supply Index (SWSI). Other indicators include groundwater levels, reservoir storage and pasture/range conditions. The following drought severity (Table 2.1) adopted by US Drought Monitoring Cell shows the range of indicators for each dryness level. Though the ranges of various indicators often tally, the final drought category is computed based on the majority of the indicators. Also weightage to indices is given based on how well they performed in various parts of the country during different times of the year.

Table 2.1 Ranges of various drought indicators for different drought categories in USA

Drought category	Palmer Drought Index	CPC Soil Moisture Model (percentile)	USGS Weekly Stream Flow (percentile)	Standardized Precipitation Index (SPI)	Short and Long-term Drought Indicator blends (percentile)
D_0	-1.0 to 1.9	21-30	21-30	-0.5 to -0.7	21-30
D_1	-2.0 to -2.9	11-20	11-.20	-0.8 to -1.2	11-20
D_2	-3.0 to -3.9	6-10	6-10	-1.3 to -1.5	6-10
D_3	-4.0 to -4.9	3-5	3-5	-1.6 to -1.9	3-5
D_4	-5.0 or less	0-2	0-2	-2.0 or less	0-2

D_0 = Abnormally dry; D_1= First stage drought; D_2 = Severe drought; D3 = Extreme drought; D_4 = Exceptional drought

(*Source* www.droughtmonitor.unl.edu/classify.htm)

China monitors its drought at 10-day interval by computing a comprehensive index (CI) which is a function of the SPI values and the corresponding potential evapotranspiration values. Based on the CI and the soil moisture monitoring from Agro met Stations and Remote Sensing indices, a number of drought monitoring products were developed and are in use to monitor droughts.

South African Weather Service (SAWS) uses SPI in assuming the severity of drought. SPI values less than zero indicate the severity of drought conditions. Palmer Drought Severity Index (PDSI) is used to characterize the droughts in Portugal. Many of the nations located in South and South East Asian uses the rainfall departures from its normal as criteria for characterizing the droughts.

Agricultural Droughts

For agricultural planning purposes, a knowledge on probability of occurrence of agricultural droughts over a region is of paramount importance. It can be defined as situation when both soil moisture and rainfall are inadequate to support a healthy crop. Crop production under rainfed conditions is influenced by various intensities of drought experienced at different crop growth stages. Therefore, knowledge on the frequency of agricultural droughts of varying intensities is essential for developing suitable crop management technologies for increased and sustainable production. The complexity in defining the magnitude of agricultural drought is due to (1) diversity of crops and cultivars grown in a given location,(2)Wide variability in sowing time (3) variability in soil and its fertility, (4) spatial variability of rainfall, (5) delay in timely operations of agriculture and (6) diversity in crop management practices.

The definition of Agricultural droughts as per National Commission on Agriculture (1976) is that a period in which at least four consecutive weeks receiving less than half of normal rainfall during *kharif* season and six such consecutive weeks in *rabi* season. The normal rainfall of the week should be above 5 mm for consideration of the rainy season.

For this purpose, long term weekly rainfall data (52 weeks) is essential to compute the mean weekly rainfall values and then workout half of the mean values for the station. Using a simple computer program or in excel spread sheet, the occurrence of consecutive periods of weeks whose rainfall less than half of normal values can be identified and the periods of agricultural droughts coinciding with crop stages can be worked. This computation should be restricted only to cropping season. In India, the crop season commences with the onset of rains (June) and normally ends with the withdrawal of monsoon (September). In the southern parts of country, the season extends up to end of November due to occurrence of NE monsoon. Within a year there is a possibility that agricultural drought could occur more than once. A simple statistical analysis can be carried out to compute the frequencies and their probabilities of occurrences of droughts at different growth stages.

Various approaches to quantify the intensity of agricultural drought have been attempted by many researchers. Apparao *et al* (1981) utilized the departures of weekly aridity index week by week as criteria for classifying agricultural droughts. Victor *et al* (1988) indicated the possibility of using Water Requirement Satisfaction Index (WRSI) for drought monitoring of field crops.

The impact of drought depends on the phenological stage of crop as the water requirement of a crop varies during different phenophases. Hence, a novel method of classification of agricultural droughts was attempted at Central Arid zone Research Institute (CAZRI), Jodhpur (Sastri *et al*, 1981; Ramana Rao *et al*. 1981; Sastri *et al*., 1982) by considering the weekly values of AE/PE which is termed as Moisture Adequacy Index (MAI) derived from computations of weekly water balance of Thornthwait and Mather (1955) method during different phenophases of crop growth as follows:

AE/PE (%) during different phenophases	Droughtintensity	Code		
		Seedling	Vegctative	Reproductive
76 to 100	Nil	S_0	V_0	R_0
51 to 75	Mild	S_1	V_1	R_1
26 to 50	Moderate	S_2	V_2	R_2
25 or less	Severe	S_3	V_3	R_3

(*Source*: Ramana Rao *et al*., 1981)

The suffix associated with crop stage denotes the intensity of drought. Depending upon the average values AE/PE during different phenophases, the drought code varies. This is a generalized classification without specification of any crop. With the combination of occurrence of different intensities, a unified drought code viz., A_0, A_1, A_2, A_3 representing nil, mild, moderate and severe drought respectively was proposed and a ready reckoner table showing the intensities of droughts at Seedling (S), Vegetative (V) and Reproductive (R) as reported by CAZRI as follows:

Table 2.2: Intensity of Agricultural Drought based on Moisture Adequacy Index

No drought (A_0)	Mild (A_1)	Moderate (A_2)	Severe (A_3)
$S_0V_0R_0$	$S_0V_0R_2$	$S_0V_0R_3$	$S_0V_2R_3$
$S_0V_0R_1$	$S_0V_1R_2$	$S_0V_1R_3$	$S_0V_3R_2$
$S_0V_1R_1$	$S_0V_2R_0$	$S_0V_2R_2$	$S_0V_3R_3$
$S_1V_0R_0$	$S_0V_2R_1$	$S_0V_3R_0$	$S_1V_2R_3$
$S_1V_0R_1$	$S_1V_0R_2$	$S_0V_3R_1$	$S_1V_3R_3$
$S_1V_1R_0$	$S_1V_1R_1$	$S_1V_0R_3$	$S_0V_2R_3$
$S_0V_1R_1$	$S_1V_1R_2$	$S_1V_1R_3$	$S_2V_0R_3$

(*Source*: CAZRI)

Rao, G.G.S.N. (1986) slightly modified the above criteria to be used for the natural grasslands in arid parts of NW India in which the life cycle duration for grasses varies from 10, 12 and 14 weeks in the rainfall zone of less than 200, 200-400 and > 400 mm per year. These grasslands supports large livestock population and the economy of the animal husbandry from these arid parts are influenced by the productivity of natural grasslands. The water demands of the grasses are comparatively less over crops and they could survive also under low water availability condition. The following is the drought intensity criteria for grasslands in NW India.

$\frac{AE}{PE}$ (%) at different growth stages	Drought intensity	Code			Drought Code
		Sprouting (S)	Vegetative (V)	Reproductive (R)	
> 50	No	S_0	V_0	R_0	A_0
26-50	Mild	S_1	V_1	R_1	A_1
16-25	Moderate	S_2	V_2	R_2	A_2
< 15	Severe	S_3	V_3	R_3	A_3

Similar to the combinations of occurrence of drought intensities at different growth stages as mentioned in Table 2.2 a unified drought code viz, A_0, A_1, A_2 and A_3 can also be worked out for forage grasses.

Hydrological Droughts

Hydrological droughts are measured by analyzing the minimum flow of rivers. Benzvi (1987) defined drought index as annual runoff volume less than one standard deviation below the mean, i.e, SAI <-1, where the annual runoff remains below the mean value for four years or more.

Droughts are stochastic in nature and forecasting accurately is not possible. Therefore, stochastic and probability theories are used in describing the drought characteristics and evaluation of uncertainties. Though there are number of methods, *viz.*, probability analysis, cluster analysis, periodic stochastic modeling auto regressive models, Fuzzy rule-based modeling and etc. it may not be possible to provide the procedure for computation as it is beyond the scope of the present work. However, as an example, a simple probability analysis as reported by Nagarajan (2003) is presented here. The readers are advised to refer Nagarajan (2003) for further reading on this topic.

Probability Analysis

The monthly stream flows sequence is given as by Q11.......Q1s, Q21......Q2s....QN1......QNs. where S the number of months =12 and N is number of years of data. The drought characteristics are to be found for a return period of T years ((S*T) months) and corresponding to a truncation level of (TL), where monthly truncation level is expressed as percent of mean monthly flow. For a given sequence of length (N*S), which is split into number of overlapping subsequences (Nos) for each of length (S*T) months, the critical drought characteristics of each of these overlapping sequences are determined corresponding to the pre-selected TL and T then the cumulated and averaged

$$DD_{ave=} \frac{\left(\sum_{I=1}^{Nos} (DD)_1\right)}{Nos}$$

$$DS_{ave=} \frac{\left(\sum_{I=1}^{Nos} (DS)_1\right)}{Nos}$$

Where DD_{ave} and DS_{ave} are the average critical drought duration and average critical deficit sum respectively.

Commencement of drought due to deficit conditions or surplus inflows could be determined from the weather balance algorithm of Herbst *et al* (1966).

2.4 Data Requirements in Drought Analysis

Primarily drought analysis is carried out with rainfall data alone. Majority of the drought indices developed in the world use rainfall accumulated for a specific period (weekly, monthly, seasonal or annual) and analyze the deviation of the totals with respect to Long Period Averages (LPA). Some indices use the ratio between rainfall and the potential evapotranspiration (PET). Hence, for computation of PET, temperature and other weather parameters like vapour pressure, wind speed, radiation, and sunshine hours are required. Therefore, for computation of Meteorological droughts, preferably the daily values of rainfall along with other meteorological parameter recorded for a long period (minimum of 30 yrs) is essential.

Agricultural drought analysis involve computations of water balance to derive parameters like Soil Moisture Index (SMI), Moisture Adequacy Index (MAI) and Aridity Index (AI) for identification drought periods and their intensity at different crop growth stages. The water balance computations of Thornthwaite and Mather (1955) and Frere and Popov (1979) which is popularly known as FAO Method require information on rainfall and PET, soil water holding capacity and the cropping pattern grown in the study area. As water balance computation for drought studies are done at weekly intervals, data on weekly total rainfall, PET along with initial soil moisture level at the beginning of year is required in the data series. If droughts are to be computed for specific crop, then the crop coefficient (Kc) values are essential to convert the PET values to ETo (Crop Specific evapotranspiration values) which would work as input in the water balance models.

To analyze the impact of droughts on agricultural crop production, crop statistics (area, production and productivity) along with the doses of fertilizer and irrigation frequency and amount are necessary in analyzing the impacts critically.

Water level data in reservoirs and dams, rainfall in the upper catchments areas, ground water status, snowfall in higher ranges, snow melt data are necessary for evaluating the Hydrological drought conditions. For regulating the water supply to crops during drought conditions, information on planting schedule and area of water stress of major crops raised under main canal command is desirable. Also the daily rainfall values received in the command area is needed for regulating the water supply at distributor level. Similarly, the drinking and other water requirements data of urban population is also essential to be collected for management of droughts.

As all the droughts affect ultimately the human population directly or indirectly, for assessing the drought on socio-economic front, data on the following parameters are needed.

- Land use Statistics
- Status of farmer (Small, marginal and big)
- Live stock population and fodder requirements
- Literacy status

- Infra structure developments
- Availability of financial institutions
- Developmental plans initiated by Governments (Watershed programs)
- Adaptation level to improved crop management practices
- Ground water availability and its utilization.
- Health and nutritional status of rain fed farmers.

2.5 Drought Indices

For monitoring and assessment of various types of droughts conditions at different scales, most important is the identification of suitable methodology and computation procedure based on the availability of data and its applicability over a large area.

Various methods are available in literature for estimation of droughts from rainfall data using standard statistical procedures. WMO Bulletin No.138 (1975) reviewed the statistical methods, *viz.*, normal rainfall, rainfall variability, frequency distribution deciles of rainfall, frequency of droughts in sequences, persistence, extreme values, and spatial correlation methods.

Unreliable data points, lack of standard procedures to compute indices of drought prevalence and intensity could lead to inefficient management of droughts. Therefore, there is need to identify and classify various indices available in the literature to compute the droughts. Drought indices like Percent of Normal (PN), Deciles, Standard Precipitation Index (SPI), Effective Drought Index (EDI) and Bhalme-Mooley Drought Index (BMDI) use only rainfall. These indices are not only simple to compute and perform better over other complex indices. The climate derivatives such as Soil Moisture Index (SMI), Moisture Adequacy Index (MAI) and Aridity Index (Ia) and remote sensing derivatives, *viz.*, NDVI, etc. have also been developed and are extensively being in use in drought monitoring over the years.

2.5.1 Rainfall-based Indices

Rainfall is the single most important factor influencing the incidence of drought and many definitions available in literature use this variable singly or in combination with other weather variables. Murthy and Seshasai (2010) identified 11 such indices developed by research workers around the world that uses weather parameters (Table 2.3).

Table 2.3: Meteorological Drought indices

Year of Development		Name of Index
1916	→	Munger's
1919	→	Kincer
1930	→	Morkowitch
1942	→	Blumenstock
1954	→	Antecedent Precipitation
1957	→	Moisture Adequacy
1965	→	Palmer's (PDSI< PHDI)
1968	→	Crop Moisture
1968	→	Keetch Byram Drought
1981	→	Surface Water Supply
1993	→	Standard Precipitation

(*Source*: Murthy and Seshasai, 2010)

Sikka (1999) reviewed various drought indices developed by research group of IMD and observed that Blanford in 1880 was the first to assess the drought years on the basis of annual rainfall followed by Eliot (1902) and Walker (1910, 1923 & 1924). Ramdas (1950) defined monsoon drought on sub-region scale over India if the rainfall over an area was deficient by 25 percent or more of the normal during the season. Chowdhary *et al* (1989) devised a drought index (DI) based on the area receiving less than 75 percent of the normal rainfall.

$$DI=\frac{\text{Drought area-Meandrought area}}{\text{Standard deviation of drought area}}$$

And had categorized drought as

DI > 3.0	Calamitious
DI > 1.0 and < 3.0	Moderate
DI > 0.5 and < 1.0	Slight

They computed droughts for the period 1875 to 1987 using the above criteria for India and concluded that droughts occur over 20 percent of area in the country once in 5 years, 33 percent once in 10 years and higher than 50 percent area only once in 50 years.

Parthasarathy *et.al* (1995) has identified major droughts in India from the standard deviation values of annual rainfall deficiencies is given below.

Rainfall deficiency		Drought classification
Up to 1.0 σ	→	Drought
1.0 to 1.24 σ	→	Mild
1.25 to 1.49 σ	→	Moderate
1.5 to 2.0 σ	→	Severe
> 2.0 σ	→	Phenomenal

Van Rooy (1965) developed a drought anomaly index based on rainfall departure and mean of lowest ten values of rainfall series.

$$I = -3 + \frac{P - \bar{P}}{\bar{M} - \bar{P}}$$

Where I is an anomaly index, P is actual precipitation $\bar{P}$ in the normal precipitation; M is the mean of the ten lowest values of P on record.

Percent of Normal (PN)

It is calculated by dividing the actual rainfall over a season by its Long-Term Average (LPA) or normal rainfall and multiplied by 100 percent (Pai DS *et al*, 2010).

- The seasonal rainfall over the country as a whole is termed as deficient when rainfall is less than 90 percent of LPA.
- When rainfall is in between 90-110 percent of LPA, it is termed as normal.
- When the seasonal rainfall is deficient (<90% of LPA) and 21 to 40 percent of area of country is under drought, it is termed as All India drought.
- When the seasonal rainfall over the country is deficient (<90% of LPA) and more than 40 percent of area of the country is under drought, then it is called as All India severe drought.

Bhalme and Mooly Index

It is popularly termed as accumulated Negative Moisture Index (NMI). The values vary between normal to extreme drought conditions which is similar to Palmer's drought intensity classification.

$$NMI(m) = 100\,(\mathbf{P}_{m\,tot} - \mathbf{P}_{m\,mean})/e$$

Where $\mathbf{P}_{m\,mean}$ is monthly mean rainfall

$\mathbf{P}_{m\,tot}$ is total monthly rainfall of month m

e is the standard deviation

The negative values of NMI indicate the drought condition. The intensity of drought based on NMI value is as follows.

NMI Values		Drought condition
0.99 to -0.99	-	Near normal
-1.00 to -1.99	-	Mild
-2.00 to -2.99	-	Moderate
-3.00 to -3.99	-	Severe
-4.00 or less	-	Extreme

Standardized Precipitation Index (SPI)

SPI is an index developed by McKee *et al* (1993) based on the rainfall for the time scale of interest and is relatively less complex to compute. The versatility allows the SPI to monitor short term water supplies, such as soil moisture, ground water supplies, stream flows, lakes and reservoir levels. Fór computations of SPI for any location, long-term rainfall data are desired. This data is fitted to a probability distribution which is then transformed into a standardized normal distribution so that the mean SPI for the location and desired period is zero.

The SPI represents the total difference of precipitation for a given period of time from its climatological mean and then normalized by the standard deviation of precipitation for the same period computed using the data over the entire period. It is written as

$$SPI = \frac{a\text{-}A}{Sd}$$

Where a = current precipitation for a given period (Week, month)

A = long term normal of precipitation for the same period

Sd = standard deviation of precipitation for the given period

To compute SPI, long term precipitation data is fitted to a probability distribution which is then transformed into normal distribution, so that mean SPI for the location and for specific period is zero. Positive SPI values indicate greater than median precipitation while negative values indicate less than median values. McKee *et.al,* (1993) defined the criteria of a drought as an event that occurs whenever the SPI is continuously negative and reaches an intensity where SPI is -1.0 or less and the event ends when the SPI becomes positive. Therefore, using SPI, the duration and intensity of drought can be defined precisely.

As SPI is a normalized value, the wetter and drier climates can be represented in the same way. McKee *et.al,* (1993) used the following classification system for SPI value to identify drought intensities as follows:

S.No	SPI values	Intensity of dry/wet spells
1.	2.0 and above	Extremely wet
2.	1.5 to 1.99	Very wet
3.	1.0 to 1.49	Moderately wet
4.	-0.99 to +0.99	Near normal
5.	-1.0 to -1.49	Moderately dry
6.	-1.5 to -1.99	Severely dry
7.	-2.0 and less	Extremely dry

As the theory behind transformation and computational procedures are lengthy and time consuming, computer software for computation of SPI were developed by many Scientists. The readers are advised to download SPI program from the software http://drought.unl.edu/monitoringtools and compute the SPI values for identification of drought intensities

Deciles Method

Deciles of rainfall are widely used in Australia and are more popular. This is an example of cumulative frequency distribution in which, the limits of each deciles of the distribution are calculated from the cumulated frequency curve or an array of data. The first decile is that rainfall amount, that would not exceed by the lowest 10 per cent of totals. The second decile is that rainfall amount that would not exceed by 20 percent of totals and so on. The decile ranges are the ranges of values between deciles, for example, the eight decile range represents the values between seven and eight deciles. The decile values provide a comprehensive picture of rainfall distribution at a place and knowledge of decile range into which a particular total falls gives useful information on departing from normal. The following are(Table 2.4) the decile ranges in classifying rainfall.

Table 2.4: Rainfall distribution under different Decile ranges

Rainfall departure	Decile class (%)	Decile range
Very much above normal	Highest 10	10
Much above normal	80-90	9
Above normal	70-80	8
Slightly above normal	60-70	7
Normal	40-60	6 & 5
Slightly below normal	30-40	4
Below normal	20-30	3
Much below normal	10-20	2
Very much below normal	Lowest 10	1

Spatial Correlation

For using this statistical technique, it is essential first to test the normality of rainfall data series to be correlated and if these are significantly different from normal, a transformation procedure should be applied. Maher (1968) studied the spatial association of rainfall over Australia using 100 stations data for the period 1900-1964. He has selected a master station and correlation coefficients were calculated for annual rainfall values between master station and each of the other stations in the region. Taking hypothesis that the population value of correlation coefficients *r* is not zero, then at 5 per cent significance the value of *r* were plotted and isopleths at 0.25 of *r* values were drawn relative to each station (Figure 2.3).

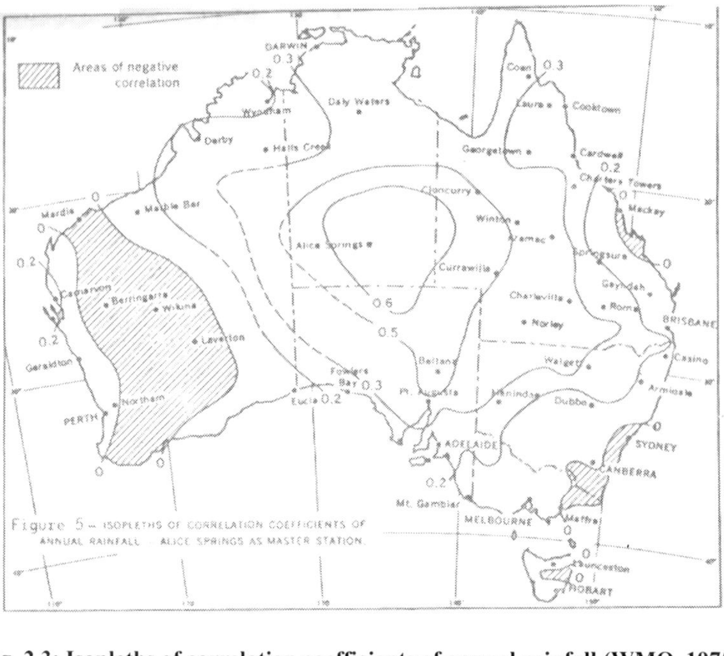

Fig. 2.3: Isopleths of correlation coefficients of annual rainfall (WMO, 1975)

From the above figure a significant correlation observed between annual rainfall among the stations in the climatic zone centered on the master station (Alice spring). However, no correlation exists between master station with rainfall stations located in the eastern and western parts of the continent. Maher (1968) further suggested the computation of *r* values for two years when the annual rainfall was below the median value. By this, it would be possible to evaluate the degree and pattern of rainfall association when conditions are dry at the master station.

Palmer Drought Severity Index (PDSI)

Palmer (1965) developed an index to measure the departure of the moisture supply. This index was developed on the concept of supply and demand of the water balance equation taking into account more than just the precipitation deficit at specific locations. It indicates the standardized moisture conditions and allows comparison to be made between locations and periods. The PDSI values are normally computed on monthly basis.

To compute the Palmer Index, long term climatological data is required to derive five constants which define the moisture characteristics of the climate of the region under study. Steps in computating PDSI are as follows. The first step in the computation is the monthly water balance accounting for long period (30 years more). A two layer soil model was used and PE is estimated by any standard method. The upper layer is assumed to have 25 mm of available water at field capacity. The loss from the underlying layer depends on the initial soil moisture content as well as on PE and available water capacity (AWC) of the soil system. Runoff is assumed to occur only if both layers reach their combined moisture capacity. In addition to PE, three more additional potential terms such as Potential Recharge (PR), Potential Loss (PL) and Potential Runoff (PRo) are introduced. Potential Recharge is defined as the amount of moisture required to bring the soil in to its water holding capacity. Potential Loss is defined as the amount of moisture that could be lost from the soil by the evapotranspiration during rainless period. The Potential Runoff is defined as the difference between the potential precipitation and potential recharge.

From the results of summarized water balance computations, five constants for each of 12 months have to be generated. The constants are as follows;

1. α is the coefficient of evapotranspiration

$$\frac{Mean\ of\ \ monthly\ evapotranspiration(ET)}{Mean\ monthly\ PE}$$

It varies from 1.0 in humid climate to approaching 0 in extreme arid regions.

2. β is the coefficient of Recharge

$$\frac{Mean\ monthly\ moisture\ gain(R)}{Mean\ maximum\ possible\ gain(PR)}$$

3. δ is the coefficient of loss

$$\frac{Mean\ moisture\ loss(L)}{Mean\ potential\ loss(PL)}$$

4. γ is the coefficient of runoff

$$\frac{Computed\ mean\ runoff\,(Ro)}{Mean\ potential\ runoff\,(PRo)}$$

5. The constant K is an empirically derived weighting factor for each month that depends on number of measures of the moisture supply and demand characteristic of the climate in question

$$K = 448.8\ K'/DK'$$

Where $D=\left|P-\hat{P}\right|$

$\hat{P}$ is the adjusted value of precipitation which is climatically appropriate for existing conditions

$$K'=1.5\log_{10}\left[\left(\frac{\overline{PE}+\overline{R}+\overline{R}_O+2.80}{\overline{P}+\overline{L}}\right)\frac{25.4}{\overline{D}}\right]+0.50$$

Where values are expressed in mm.

It is possible to compute the amount of precipitation ($\hat{P}$) that should have occurred in a particular month to sustain the evapotranspiration, runoff and moisture storage that could be considered as normal.

$$\hat{P} = \alpha PE + \beta PR + PRo + S_{PL}$$

This computed precipitation is an adjusted normal precipitation, which is dependent on antecedent weather as reflected by computed moisture storage and an anomaly of PE during the month. Over a long period, the mean of the computed precipitation is equal to the mean of actual precipitation. For a particular month, the difference between actual precipitation and computed precipitation provides a measure (d) of the degree to which the month was abnormally wet or dry. The moisture anomaly index Z = k*d provides a measure that is comparable in space and time.

The drought index (x) depends on the sequence of z values which was derived empirically is as follows

$$X_i = X_{i-1} + \frac{z}{3.0} - 0.103X_{i-1}$$

The values vary from -4.0 (extreme drought) to +4.0 (extreme wet)

Positive values show wet conditions and negative values shows dry conditions. The PDSI values between -2 and +2 would indicate normal. However, the sub-range of -1 to -2 is treated as mild drought. If the values range -2 to -3, it is termed as moderate drought and -3 to -4 indicate severe drought and the values below -4 are considered as extreme drought. A drought sequence is considered as a sequence of 3 or more consecutive months with PDSI value $\leq$ 2.0. A sequence of 6 or more months where the PDSI values are less than -2.0 is termed as a major drought event and the end of the drought sequence is taken where the PDSI value of the last month is $\leq$ - 2.0.

2.5.2 Climate Derivate Based Indices

Researchers have evolved methods to estimate drought conditions using climate derivatives instead measured parameters, which provides the actual field conditions experienced by agricultural crops. These methods are quite useful in crop management and agricultural planning.

Moisture Adequacy Index (MAI)

Weekly Moisture Adequacy Index which is defined as the ratio of AE by PE is computed from water balance procedure (Thornthwaite & Mather, 1955). Drought impact is related to moisture availability at certain crop growth stages, which is described in earlier section under agricultural droughts.

Soil Moisture Index (SMI)

The soil moisture index is defined as the ratio of actual soil moisture to that of the available soil water of that particular type of soil.

$$\text{SMI} = \frac{\text{Actual Soil Moisture}}{\text{Available Soil Water}}$$

Actual soil moisture either can be measured or estimated using weekly water balance approach of Thornthwaite and Mather (1955). These values are used in identifying the water stress period.

During the crop growing season, the weeks that record SMI above 0.75 are considered as drought-free. If SMI values vary between 0.50 and 0.75, the crop in these weeks is subjected to moderate drought condition and if SMI is in between 0.25 and 0.50, the crop is subjected to severe droughts and if SMI is less than 0.25, the crop is subjected to extreme drought conditions.

Crop Moisture Index (CMI)

The Crop Moisture Index (CMI) compliments the PDSI developed by Palmer (1965). It measures the degree to which crop water requirements are met, which is more applicable to short-term changes in moisture conditions and is not meant to assess long-term droughts. It is normally computed on weekly basis using the mean temperature, total weekly rainfall and the CMI value from the previous week. A simple water budgeting would estimates the CMI for each week. It is possible to use a combination of PDSI and CMI for drought monitoring. PDSI serve as a long-term drought monitoring tool and CMI may provide the progression of seasonal water shortages during crop growing stage.

Effective Drought Index (EDI)

Effective Drought Index is a function of rainfall needed for a return to normal conditions (PRN) (Buyan and Wilhite, 1999). PRN is the precipitation, which is necessary for the recovery from the accumulated deficit since the beginning of drought. PRN is computed from daily effective precipitation (EP) and its deviation from the mean for each day. Like SPI, the EDI values are standardized allowing it to compare the values at two or more locations. The range of EDI values vary from -2 to 2. The value -2 EDI indicates extremely dry, at -1.5 <EDI<-1.99 severe drought and at -1<EDI<1.49 moderate drought and near normal at -0.99<EDI< 0.99.

Soil Water Index (SWI)

The Veg ET model (Senary, 2008) conducts daily water balance on pixel by pixel basis and produces several agro hydrological products viz. Soil Water Index (SWI), Water Requirement Satisfaction Index (WRSI) anomaly and cumulative Eta (sum of daily actual evapotranspiration) forecast anomaly. The description of Soil Water Index (SWI) is given below and rest of products can be obtained from http://earlywarning.usgs.gove.gov/usewem

The SWI is the Water Stored in the vegetation root zone depth as a percentage of Water Holding Capacity (WHC) of the soil at the end of a particular day 'i'.

Where $$SWI = \frac{S_{wi}}{WHC} \times 100$$

SWI is the Soil Water content and WHC are Water Holding Capacity of the soil which can be obtained from STATSGO database. The soil water content (S_{wi}) is obtained through a simple mass balance equation as given below.

$$S_{Wi} = S_{Wi-1} + PPT_i - ET_{ai}$$

Where PPTi and ETai are rainfall and actual evaporation of current day and S_{Wi-1} is previous day's soil water

This index is an indicator of soil moisture status at the end of particular day and represented in four broad qualitative categories, such as

S.No	SWI values (%)	Status		
1.	90 to 100	Sufficient	←	(supports crop next 7 to 10 days)
2.	50-90	Satisfactory	←	(enough to support crop for a few days)
3.	10-50	Stress	←	(likely to experience severe to moderate stress if no rain in next few days)
4.	0-10	Wilting	←	(Continued drought results in wilting of crop)

Spatial distribution of these classes can be used to identify areas that are experiencing moisture stress. This index can be potentially used along with forecasted weather for preparation of weather based agro advisories.

Water Balance Methodology

The Water balance equation at any location can be written as

$$P\text{-}Q\text{-}U\text{-}E - \Delta W = 0$$

Where P = precipitation, Q = runoff, U = deep drainage,

E = Evapotranspiration, ΔW = change in soil water storage

All water balance methods developed over years use the above basic equation and many models ignore the details of infiltration process and assume that soil water storage is always replenished provided adequate water is available at the surface. Also in the absence of knowledge about the rate of loss of water during a drying cycle of soil, errors in estimation of actual evapotranspiration may occur. In India, largely the scientific group uses the Book keeping method of Thornthwaite and Mather (1955) which is described briefly here.

Computations of Weekly Water Balance by Thronthwaite and Mather's method (1955):

To compute the weekly water balance by the Thronthwaite and Mather's method (1955), the following information at a place is required.

- Weekly rainfall (in mm)

- Weekly potential evapotranspiration (in mm)
- Available water holding capacity of the soil (in mm)
- Initial soil moisture
- Soil moisture tables for different WHC

Steps involved in the calculation of weekly water balance are given below:

Step-1: Enter weekly rainfall (P)

Step-2: Enter weekly potential evapotranspiration (PET)

Step-3: Enter available water holding capacity of the soil (AWC)

Step-4: Compute (P-PET) for different weeks

The difference between actual rainfall and potential evapotranspiration expresses whether the rainfall is adequate to meet the atmospheric demand, without taking into account the soil moisture stored in the soil. The negative value indicates that the atmospheric demand is not met by the rainfall. Therefore, the demand has to be met from the stored soil moisture if available in the soil and from the rainfall. A positive value indicates that the water supply (rainfall) is more than the atmospheric demand. The excess water retained after meeting demand goes to recharge the soil up to its field capacity. The excess amount beyond the soil capacity goes as surplus.

Step-5: Surplus (SPL)

Surplus refers to quantity of water whenever the soil moisture reserve exceeds available water capacity of the soil

$SMi = SM_{i\text{-}1} + (P_i\text{-}PET_i)$

If SM_i is greater than AWC, surplus occurs

Then $SPL_i = SM_i - AWC$

$SM_i = AWC$ when

$AET_i = PET_i$

SM_i = Soil Moisture at i^{th} week

$SM_{i\text{-}1}$ = Soil Moisture at one week prior to i^{th} week

P_i and PET_i are rainfall and PET at i^{th} week

AWC = Available Water Capacity

SPL = Surplus

Step-6: Deficit (D)

When the difference between (P-PET) is negative, it indicates water deficit in that week. This refers to short fall in the atmospheric demand. To get the Deficit values, one has to compute the Accumulated potential water loss (APWL) which is the cumulative negative values of P-PET up to i[th] week.

$$\text{APWL}_i = \sum (P_i - PET_i)$$

and $$\text{SM}_i = \text{AWC} * \exp($$

$$\text{AET}_i = \text{P}_i + \Delta\text{SM}_i$$

$$\text{D}i = \text{PET}_i - \text{AET}_i$$

Where, ΔSM is change in the soil moisture storage

$$\Delta\text{SM}_i = \text{SM}_{i-1} - \text{SM}i$$

AETi is actual evapotranspiration at in i[th] week

SMi is the soil moisture reserve at in i[th] week

Di is Water deficit at i[th] week $\left(\frac{-APWL}{AWC}\right)$

2.5.3 Remote Sensing Based Indices

Normalized Difference Vegetative Index (NDVI)

The index is based on the concept that vegetative vigour is an indicator of water availability. The crop/vegetation reflects high in near infrared due to its crop geometry and the health of crop and absorbs high in red reflected radiance due to its biomass and accumulated photosynthesis. (Murthy and Seshasai, 2010) Using these characteristics of near infrared, red and middle infrared band, which indicates both the health and condition of crops, different types of vegetative indices have been developed. Among these, NDVI is a popular index due to its simplicity in calculation and easy to interpret. It is widely used in Remote Sensing Application Studies. NDVI is a transformation of reflected radiation in the visible and near infrared bands of NOAA AVHRR and is a function of green leaf area and biomass.

$$\text{NDVI} = \frac{\text{NIR} - \text{RED}}{\text{NIR} + \text{RED}}$$

Where, Near Infrared (**NIR**) and **RED** are reflected radiations in these two spectral bands

Water, clouds and snow have higher reflected radiation value in the visible region consequently NDVI assumes negative values for these features. Bare soil and rocks exhibit similar value in both visible and near IR regions and the index values are near zero. The NDVI values for vegetation generally range from 0.1 to 0.6. The higher index values are associated with greater green leaf area and biomass where as lower values of index show moisture stress under ideal climate conditions (Tucker, 1979). The severity of drought situation can be assessed by the extent of NDVI deviation from its long term mean value (Burgan and Hartford, 1996) which is given by

$$NDVI_{dev} = NDVI_i - NDVI_m$$

Where $NDVI_i$ is the ith monthly NDVI value

$NDVI_m$ is long term average for the same month.

Maps producing standard $NDVI_{dev}$ using relative greenness are useful to assess drought situation and is also being widely used (Johnson *et.al*, 1993). Peters *et.al.* (2002) developed standardized vegetation index based on NDVI, which indicates the probability of vegetation condition's deviation from normal using 2 scores of NDVI distribution derived from long term historic NDVI data sets. Currently, NDVI products can be generated from the data recorded by most of the satellite sensor systems. The MODIS NDVI of 250 m and 1000 m, SPOT VGT NDVI of 1000 m, NOAAAVHRR NDVI of 1000 m, IRS WiFS NDVI of 188 m and IRS AWiFS NDVI of 56 m are widely used in India for drought monitoring purpose because of the advantages of spatial and temporal coverage of these products. The NDVI values can be contaminated by high cloud cover during crop season due to visible band cloud masking or off-nadir viewing or insufficient cloud free passes in compositing period require validation before use in drought assessment. Heavy rains during crop maturity reduce the yields which are not detected by satellites. Therefore NDVI can be used as an indicator of crop development only after recording of significant spectral emergence of crops, which has a lag of 2-3 weeks.

The other commonly used vegetation indices as reviewed by Kasturi Rangan (1996) are as follows (Table 2.5).

Table 2.5: Commonly used Vegetation Indices

Vegetation Indices	Formula
Difference Vegetation Index (DVI)	NIR–RED
Ratio Vegetation Index (RVI)	NIR/RED
Normalized Difference Vegetation Index (NDVI)	(NIR-RED)/(NIR + RED)
Perpendicular Vegetation Index (PVI)	$(1/(a^2+b^2))*$ (NIR – aRED-b)
Soil Adjusted Vegetation Index (SAVI)	{((NIR-RED)/(1+L)/(NIR+RED+L) L = 0.5
Transformed Soil Adopted Vegetation Index (TSAVI)	(aNIR-bRED +b)/{(aNIR+RED–ab+X $(1+a^2)$}
Weighted Difference Vegetation Index (WDVI)	NIR–C*RED: C=NIR (bare soil)/RED (bare soil)
Vegetation Condition Index (VCI)	$(NDVI_{max}$-NDVI)/$NDVI_{max}$-$NDVI_{min})$
Transformed Vegetation Index (TCI)	(NDVI + 0.5)**0.5
Green Vegetation Index (GVI)	-0.29 MSS 1-0.5622 MSS 2 +0.6 MSS 3+0.491 MSS 4
Stress related TM based Vegetation Index	STV11 = (TM5 * TM3) / TM4 STV12 = TM4/TM3+TM7 STV13 = TM4/TM3+TM5
Mid Infra-red stress related Vegetation Index	MSV11 = TM4/TM5 MSV12 = TM4/TM7 MSV13 = TM4/TM5 + TM7
Stress Degree Days (SDD)	$\Sigma (T_c - T_a)$

(Kasturirangan, 1996)

Normalized Difference Water Index (NDWI)

Short wave infrared (SWIR) band is sensitive to moisture availability in the soil as well in crop canopy. Initially during crop season, soil background is dominant and hence SWIR is sensitive to soil moisture in the top 1-2 cm layer. With the development of crop, SWIR becomes sensitive to leaf moisture content. NDWI using SWIR can complement NDVI for drought assessment particularly in the beginning of the season (Murthy and Sesha Sai, 2010).

$$\text{NDWI} = \frac{(\text{NIR} - \text{SWIR})}{(\text{NIR} + \text{SWIR})}$$

where Near Infrared and SWIR are the reflected radiations in these two spectral bands. Higher values of NDWI signify surface wetness and lower values indicate leaf water stress.

2.5.4 Process-based Indicators

A process-based indicator known as EF is derived by modeling the energy and matter transfer between atmosphere and surface. EF is derived as under:

$$Rn = G + H + LE$$

Where,

$Rn =$ Net Radiation

$G =$ Soil heat flux

$H =$ Sensible heat flux

And $LE =$ Latent heat flux

$LE = (Rn\text{-}G) - c_p * \gamma_{air} * 1/r_a * (T_s\text{-}T_x)$ where

$c_p =$ Sp. Heat of air

$\gamma_{air} =$ Air density

$r_a =$ Surface roughness

$T_s =$ Surface temperature

$T_x =$ Maximum temperature

$EF = LE / (Rn - G)$

EF is Evaporative fraction

As long as adequate soil moisture is available, energy will be used for evapotranspiration and EF will be closer to one (no water stress). In the absence of soil moisture, all available energy will be directed towards heating of surface and the ambient air and EF will approach zero indicating serious water stress (Bastiaanssen, 1998). Although there are many vegetation indices and derived indices are available, still Vegetation Condition Index (VCI) and NDVI deviation from historic NDVI are widely used on operational basis to assess drought situation (Kogan, 1995; Burgan *et al.*, 1996; Hayes and Decker, 1998).

2.6 Assessment of Drought Prone Regions in India

The vulnerability of an area to drought proneness is dependent on the rainfall characteristics, length of crop growing season, cropping pattern and types of crops and varieties, livestock management, socio-economic status of farmers, infrastructure facilities, literacy status, and etc. Among all the above, the key indicator for assessing the drought proneness is the water availability through rainfall, surface runoff, irrigation and ground water. Rainfall being variable annually, limits crop production and often threatens the food and social security of the human population. Variability (coefficient of variation) in annual rainfall in a region over 40 percent is considered as risk prone region for droughts. The variability is above 50 percent in NW part of India and in the adjoining arid zone covering Rajasthan and Gujarat States and touches above 70 percent in extreme regions of West Rajasthan where annual rainfall is extremely low (less than 300 mm/year).

The Second Irrigation Commission of India requested the State Governments to identify drought affected areas by the criteria, *viz*., (1) meteorological data, (2) revenue remission, (3) frequency of famine / scarcity of food, and (4) availability of irrigation resources. However, identification of drought prone regions using the above criteria varied considerably among the states and the areas were magnified. For example, Tamil Nadu included all the talukas which receive rainfall less than 900 mm/year or areas that have less than 35 percent of area under irrigation as drought prone regious. Rajasthan Government considers a drought area if the ratio of good crop years to scarcity years be 2:1 or less and 50 percent of the villages of that area should be affected by drought. The Karnataka Government considers those areas that receive less than 400 mm rain during *kharif* season and less than 150 mm in *rabi* season with a CV more than 30 percent during each crop season and rainfall deficiency of more than 20 percent during crucial crop stages. The Irrigation Commission in 1972 finally adopted the meteorological criterion for defining drought prone areas. It defined an area prone to drought if meteorological drought (75% of normal) occurred in 20 percent of the years in data series. If it occurred more than 40 percent of years, it is termed as chronically drought prone.

So far, the assessment of rainfall variability in a region is carried out with available rainfall data collected at the district headquarter in the country. Perhaps, analyzing the rainfall data from sub-district units along with the soil fertility map and ground water resources shall help in identifying the regions that are prone to droughts for overall planning and development. The assessment of length of crop growing period (LGP) under different agro climatic regions is very much essential to develop suitable cropping plans for sustainable crop production. It could be used as a criterion for assessing the drought prone areas and also help in introducing new crops or alternate crops. For example, the farmer in a traditionally groundnut grown district like Anantapur in South India was asked to grow cluster beans, equally remunerative crop that can withstand water stress and fits in with the rainfall distribution (CRIDA, 2012). Characteristics of LGP at selected stations in India are given in Table 2.6.

The mean LGP for different soil water holding capacity vary from 20 to 30 weeks at Hyderabad and for Jodhpur it is around 10 weeks only. For higher rainfall zones and better water holding capacities, the LGP vary around 32 to 36 weeks. The percent probability of length of available growing season decreases when expected period of duration increased.

Table 2.6: Characteristics of length of growing season at selected stations

Station	Mean annual rainfall (mm)	Available waterholding capacity (mm)	Start of growing season (weeks)			LGP (weeks)			Probability (%) for growing season to exceed			
			Mean	Extremes		Mean	Extreme		Weeks			
				Early	Late		Low	High	10	16	22	30
Jodhpur	369	100	28	24	40	10	0	20	63	10	—	—
Hyderabad	767	50	25	19	33	20	12	29	97	79	49	4
	767	150	25	19	33	24	15	36	97	86	62	23
	767	250	25	19	33	30	16	48	100	98	89	52
Indore	963	300	25	20	29	32	16	43	100	100	99	73
Bhubaneswar	1525	200	23	17	25	36	29	51	100	100	100	100

(*Source*: Ramakrishna et.al. 2000)

Traditionally, the dependence of farmers on livestock is high in drought prone regions as it supplements the income from agriculture under normal situations. The availability of fodder to animals during drought period varies considerably. Therefore, improving fodder resources both through improved fodder varieties and controlled grazing methods may reduce drought impacts and conserve the natural resources. A severe shortage of fodder may lead to migration to green pastures in the neighbouring areas. Assessment of social structure and literacy percentage and the infrastructure such as roads, transportation, safe drinking water and power would provide necessary information on the fragile nature of the drought prone region. The social structures of the farmers in drought prone regions are poor and lives on subsistence farming. The low literacy rates in addition to physical and climate constraints inhibit the community to adapt to new methods for improving their livelihoods.

During the last six decades, development of drought prone regions was taken up through various Governmental plans with liberal funding. Transportation of surface waters from long distances in to the drought prone areas during 1970-1980 such as Indira Gandhi Canal (in to arid parts of Rajasthan), Tungabhadra Canal (KC Canal into the southern arid districts like Kadapa, Bellary, etc.) had totally changed the agricultural and economic scenario considerably over the last decades. Watershed development, soil conservation measures and mining of ground water, good road network, fodder banks, improved public distribution systems, and better financial facilities considerably improved the living standards in traditionally drought prone areas in the country. These measures yielded the results and has not allowed the droughts to escalate to famine conditions in spite of the country having witnessed phenomenal droughts within a decade viz in 2002 and again in 2009.

2.7 Indicators for Drought Declaration in India

There are number of indicators that are used for monitoring droughts and every indicator has some limitation due to non-availability of required data to compute indicators. Also the reliability of data is one of the serious limitations in this regard. The Manual for Drought Management (2009) bought out by Government of India provided guidelines to declare drought on a regional scale with a combination of important indicators which were identified based on wide ranging consultations with meteorologists and agricultural scientists. The guidelines reported in the manual are as follows:

Rainfall deficiency, the extent of area sown, NDVI and MAI are recommended as four standard monitoring tools which could be applied in combination for drought declaration. Since information on these indicators is available at the level of Taluka / Tehsil / Block, drought may be declared by the State

Government at the level of these administrative units on the basis of observed deficiencies. It recommended that at least three out of four indicators must satisfy the criteria for drought declaration. The limits set for four indicators, as per Manual for Drought Management (2009) are as follows:

2.7.1 Rainfall deficiency

- The State Government could consider declaring a drought if the total rainfall received during the months of June and July is less than 50 percent of the average rainfall for these two months and if there is an adverse impact on vegetation and soil moisture as measured by the vegetation index and soil moisture index. Such a rainfall deficit would cause so much damage to agriculture crops that it would be difficult to revive crops.
- The State Government could consider declaring a drought if the total rainfall for the entire duration of the rainy season of the state, from June to September (the south-west monsoon) and or from December to March (north-east monsoon), is less than 75 percent of the average rainfall for the season and there is an adverse impact on vegetation and soil moisture, as measured by the vegetation index and soil moisture index.

Rainfall data are easily available through the weather observatory net work of the IMD and State Government and the procedure is simple to compute and analyse the rainfall departures. However, it is necessary for the State Government to standardize rainfall data with respect to its measurements and recording for all the Talukas/Tehsils/Blocks for each month and the entire rainy season.

2.7.2 Area Under Sowing

Sowing is another important indicator of the spread and severity of drought. Sowing operations are linked to rainfall and availability of water. Mostly, farmers undertake sowing operations at the commencement of the monsoon. If sowing fails due to water stress, farmers sow a second or even a third time. Therefore the *area under sowing* provides reliable information on the availability of water for agricultural operations.

Drought conditions could be said to exist if the total sowing area of *kharif* crops is less than 50 percent of the total cultivable area by the end of July / August, depending upon the schedule of sowing in individual states. In such situations, even if rainfall revives in the subsequent months, reduction in the area under sowing cannot be compensated for and the agricultural production would be substantially reduced. The State Government should, therefore, consider declaring a drought if, the total area sown by the end of July / August is less than 50 percent of the total cultivable area along with the other indicators.

In case of *rabi* crops, the declaration of drought could be linked to the area of sowing being less than 50 percent of the total cultivable area by the end of November / December along with the other indicators.

The Agriculture Department(s) in the State Government(s) provides data on crop-wise sowing operations; generally available for all the Talukas / Tehsils / Blocks and is compiled for all the districts in a State.

2.7.3 Normalized Difference Vegetation Index

The National Agricultural Drought Assessment and Monitoring System (NADAMS), operated by the National Remote Sensing Centre (NRSC), Hyderabad, an institution under ISRO issues a bi-weekly drought bulletin and monthly reports on detailed crop and seasonal condition during the *kharif* season. These reports present the Normalized Difference Vegetation Index NDVI) and Normalized Difference Wetness Index (NDWI) maps from the data obtained from National Oceanic and Atmospheric Administration – Advanced Very High Resolution Radiometer (NOAA-AVHRR) and Indian Remote Sensing (IRS) satellite (Wide Field Sensor (WiFS) data. They provide quantitative information on sowings, surface water spread and District / Tehsil / Taluka / Block level crop condition assessment along with spatial variation in terms of maps. At present, 11 agriculturally important and drought-vulnerable States, *viz*., Andhra Pradesh, Bihar, Gujarat, Haryana, Karnataka, Maharashtra, Madhya Pradesh, Orissa, Rajasthan, Tamil Nadu and Uttar Pradesh are covered through these reports.

The values obtained for a given NDVI always ranges from -1 to +1. A negative number or a number close to zero means no vegetation and a number close to +1 (0.8-0.7) represents luxurious vegetation shown in the Figure 2.4.

For declaring drought, States need to obtain NDVI values through the NADAMS. All the above-mentioned States receive NADAMS reports on a regular basis. Those States which do not receive the report could approach the NRSC for receiving the information. It is necessary that the States declare drought only when the deviation of NDVI value from the normal is 0.4 or less. However, the NDVI value needs to be applied in conjunction with other indicators and values. For the declaration of drought, NDVI must not be invoked in isolation from the other two key indicators.

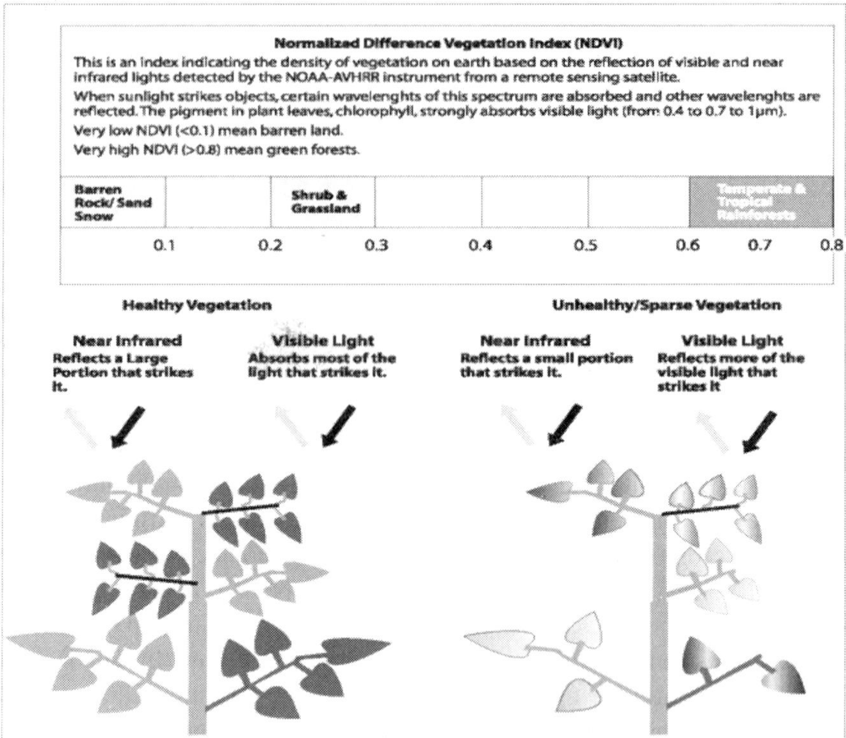

(*Source*: Manual for drought management-2009)

Fig. 2.4: Normalized Difference Vegetation Index (NDVI)

2.8 Drought Management Procedures in India

After the debacle of 2002 All India Drought that affected agricultural production by about 29 MT compared to its previous year, the Indian Government gave a serious thought to identify and streamline the procedures to adopt a uniform drought code for the whole country. Samra (2004) reviewed in detail the drought, declaration and management procedure adopted in India. He has schematically represented the drought management cycle consisting of Monitoring, Declaration, Response, Mitigation, and Preparedness etc. (Fig 2.5). It also includes Preventive rather providing only short term relief measures.

At national level, India Meteorological Department (IMD) has evolved and standardized techniques to forecast the onset of SW monsoon and subsequent rainfall and the expected departures of regional rainfall over the normal value. It prepares weekly aridity anomaly maps and compiles weekly rainfall departures at sub-divisional as well at district level. The onset and the progress of SW monsoons are continuously monitored throughout the season. It also

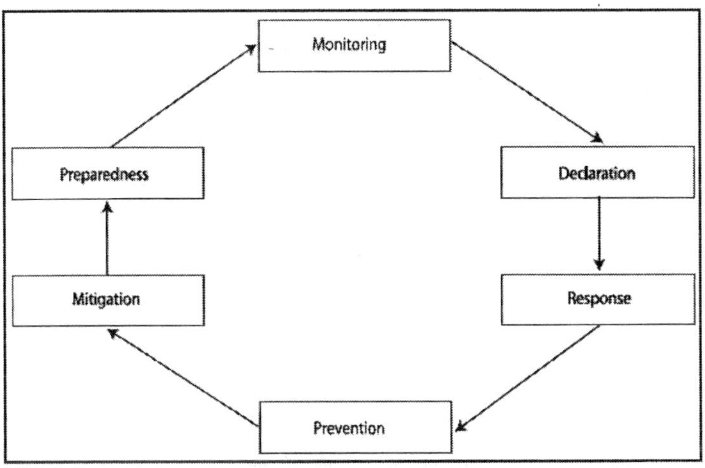

(*Source*: Samra; 2004)

Fig. 2.5: A schematic description of a drought management cycle in India

identifies and monitors the regions with rainfall deficits. The weekly and special reports on monsoon forms a basis for an early signal on the initiation of drought conditions at different parts of the country. At the regional scale, the State Agricultural Department in the respective states closely monitors the progression of agricultural activities like progress in sowing, irrigation, seeds and fertilizers availability etc, and hydrological information at weekly intervals along with the daily distribution of monsoonal rains, dry spells and their duration over the state. Week-wise information is compiled at each state and submitted to Crop Weather Watch Group (CWWG), which is nodal agency under the Ministry of Agriculture to monitor the drought over the country and prepare contingency plans to effectively manage the drought (Table 2.7). The group consists of the following members from various departments.

Table 2.7 : Composition and role of CWWG of the Ministry of Agriculture, Government of India

Partners		Tasks
Additional Secretary, Ministry of Agriculture	→	Chairperson of the Group; promotes overall coordination
Economic & Statistical Adviser, MOA	→	Report behavior of agro climatic and market indicators
India Meteorological Department	→	Rainfall forecast and progress of monsoonal conditions
Central Water Commission	→	Water availability monitoring in major reservoirs
Plant Protection Division	→	Watch pests and diseases outbreak
Crop Specialists	→	Crop conditions and production
Agricultural Input Supply Divisions	→	Supply and demand of agricultural inputs
Agricultural Extension Specialists	→	Report on field-level farm operations
Ministry of Power	→	Manage electrical power for groundwater extraction
Indian Council of Agricultural Research	→	Technical input and contingency planning
National Centre for Medium Range Weather Forecast	→	Provide medium-term forecasts

(*Source*: Samra, 2004)

This group meets on every Monday and evaluates information and data furnished by IMD and other scientific and technical bodies. The exchange of information with similar kind of groups at state level serves as triggering mechanism to activate drought response system.

Monitoring and information management system at different levels and at different periods viz., daily, weekly, fortnightly, monthly and seasonally by CWWG are summarized as below in table 2.8.

Drought declaration is a state subject while the Central Government only facilitates the financial and institutional process. Traditionally in India, the District Administrator (Collector/Dist Magistrate) recommends the declaration of drought only after crop production estimates are available through Annawari / Paisawari / Gidawari system. Generally, area with less than 50 percent of the above system are considered to be affected by the droughts. Final figures on crop losses are only available in December and the Government of India sends a team of experts to the drought affected regions to assess the drought status after its declaration. Based on this report, State Government submits the requirement for relief and release of funds.

Table 2.8 : Details of CWWG monitoring and information management

Parameters	Agency				Communication mode
	National	State	District	Field	
A. Meteorological					
Delay in the onset of monsoon	W	W	D	D	Wireless/Fax/Telephone/e-mail
Dry spells during sowing	W	W	D	D	Wireless/Fax/Telephone/e-mail
Dry spells during critical crop growth periods	W	W	D	D	Wireless/Fax/Telephone/e-mail
B. Hydrological					
Water availability in reservoirs	W	W	D	D	Wireless/Fax/Telephone/e-mail / Written Reports
Water availability in tanks / lakes	F	F	F	W	Written Reports
Stream flow	F	F	F	W	Written Reports
Groundwater level	S				Written Reports
Soil moisture level	F	F	F	F	Written Reports
C. Agricultural					
Delay in sowing	W	W	W	W	Wireless/Fax/Telephone/e-mail
Sown area	W	W	W	W	Wireless/Fax/Telephone/e-mail
Crop vigor	F	F	F	W	Written Reports
Change in cropping pattern	W	W	W	W	Wireless/Fax/Telephone/e-mail
Supply and demand of agricultural inputs		W	W	W	Wireless/Fax/Telephone/e-mail

D=Daily; W=Weekly; F=Fortnightly; M=Monthly; S=Seasonal (Pre and Post-rains)
(*Source:* Samra, 2004)

2.9 Drought Code Adopted by Local Governments

The criteria adopted in different states to identify drought prone regions and to declare drought depending upon rainfall and the crops grown in that region was reviewed by Ramakrishna *et al*, 2000. Many states in the country follow the Annewari System, wherein crop conditions are assessed through visual estimates. The criteria are as follows:

Production		Criteria
Above 75% of normal	→	No drought
50-75% of normal	→	Moderate
25-50% of normal	→	Severe
Less than 25% of normal	→	Disastrous

2.10 Spatial Variability of Drought and Their Frequency of Occurrence in India

Spatial variability of droughts closely follows the rainfall variability (CV). The higher the CV values, drought proneness is also higher. As per the drought research work conducted by many organizations, it was estimated that about 68 percent of the total geographical area in the country is prone to different intensities of droughts. The area which receives annual rainfall less than 500mm (arid zone) and in the range of 500-750 and 750-1125 mm/year (semi-arid to sub-humid climates) is considered to be drought prone (Manual for Drought Management, 2009). The cropped area falling under various ranges of rainfall in India is as follows (Table 2.9)

Table 2.9: Cropped area falling under various ranges of rainfall in India

No.	Rainfall ranges	Classification of rainfall	Percent of total cropped area
1.	Less than 750 mm	Low	33
2.	750-1125 mm	Medium	35
3.	1126-2000 mm	High	24
4.	Above 2000 mm	Very high	8

(*Source*: Manual for Drought Management, 2009)

It can be observed that about 59 per cent of the total cropped area of the country is covered under the rainfall ranges of 750-2000mm /year. In spite of climate risks, considerable area (33 per cent) is under cultivation in low rainfall region.

The Drought Research Unit of IMD at Pune have categorized drought on the basis of rainfall departures. It further classified the regions as drought areas and chronic drought areas depending upon the frequency of occurrence of deficit rainfall conditions. From the analysis of annual and southwest monsoon rainfall departures for about 500 stations in the country, the distribution of drought prone areas have been identified (Table 2.10).

Table 2.10: Drought prone regions in the country

Drought prone areas	→	Gujarat, Rajasthan, Punjab, Haryana, West Uttar Pradesh and West Madhya Pradesh, Madhya Maharashtra, Interior Karnataka, Rayalaseema, South Telangana and parts of Tamil Nadu, a small portion of NW Bihar and adjoining east Uttar Pradesh, a small portion of NE Bihar and adjoining portion of West Bengal.
Chronically drought affected	→	West Rajasthan and Kutchh areas

Source: Apparao et al. (1991)

Most of the drought prone areas identified in the above fall either in arid or in semi-arid region where the droughts occur more frequently. Therefore the probabilities of occurrence of droughts in different meteorological sub-divisions have been worked out by Apparao *et al.* (1991) and shown in the following Table 2.11.

As expected, the probabilities are high in the arid region of west Rajasthan compared with other sub-divisions. The extent to which the monsoon rainfall occurred over the country during major drought years (Table 2.12) compared to average annual rainfall for the years 1880 to 1986 were analysed by IMD and it showed that lowest amounts of rainfall occurred during the year 1899, 1918 and 1972.

Table 2.11: Probability of occurrence of drought in different meteorological sub-divisions

Meteorological sub-division	Frequency of deficient rainfall (75% of normal or less)
Assam	Very rare, once in 15 years
West Bengal, Madhya Pradesh, Konkan, Bihar and Orissa	Once in 5 years
South Interior Karnataka, eastern Uttar Pradesh and Vidarbha	Once in 4 years
Gujarat, east Rajasthan, western Uttar Pradesh	Once in 3 years
Tamil Nadu, Jammu & Kashmir and Telengana West Rajasthan	Once in 2.5 years

Source: Apparao *et al.* (1991)

Region-wise percentage departure of rainfall in these phenomenal drought years in various regions of the country are also given in the Table 2.12. Large departures were noticed in NW region.

Table 2.12 : Region-wise percentage of departure of rainfall from normal long-term average during SW monsoon in major drought years

Region	1918	1965	1972	1979	1987	2002	2009
All India	-26	-18	-25	-19	-19	-19	-23
North-West	-36	-28	-36	-18	-46	-48	-36
Central	NA	NA	NA	-31	-28	-31	-20
East	-20	-30	-22	-23	-14	+2	-27
Peninsular	-32	-22	-31	-14	-27	-27	-4

(*Source*: India Meteorological Department)

Droughts occur at random and no periodicity has been noticed. Severe droughts that occurred over the past 100 years in different parts of the country as reported by Upadhyay (1994) are shown in Table 2.13.

Table 2.13: Severe droughts over India in the past 100 years

Year of drought	Regions affected	Per cent of country's area affected	Per cent Departure of monsoon rainfall from normal	
			Over India	Over drought region
1877	Sub-Himalayan West Bengal, Bihar Plains, Uttar Pradesh, Haryana, Himachal Pradesh, Madhya Pradesh, Rajasthan, Gujarat, Konkan, Marathwada, Vidarbha, coastal Andhra Pradesh, Telangana, Jammu and Kashmir	64	-32	NA-
1899	Haryana, Punjab, Himachal Pradesh, Madhya Pradesh, Rajasthan, Gujarat, Andhra Pradesh, Maharashtra, Tamil Nadu and Kerala	65	-28	NA-
1918	Haryana, Punjab, Himachal Pradesh, Rajasthan, western Madhya Pradesh, Gujarat, Maharashtra, Telangana, Jammu & Kashmir, Rayalaseema, Karnataka, Tamil Nadu and Kerala	71	-26	-49
1965	Eastern Uttar Pradesh, Haryana, Punjab, Himachal Pradesh, eastern Rajasthan, Madhya Pradesh, Gujarat, Vidarbha, Kerala, Jammu and Kashmir	41	-17	-36

(*Contd.*)

Year of drought	Regions affected	Per cent of country's area affected	Per cent Departure of monsoon rainfall from normal	
			Over India	Over drought region
1972	Parts of north-east India, Bihar Plain, Punjab, eastern Rajasthan, Gujarat, Maharashtra, Kerala, Andhra Pradesh, Jammu and Kashmir and south Karnataka	47	-25	-35
1979	Bihar Plateau, Uttar Pradesh, Himachal Pradesh, eastern Rajasthan, Jammu and Kashmir, Konkan and Goa, coastal Andhra Pradesh	45	-21	-38
1987	Nagaland, Manipur and Tripura, Jammu and Kashmir, Himachal Pradesh, Uttar Pradesh, Haryana, Punjab, Rajasthan, Gujarat, Vidarbha, Orissa, coastal Andhra Pradesh and Kerala	50	-18	-45

Source: Upadhyay (1994)

Largest area affected by drought was observed during 1918, where as highest depateses in rainfall was noticed in 1877.

Bagchi (1991) identified 100 drought prone districts in 13 states in the country based on the deficiency of rainfall during critical periods of crop growth is given in the following Table 2.14.

Table 2.14 : Number of drought prone districts in different states identified

State	Total No. of districts	Drought prone districts	Area cover (%)	Population (%)
Andhra Pradesh	23	8	42	35
Bihar	25	7	25.5	22
Gujarat	19	12	79	68
Haryana	12	4	33	32
Jammu and Kashmir	9	2	-	-
Karnataka	20	14	75	-
Madhya Pradesh	45	11	-	-
Maharashtra	26	9	40	36
Orissa	13	2	14.6	-
Rajasthan	26	13	63	45
Tamil Nadu	16	9	64	-
Uttar Pradesh	56	6	15	12
West Bengal	16	3	31	20

Source: Bagchi (1991)

It is noticed that highest number of districts prone to deficit rainfall conditions during critical periods of crop growth are observed in Gujarat where 12 districts out of 19 covering 79 per cent of area is prone to droughts followed by Karnataka (75 per cent), Tamil Nadu (65 per cent), Rajasthan (63 per cent) and 42 per cent in Andhra Pradesh. With improved irrigation facilities and bifurcation of districts for administrative reasons, the number of drought prone districts might have changed over the period.

In view of the importance of *kharif* season rainfall for dry land crops, the probabilities of meteorological droughts (negative departures of *kharif* seasonal rainfall) in various regions of the country have been worked out by many scientists. Using the sub-divisional annual rainfall data for the period 1871-1999, CRIDA Hyderabad has computed the frequency of droughts across different meteorological sub-divisions (Fig 2.6) using the criteria adopted by IMD for identification of a drought year.

As expected, highest frequency of droughts of about 36-42 occurred in Saurashtra and West Rajasthan over 128 year period. In the semi-arid region covering, Haryana, East Rajasthan , North Gujarat, Central Maharashtra, Vidarbha, Telangana, Rayalaseema, North Interior Karnataka, Tamil Nadu about 15-28 droughts occurred over the period 1871-1999. Lowest values were observed in all the sub-division of NE India, Orissa, West Bengal, Jharkhand, Konkan and Coastal Karnataka region.

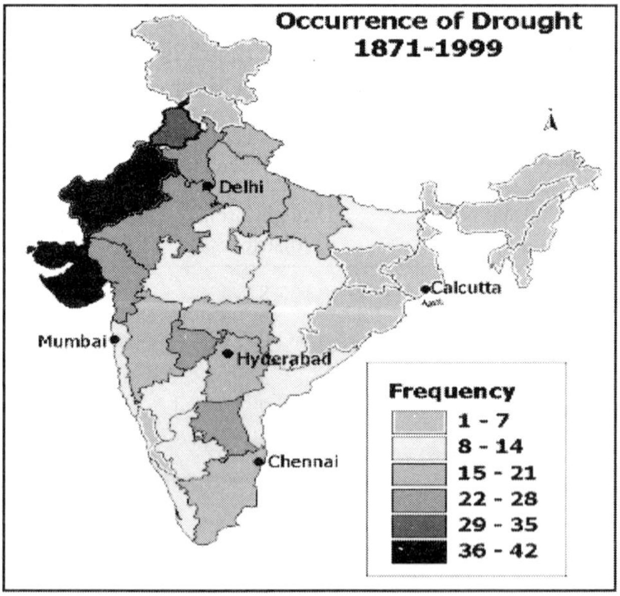

(*Source*: CRIDA)

Fig. 2.6: Occurrence of drought during 1871-1999

The spatial distribution of probability of occurrence of various categories of meteorological droughts viz., moderate and severe were computed using more than 1100 stations rainfall data across India by CRIDA and is shown in the Figure 2.7. At national level, irrespective of the kind of drought, major part of the country can experience 3 to 4 drought years out of every ten years. The probability of occurrence of moderate drought is about 10 to 20 per cent indicating the chances of one to two droughts for every 10 year period over large part of the area in the country. The probability of experiencing severe drought is very low (less than 10 percent) and mostly confined to northwestern parts adjacent to Indian desert.

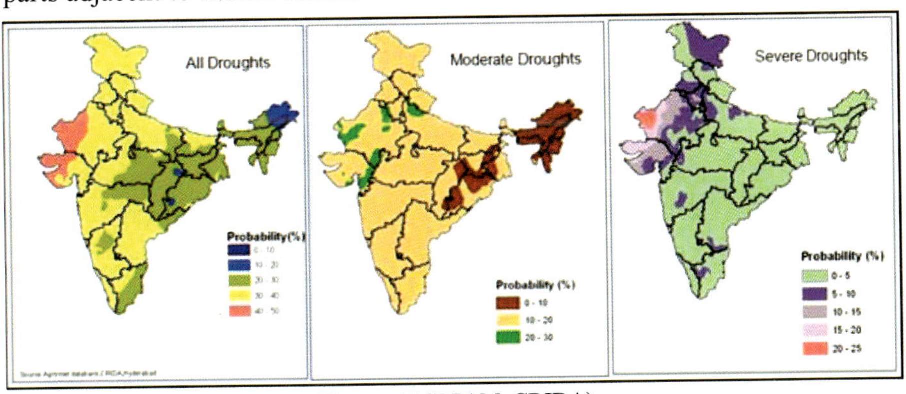

(*Source*: AICRPAM, CRIDA)

Fig. 2.7: Probability of drought of different intensities in India

The district wise probabilities of moderate and severe drought computed by IMD 2010 for the period 1901-2003 is shown below

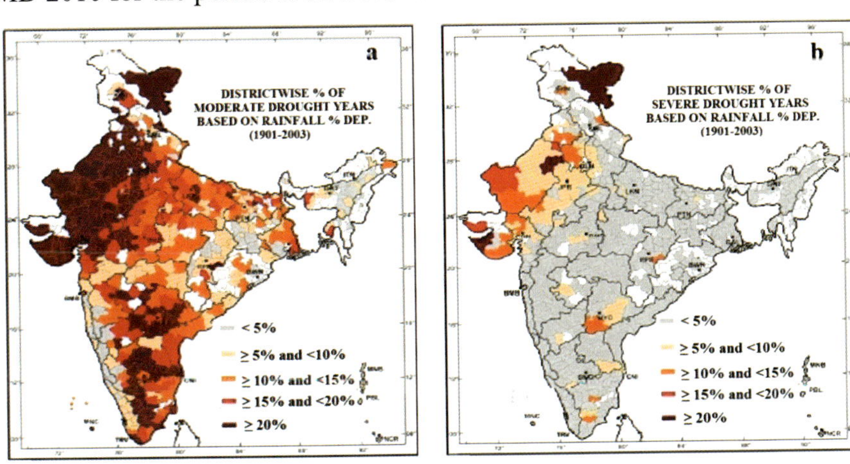

(*Source*: IMD, 2010)

Fig. 2.8: District-wise percentage of incidences (probability) of drought of (a) moderate intensity and (b) severe intensity during the southwest monsoon season for the period 1901-2003

It is observed that the districts vulnerable to record two droughts in a decade of moderate intensity are seen in NW parts covering, Rajasthan, Gujarat, Haryana, Punjab and in Southern peninsular covering the states of Maharashtra, Karnataka, Andhra Pradesh and Tamil Nadu. Similarly two severe droughts in a decade could be expected in some districts in Saurashtra, West Rajasthan and at higher altitude regions in Himachal Pradesh. One severe drought year in a decade is expected in districts Rajasthan, Gujarat, Rayalaseema in Andhra Pradesh and southern districts in Tamil Nadu.

Gore *et al* (2010) published probability maps of moderate and severe meteorological droughts over the country by considering the daily rainfall data of various locations in the country for the period 1901-2000 (Figure 2.9). They have also worked out the probabilities of moderate and severe drought for different districts in subdivision which were grouped in to six homogenous regions, *viz*., Northwest India, West Central India, Peninsular India, Central Northeast India and Northeast India and hill regions.

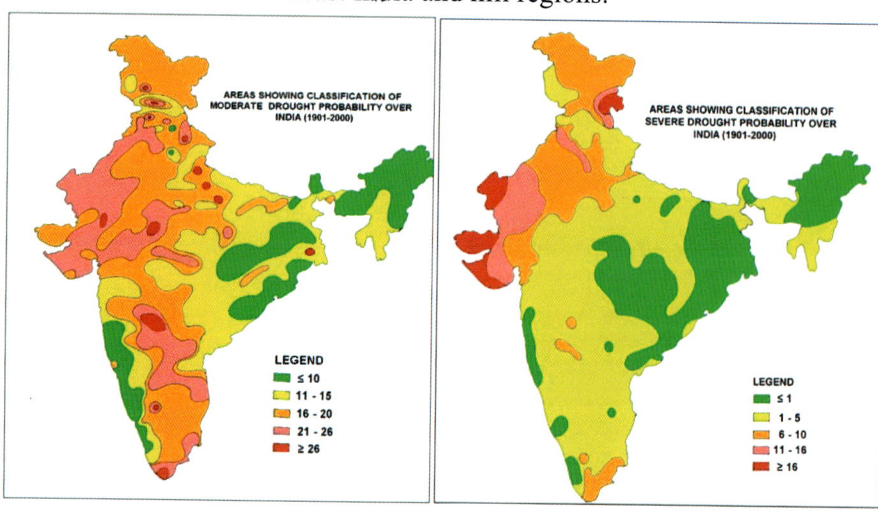

Fig. 2.9: Moderate and severe drought probability maps of India (Gore *et al.*, 2010)

The results of the study by Gore et al., 2010 in given below

- In general, in most parts of India, probabilities of moderate drought are in the range 11 to 20%. Major parts of India show probabilities of severe drought in the range 1 to 5%. In some West Central, Central, East and Northeast region of India, no severe drought is experienced
- In the Northwest region of India, the probability of moderate drought varies from 12 to 30% and that for severe drought varies from 1 to 20% in most of the parts and about 20-30% in the extreme northwestern parts.

- In West Central India, the probability of moderate drought varies from 5 to26% and that for severe drought varies from 1 to 8%.
- In the Peninsular region, the probability of moderate drought varies from 3 to 27%, and that for severe drought varies from 1 to 9% in major parts.
- In the Central Northeast region, the probability of moderate drought varies from 6 to 37% and that for severe drought varies from 1 to 10%.
- In the Northeast region, the probability of moderate drought varies from 1 to 26% and that for severe drought varies from 1 to 3.
- In the hilly region, the probability of moderate drought varies from 9 to 31% and that for severe drought varies from 1 to 12% except in Leh and Lahul & Spiti.

2.11 Drought Vulnerability Assessment

Vulnerability can be defined as the degree of susceptibility of a system to a hazard either as a result of varying exposure to that hazard or because of variations in the ability to cope with its impacts (Nagarajan 2003). These studies would help in identifying socio-economic and environmental causes of drought impacts and reduce the gap between impact assessment and policy formulations. The vulnerability profile of a region provides necessary input for development of Early Warning Systems (EWS), for managing drought effectively.

The vulnerability of a region to drought depends on the extent to which physical and climate conditions play an adverse role in crop production, water resources and the socio-economic conditions of the farmers. It is the frequency of occurrence of drought that demarcates regions of droughts with chronically drought prone region. Therefore, for the classification of a region as drought prone according to the Irrigation Commission Method (1972), the rainfall data must be collected and analyzed as per the procedure laid out in the report and declare drought prone areas. Government of India thus has identified 120 districts as drought prone. Hanumantha Rao (1994) Technical Committee using the criteria of percent irrigated area in different climatic zones had found 185 districts in 13 states occupying an area of 120 Mha as drought prone. Rapidly changing agricultural scenario with more area being brought under major and minor irrigation system (ground water) indicates the need for development of new set of criteria for delineation of drought prone regions.

As district being a large area and analyzing the rainfall collected at single station or at a few places may not represent the actual conditions existing in the district. Hence, CRIDA during 2006 carried out studies to delineate drought prone blocks in Mahabubnagar and Anantapur districts in Andhra Pradesh which were already delineated as drought prone districts category. Based on the percent of irrigated area and rainfall, a composite drought index was developed with weighted drought indices derived from the above components. The procedure needs to be fine tuned for its wider applicability. Murthy *et.al* (2008) identified drought proneness at block level for Mahabubnagar and Anantapur using a set of NDVI based indicators, such as, NDVI differences, Inter annual NDVI variability, and probability of occurrence of low NDVI etc. Blocks more vulnerable and less vulnerable to droughts had been delineated as shown below (Figure 2.10).

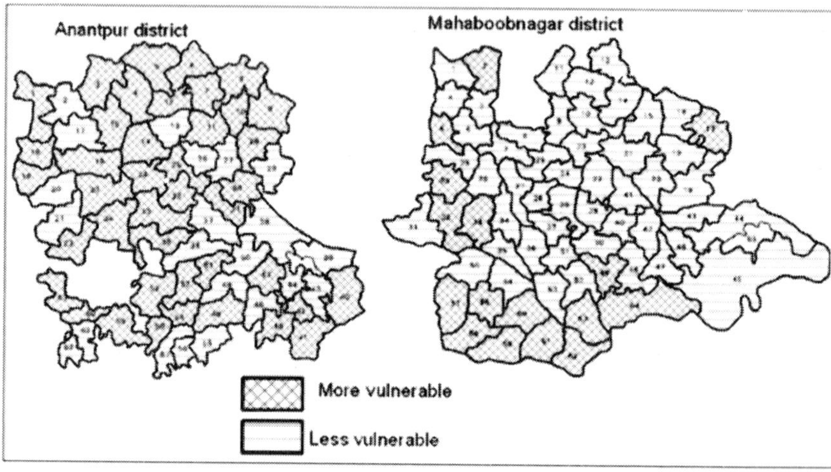

Fig. 2.10: Drought proneness assessment on the basis of historic NDVI
(Murthy and Seshasai, 2010)

Remote sensing data (RS) and GIS environment with image processing capabilities can transform the raw image into information. The information needed for drought assessment using RS and ground measurement given by Nagarajan (2002) is shown in the following Table:

Table 2.15 : Information requirement for drought assessment

Information	Remote sensing	Ground measurement	Indices
Temperature	♦	♣	**Meteorological indices:** Dryness Index, De Martonne's Index, Pluvothermic Quotient, Bhalmey and Mooley Index, Rainfall Anomaly Index, Mean Monthly Rainfall Deficit, Rainfall deciles, PDSI, Relative Drought Resistance
Rainfall - monthly and annual	♦	♣	
Evaporation		♣	
Humidity		♣	

(Contd.)

Information	Remote sensing	Ground measurement	Indices
Soil moisture	♦	♣	Crop Moisture Index, Soil Moisture content
			Remote Sensing Indices
Vegetation cover	♦		Vegetation Condition Index, NDVI
Crop area and type	♦	♣	Soil adjusted Vegetation Index, Stress related TM-based vegetation indices, Stress degree days, Crop yield estimation, Water demand analysis
			Hydrological Indices
River flow	♦	♣	Low flow analysis, total surface water and groundwater availability
Surface water storage area/ volume	♦	♣	
Aquifer type		♣	
Groundwater level		♣	
Population		♣	**Social Indices**
Population density		♣	Food and water demand analysis
Human and livestock population		♣	
Agriculture dependant people		♣	Purchase capacity and target relief
Gross income		♣	Relief/Mitigation camp selection and functioning
Food storage facility		♣	
Medical facility		♣	
Transportation network	♦		

(*Source*: Nagarajan, 2002)

Nagarajan (2002) attempted weighing and ranking method in classifying areas that are prone to drought conditions in Maharashtra state. The methodology as given by him follows. Weight is assigned to each parameter that reflects its importance in the event of occurrence together with the rating for individual classes that denotes the event intensity. The class intervals are derived from the observed events severity of past. The parameters were grouped as, *viz.*, frequency events (historical), climate (time series of rainfall, rainfall variability, evaporation, aridity) resource availability (vegetative cover, food-crop area, cash crop area, surface and ground water), demand (population density, growth, industry) distribution loss (surface storage, medical facility, systems efficiency) and sharing of resources (proximity to infrastructure facility) etc. Though these factors are interrelated, weights for these factors were assigned based on observed deficits. The drought vulnerability is estimated as follows.

Drought Vulnerability (DV) = Assessment + Mitigation

$$DV=\sum_{i=1}^{Nos} DV=\Sigma[PD+FE+ARF+ArI+PDSI+Eva+NDVI+TCr+CaCr+CMI+SWS+PeR+AqT+GWLF+PoD+HLG+OcAg]+[Inco+Stor+MeF+TrN]$$

Where PD = period of difficulty (Time series analysis)

FE = Frequency of events

ARF = Annual rainfall

ArI = Aridity Index

PDSI = Palmer drought Index

Eva = monthly pan evaporation (in mm)

NDVI = Land vegetation cover

Tcr = Total Crop area

CaCr = Ratio of cash crop to cultivated area

CMI = Crop moisture index

SWS = surface water storage (%)

PeR = Perennial River

AqT = Type of Aquifer

GWLF = Ground water level fluctuations

PoD = Population density

HLG = Growth rate of Human and livestock population

Ocag = Agriculture dependent labour

Inco = Income

Stor = Suitable relief material storage

MeF = Trained medical fraternity

TrN = Surface transport network facilities

The drought vulnerability was estimated by (1) collection of information (2) overlaying of different map layers and deriving a polygon representing surface cover, and (3) assigning class ratings for the polygons and estimating cumulative class value. The cumulative DV values were fitted into vulnerability categories as very high, high, moderate and low.

Droughts affect all categories of people; however, the intensity of damage varies with the involvement of the society. For example, small and marginal farmers are more prone to drought effects compared to big and progressive farmers. Therefore, social and economic status, land holdings, family size, live stock population, income other than agricultural sources are some of the indicators that need to be assessed to identify the society vulnerability to droughts. The common variables used by USAID (1995) in vulnerability assessment studies for Kenya are drought risks (CV of NDVI for the past 3 years, average NDVI for the three years), physical insecurity (pastoral insensitivity, conflicts, wildlife) and household entitlements base (livestock unit per capita, share of income from non agricultural activities, income from cash crop, fertile land per capita, access to urban infrastructure).

People have lived with droughts in the past also and must have evolved some methods where the society survived even after devastating droughts/famines that killed thousands of people. Hence there is a need to dig out the traditional knowledge for managing drought effectively. A critical analysis is also required whether the hidden knowledge is still applicable in the fast changing world and also of the climate. With the increasing population of both human and livestock, the pressure on the water resources has increased many folds in the recent past. Availability of fresh water for drinking and irrigation decreased and the situation becomes worst with high variable climates. Therefore there is need to assess the existing water bodies with respect to the demand from various sectors for conservation and judicious management of water resources through improved water management and soil conservation methods

A micro-level assessment of drought vulnerability need to be carried out as social and physical structure variability among the sub district regions (Blocks/Tehsil/mandals/hobli) is large in drought prone districts.

2.12 Vulnerability of Society to Drought

Drought does not appear all of a sudden. It is characterized as creeping phenomenon that affects larger area. Its impacts are felt at all the levels of human and livestock population. The impacts of droughts on society can be divided into two categories viz., the economic and social. On the economic front, decrease in agricultural and fodder productions, reduced income from farm lands, loss of livestock, fall in industrial production, unemployment, shortage of essential commodities, inflation and price rise, poor purchasing power, increasing debts etc. On the social front, it is the malnutrition, poor hygiene, ill health, migration, social liabilities like marriage and education of children, supporting the old age parents etc. The major implications of drought on the society are the starvation due to poor purchasing power resulting in loss of human life. Fortunately from the second half of 20 century, deaths due to drought / famine were not reported in India. With improved stock in food grains and better infrastructure and transport facilities, human loss was almost nil or negligible in the event of droughts. The recent droughts in 1987 and 2002 affected about 285 and 300 million people respectively in the country. During 1987, the GOI has to spend about Rs.17,357.6 million as assistance to manage drought (Kulshreshta, 1997). Similarly, loss due to 2002 drought was estimated as US $ 910.72 million. The impacts of droughts on the number of human population affected, reduction in cropped area and the cattle population during the drought years for the period (1979-1989) prepared by Indu Prakash (1994) and reported by Kulshreshta-1997 is given in Figure 2.11.

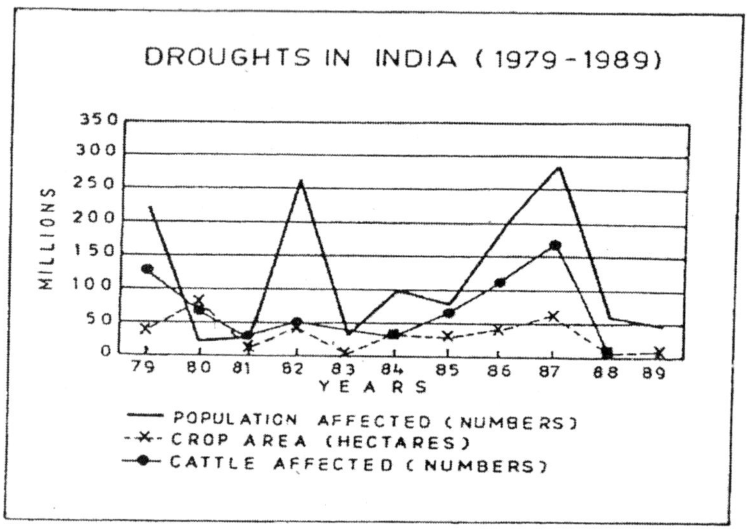

(*Source*: Kulshrestha, 1997)

Fig. 2.11: Drought effects in India (1979-1989)

It is observed that during the period 1979-1989 three drought years viz., 1979, 1982 and 1987 were noticed. Among these drought years, the maximum cropped area and highest number of population both human and livestock were affected during the 1987 drought year.

For identification and establishment of a region affected by drought which has affected the socio-economic structure of the society, the following indicators both physical and of social nature which need to be surveyed as compiled by Nagarajan (2003) is given the following (Table 2.16).

Table 2.16: Physical and social indicators for establishing droughts

- **Revenue units**

  Number of villages chronically under drought

  Marginal villages that are affected by drought

  Population within the chronically and occasional drought areas

  Livestock within the susceptible drought villages

  Spatial distribution of villages

  Estimate of crop yield and grain requirement

- **Scarcity of essential resources**

  Scarcity of possible food, water, fodder, finance etc

  Crop production and supply augmentation

  Fodder availability and supply augmentation

- **Water scarcity**

  Surface water area reduction in perennial reservoir used for public distribution

  Storage level above the critical level for pumping

  Number of reservoirs in critical storage condition

  Reduced flow from natural springs

  Dryness status of village supply tanks/ponds

  Number of village supply tanks under critical condition

  Ground water level reduction-bore/open dug well

  Number of supply wells that are beyond drawing capacity

  Number of villages where source wells are dry

  Spatial distribution hamlets/village/urban areas and their level of deficiency

- **Crop production**

  Crop area

  Sown area of crops

  Loss in critical crop area

  Anticipated crop yield and deficit

  Annual and perennial crop loss

  Damage to crop quality and quantity

  Insect infestation

  Wildlife damage

- **Energy requirement**

  Short supply in energy requirement for agriculture

  Frequent power supply disruption

  Hydro-power generation unit breakdown due to water shortage

  Enhanced price hike in alternate fuels

(Contd.)

- **Health condition and mortality**

  Increase in malnutrition related diseases

  Weaker physical condition of labourers and poor turn-around of work

  Enhanced percentage of spending on nutritional substitute medicines

  Consumption of non-potable quality water

- **Occupational diversification**

  Switch over from agro-based activities to labour for subsistence

  Participation of entire family in relief outside their villages

- **Changes in social values**

  Curtailment of expenditure on social functions

  Increased tree felling and disposal of livestock

  Enhanced anti-social activities

  Increased conflicts -water use, management, political and social

  Disruption in cultural belief system and social values

  Public dissatisfaction over government machinery

- **Consumption pattern and level of self-sufficiency**

  Intake of non-potable water and food of poor nutrition value

  Intestine soaking material intake by livestock

  Loss from dairy and livestock due to:

  Non-availability/high cost of feed and water

  High maintenance cost of livestock

- **Income**

  Wider unemployment rate

  Poor wages and enhanced expenditure

  Increased budget of families towards procurement of food and drinking water

  Increased number of people enrolling for drought relief related activities

  Distress sale of moveable and immoveable holdings

  Increased financial assistance applications under mitigation schemes

- **Financial Institution**

  Strain due to foreclosure, credit risk, capital shortfall

  Poor debt recovery

  Heavy borrowing requirement from individuals, Institutions etc.

  Poor revenue and high expenditure of Institutions

  Slow down of economic growth and retardation in development

(*Contd.*)

- **Migration**

  Land use change in fodder growing area/bio-mass/pasture areas

  Reduced land price and increase in fodder/food articles

  Migration of human and livestock to other places

  Reported mortality rate during transit due to starvation

  Rural population and excess influx into urban areas

References

Appa Rao, G., Abhayankar, V.P. and Mahajan, A. 1981. Analysis of 1979 *kharif* agricultural droughts over India, Pre-published Scientific Report No. 81/2, IMD, Pune.

Bagchi,K.S. 1981. Drought Prone India: Problems & Perspectives, Vol. I& II, Agricole Publishing Academy, New Delhi.

Bastiaanssen, 1998. Remote Sensing in Water Resources Management: The state of the art, IWMI Publication, 118 p.

Ben-Zvi, A. 1987. Indices of hydrological drought in Israel. J. Hydrology, 92: 179-191.

Blumenstock, G.Jr. 1942. Drought in the US analyzed by means of the theory of probability, USDA, Tech BN1819. GPO, Washington.

British Rainfall Organization, 1936. British Rainfall, Air Ministry. Met Office. London.

Burgan, R.E, Hartford, R.A and Eidenshink, J.C. 1996. Using NDVI to assess departure from average greenness and its relation to fire business, Gen. Tech. Report INT-GTR-333, US Department of Agriculture, Forest Service, Intermountain Research Station, Ogden, Utah, 8P.

Byun, H. R. Wilhite, D. A. 1999. Objective quantification of drought severity and duration. J. Climate 12: 2747–2756.

Chowdhury, A., M.M. Bavedkar and P.S. Raut, 1989. Variability in drought incidence over India-a statistical approach. Mausam. 40: 207-214.

CRIDA, 2012. A report of the ICAR expert committee on agricultural situation in Anantapur district, Andhra Pradesh. March 2012, 70 P.

Dezman, L. E.; B. A. Shafer; H. D. Simpson; and J. A. Danielson. 1982. Development of a surface water supply index-A drought severity indicator for Colorado. In A. I. Johnson and R. A. Clark, (eds). Proc. International Symposium on Hydrometeorology. American Water Resources Association, Bethesda, Maryland.

Donald A. Wilhite, and Michael H.Glantz. 1987. Understanding the drought phenomenon. The role of definitions in planning for Drought, West View Press [Eds. Donald Wilhite, William Easterliy and Deborah A. Wood]

Eliot, J. 1902. Monthly and annual rainfall of 457 stations in India to the end of 1900 Mem.Ind.Meteorology Department 10,709.

Frere, M., Popov, G.F., 1979. Agro meteorological crop monitoring and forecasting, FAO plant production paper 17. Food and Agricultural Organization, Rome.

Gibbs, W. J.; and J. V. Maher. 1967. Rainfall deciles- as drought indicators. Bureau of Meteorology, Bulletin No. 48. Melbourne, Australia.

Gibbs, W. J.1975. Drought—its definition, delineation, and effects. In Drought: Lectures presented at the 26th session of the WMO Executive Committee. Special Environmental Report No.5. WMO, Geneva. 3-39.

Goldberg, I Ed., 1972. Agro climatic atlas of the world, Hydrometizdat, 133p.

Gore P.G, Thakur Prasad and H.R. Hatwar , 2010. Mapping of drought areas over India. NCC Research Report IMD, Pune. 21p.

Hanumantha Rao. 1994. Report of the Technical Committee on Drought Prone Areas Programme and Desert Development Programme. Ministry of Rural Development , GOI, New Delhi. 73p.

Hayes M.J and Decker, W.L. 1998. Using satellite and real-time weather data to predict maize production, International Journal of Biometeorology, 42:10-15.

Heathcote, R. L.1974. Drought in South Australia. In G.F. White, (ed). Natural Hazards: Local, National, Global. Oxford, New York. 129 p.

Herbst, P.H., Brednkamp, D.B. and Barker, M.H.G. 1966. A technique for the evaluation of drought from rainfall data. Journal of Hydrology, 4:264-272.

Hudson, H. E.; and R. Hazen. 1964. Drought and low stream flow. Chap. 18. In V. T. Chow, (ed). Handbook of Applied Hydrology. McGraw Hill, New York.

Indu Prakash. 1994. Disaster Management, Societal Vulnerability to natural calamities and Manmade Disasters-preparedness and response. Chapter-3, Rashtra Prahari Prakashan. Ghaziabad.

Ivanov, N. N. 1948. Landscape-climatic zones of the earth surface. Proc. All Soviet Geographical Conference, New Series Vol. 1. Academy of Sciences of the U.S.S.R. , Leningrad.

IMD, 2010 Climate profiel of India. Met Monograph No.Environment Meteorology-01/2010. India Meteorological Department. Ministry of Sciences, New Delhi.129p.

Johnson E, Gary, Rao Achuntuni V, Thiruvengadachari S and Kogan Felix, 1993. The role of NOAA satellite data in Drought early warning and monitoring: Selected studies, in Drought Assessment, Management and Planning: Theory and case studies, edited by Donald A. Wilhite,& Kulwer Academic Publishers.USA.

Kasturirangan,K.1996. ground based measurements for interpretation of vegetation indices over developing countries. Advanced Space Rearch,18(8),31-40.

Kogan, F.N. 1995. Droughts of late 1980s in the USA as derived from NOAA polar orbiting satellite data, Bulletin of American Meteorological Society, 76:655-668.

Kogan, F.N. 2001. Operational Space Technology for Global Vegetation Assessment, Bulletin of the American Meteorological Society, 82: 1949-1964.

Krishnan.A and Thanvi, K.P. 1971. Occurrence of droughts in Rajasthan during 1941-60. Proceedings of All India Seminar on Dryland Farming. New Delhi.

Krishnan, A. and Thanvi, K.P. 1972. Studies on duration of agricultural growing season in India under rainfed farming using climatological water budgeting approach. Annual Progress Report, Central Arid Zone Research Institute, Jodhpur.

Kulshrestha, S.M., 1997. Drought management in India and potential contribution of climate prediction, Joint COLA/CARE Technical Report No. 1, Centres of the Institute of Global Environment and Society, September 1997, 105p.

Lee,D.M. 1979. Australian Drought Watch System. In:M.T. Hincy, (ed) Bostwana Drought Symposium, Bostwane Society, Gaborone, Botswana.

Levitt,.J. 1958. Frost, drought and heat resistance. Protoplasmatologia 8(6).

Linsley, R. K., Jr., M. A. Kohler and J. L. H. Paulhus. 1975. Hydrology for Engineers, 2nd ed. McGraw-Hill, Kogukusha, Tokyo.

Maher, J.V. 1968.Drought assessment by statistical analysis of rainfall. Proc. Of the WMO Seminar on Agricultural Meteorology, Melbourne, Australia, 28 November-16 December 1966. Published by the Bureau of Meteorology, Melbourne, Australia 2:713-730.

Manual for Drought Management. 2009. Department of Agriculture and Cooperation Ministry of Agriculture, GOI, New Delhi, 192p.

McKee, T.B., Doesken, N,J. and Kleist, J. 1993. The relation of drought frequency and duration to time scales. Preprints, 8th conference on Applied Climatology, 17-22 Jan, Anaheim, CA. 179-184.

Michale I. Caughlan, 1987. Monitoring of droughts in Australia , In ; planning for drought (eds) Donal A.Wilhite and William E.Easterling with Deborah A.Wood. went view pron-London.

Murthy C.S. and Seshasai, M.V.R. 2010. Agricultural Drought Monitoring and Assessment in : Remote sensing Applications. Eds. P.S.Roy, R.S.Dwivedi, and D.Vijayan. National Remote sensing centre, Hyderabad. 303-330.

Nagarajan, R.2002. Integrated Drought Vulnerability assessment and Monitoring using RS and GIS (Personal Communication).

Nagarajan,R, 2003. Droughts Assessment, Monitoring Management and Resources Conservation Capital Publishing Company, New Delhi-312p.

National Commission on Agriculture, 1976. Government of India, Ministry of Agriculture & Irrigation, Department of Agriculture, New Delhi.

Pai D.S.Sridhar, P.GuhataKurta and H.R. Hatwar.2010. District-wise drought climatology of the Southwest monsoon season over India based on SPI. NCC Research Report.No.2/2010.

Palmer, W. C. 1957. Drought—A normal part of climate. In Weekly Weather and Crop Bulletin 44, No. la (January 10):6-8.

Palmer, W. C. 1965. Meteorological drought. Research Paper No. 45. U. S. Weather Bureau, Washington, D.C.

Parthasarathy , B. A.A.Munot and D.R. Kothwale 1995. Monthly and seasonal rainfall series for All India homogenouos regions and meteorological sub-divisions. IITM. Research Report No. RR-065.

Popov, V. P. 1948. Moisture balance in the soil and the dryness indices of climate in the Ukrainian S.S.R. Sci. Rep. of the State Univ. of Kiev, Vol.7, No. 1.

Ramakrishna, Y.S., G.G.S.N. Rao, Kesava Rao, AVR. and Vijaya Kumar, P. 2000a. Weather Resources Management, Chapter -3 in Natural Resource Management for Agricultural Production in India, Eds J.S. P. Yadav and G.B.Singh. Int. Conf. on Managing Natural Resources for sustainable Agricultural production in 21st Century. Feb 14-18, 2000. New Delhi, 245-370.

Ramakrishna, Y.S., G.G.S.N. Rao, and Kesava Rao, AVR. 2000b. Role of weather in Dryland Agricultural production, Proc.III SERC School on Agricultural Droughts: Role of crop growth modelling. CAZRI, Jodhpur.

Ramana Rao, B.V., Sastry, ASRAS and Ramakrishna, Y.S. 1981. An Integrated scheme of drought classification as applicable to the Indian arid zone. Idojaras 85: 317-322.

Ramdas ,LA 1950. Rainfall and Agriculture , Indian Journal Department of Geoph.1.262.

Ramdas, L.A. 1960. Crops and Weather in India. Indian Council of Agricultural Research, New Delhi,127p.

Rao, G.G.S.N.1986. Some studies on influences of climate on Grasslands in western Rajasthan. Ph.D thesis. Andhra University , Visakhapatnam.

Samra, J.S.-2004. Review and Analysis of Drought Monitoring, Declaration and Management in India. Working paper No.84.IWMI, Colombo.31p.

Sandford, S. 1979. Towards a definition of drought. In M. T. Hinchey, ed. Botswana Drought Symposium. Botswana Society, Gaborone, Botswana.

Sastri, ASRAS, Ramakrishna, YS, and Ramana Rao, BV. 1981. A new methods of classification of agricultural droughts. Arch. Fur. Met. Geoph. And Biokl. Ser. B. 29: 293-297.

Sastry, A.S.R.A.S., Ramakrishna, Y.S. and Ramana Rao, B.V. 1982. Droughts in Western Rajasthan. Div. Tech. Rep 1/82, CAZRI, Jodhpur, 24p.

Sikka.D.R. 1999. Monsoon drought in India, Report no.2, Joint Report of COLA/CARE Tech. Report, 270p.

Subramanyam, V.P.1964. 'Climatic water balances of the Indian arid zone'. Proceedings of Symposium on Problems of Indian Arid Zone, Jodhpur, 405-411.

Subrahmanyam, V. P. 1967. Incidence and spread of continental drought. WMO/IHD Report No.2. WMO, Geneva.

Thornthwaite, C.W. 1947. An approach towards a rational classification of climate. Geographical Review, 38(1): 55-94.

Thornthwaite, C.W, and Mather, J.R. 1955. The Water Balance. In Climatology, VIII (1) Drexel Institute of Technology, New Jersey, USA 104p.

Tucker CJ., 1979. Red photographic infrared linear combinations for monitoring vegetation, Remote Sensing of Environment, 8:243-251.

Upadhyay, D.S. 1994. Hydrometeorology in Agriculture. Indian Farming, October, Volume 44, No.7, p 41-43.

USAIDFEWS PROJEC T6, 1995, Vulnerability Assessment , February, 1995. Bureau of Africa. Disaster Response coordination USAID/FEWS, Arlington, UA.

Van Rooy, M.P. 1965. A rainfall anomaly index independent of time and space. NOTOS Weath Bur. S. Afr. 14:43-48.

Victor, U.S., Srivastava, N.N. and Ramana Rao, B.V. (1988). Quantification of crop yields using simple soil water balance model. J. Theoretical and Applied Climatology, 39: 73-80.

Walker, G.T, 1910. Correlations in seasonal variations of weather. II.Mem.Ind. Meteor.Department. 21 (Part-1) 1-21.

Walker,G.T,1923.Correlations in seasonal Weather,VIII,Mem.Ind.Meteor.Department. 24: 275-232.

Walker, G.T. 1924. Correlation of seasonal variations of weather, the local distribution of Monsoon rainfall. Mem.Ind.Meteor.Dept.21, 30-40.

Whipple, W., Jr. 1966. Regional drought frequency analysis. Proc. ASCE 92, No. 1R2 (June 1966):11-31.

Wilhite, D.A. & Glantz, M.H. 1985. Understanding the drought phenomenon: The role of definitions, Water International, 10(3): 111-120.

WMO Bulletin No.138 (1975). Drought and Agriculture, Tech note. No138, WMO No.392. Geneva.

WMO, No 1006. 2006. Drought monitoring and early warning: concepts, progress and future challenges. World Meteorological Organization WMO No.1006. WMO. Geneva.

□□□□

CHAPTER 3

Climate Change and Droughts

3.1 Factors Contributing to Climate Change

As per IPCC, the definition of climate change refers to "any change in climate over time whether due to natural variability or as a result of human activity". This usage differs from that in the framework convention on climate change where the word climate change refers to a change of climate that is attributed directly or indirectly to human activity that alters the composition of the global atmosphere and that is in addition to natural climate variability observed over comparable time periods.

3.1.1 Natural Causes to Climate Change

The earth's climate is influenced and changed due to natural causes like volcanic eruptions, ocean current, the earth's orbital changes and solar variations.

Volcanic eruptions - When a volcano erupts, it throws out large volumes of sulphur dioxide (SO_2), water vapour, dust and ash into the atmosphere. Large volumes of gases and ash can influence climatic patterns for years by increasing planetary reflectivity causing atmospheric cooling. Tiny particles called aerosols are produced by volcanoes. Because they reflect solar energy back into space they have a cooling effect on the world. The greenhouse gas carbon dioxide is also produced; however the CO_2 production is insignificant when compared to emissions created by humans.

Ocean current - The oceans are a major component of the climate system. Ocean currents move vast amounts of heat across the planet. Winds push horizontally against the sea surface and drive ocean current patterns. Interactions between the ocean and atmosphere can also produce a phenomenon such as

El Niño which occurs every 2 to 6 years. Deep ocean circulation brings in cold water from the poles towards the equator and movement of warm water from the equator back towards the poles. Without these movements the poles would be colder and the equator warmer. The oceans play an important role in determining the atmospheric concentration of $CO_2$and changes in ocean circulation may affect the climate through the movement of CO_2 into or out of the atmosphere.

Earth orbital changes - The earth makes one full orbit around the sun each year. It is tilted at an angle of 23.5° to the perpendicular plane of its orbital path. Changes in the tilt of the earth can lead to small but climatically important changes in the strength of the seasons, more tilt means warmer summers and colder winters; less tilt means cooler summers and milder winters. Slow changes in the Earth's orbit lead to small but climatically important changes in the strength of the seasons over tens of thousands of years. Climate feedbacks amplify these small changes, thereby producing ice ages.

Solar variations - The Sun is the source of energy for the Earth's climate system. Although the Sun's energy output appears constant from an everyday point of view, small changes over an extended period of time can lead to climate changes. Some scientists suspect that a portion of the warming in the first half of the 20th century was due to an increase in the output of solar energy. As the sun is the fundamental source of energy that is instrumental in our climate system it would be reasonable to assume that changes in the sun's energy output would cause the climate to change. Scientific studies demonstrated that solar variations have performed a greater role in past climate changes. For instance a decrease in solar activity was thought to have triggered the Littie Ice Age between approximately 1650 and 1850, when Greenland was largely cut off by ice from 1410 to the 1720s and glaciers advanced in the Alps.

3.1.2 Human Causes to Climate Change

The Industrial Revolution in the 19th century saw large-scale use of fossil fuels for industrial activities. Fossil fuels such as oil, coal and natural gas supply most of the energy needed to run vehicles, generate electricity for industries and households. The energy sector is responsible for about ¾ of the carbon dioxide emissions, 1/5 of the methane emissions and a large quantity of nitrous oxide (IPCC, 2001).

Carbon dioxide is undoubtedly, the most important greenhouse gas in the atmosphere. Changes in land use pattern, deforestation, land clearing, agriculture, and other activities have all led to a rise in the emission of carbon dioxide. Methane is another important greenhouse gas in the atmosphere. It is released from animals such as dairy cows, goats, pigs, buffaloes, camels, horses and sheep. Methane is also emitted during the process of oil drilling, coal mining, leaking gas pipelines, landfills and waste dumps.

The impacts of global warming can be seen on plant growth and crop yields and increased frequency of extreme weather conditions around the world. It is more clear in the dramatic changes in the area of the polar caps.

3.2 Climatic Trends

3.2.1 Global Level

Current level CO_2 Observations

Carbon dioxide is the most important anthropogenic greenhouse gas and the concentration has increased from 280 ppm (pre-industrial value) to 393.65 ppm in 2011 (Figure 3.1). The atmospheric carbon dioxide concentration in 2005 exceeds by far the natural range over the last 650,000 years (180-300 ppm as determined from Ice Cores). It was also reported that the CO_2concentration growth rates was larger during last 10 years (1.9 ppm/year) than it has been since the beginning of continuous direct atmospheric measurements (1.4 ppm).

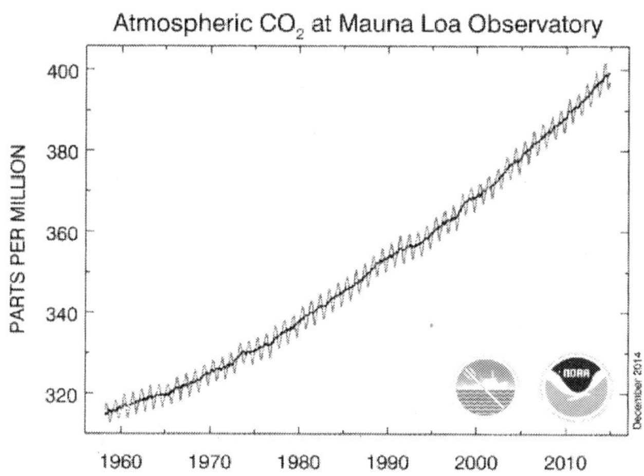

(*Source*: Scripps Institution of Oceanography NOAA Earth System Research Laboratory)

Fig. 3.1: Global CO_2 levels

Global Temperature Trends

It is observed that there are strong evidences to global warming of climate is due to increasing level of CO_2. From the observations of the global average, air and ocean temperatures, widespread melting of snow and ice, and rise in the global average sea level were noticed. Some of the observations are as follows:

- Eleven out of the 12 years (1995-2006) rank among the 12 warmest years in the instrumental record of global surface temperature (since 1850) was observed during the period 1995-2006 and the warming trend continued with years 2009 and 2010 recorded as the warmest years on record (according to GISS) since 1880. Graph below indicates the global mean land-ocean temperature index, for the period 1880 to 2010 with the base period 1951-1980 (Figure 3.2). The dotted black line is the annual mean and the solid red line is the five-year mean. The green bars show uncertainty estimates.

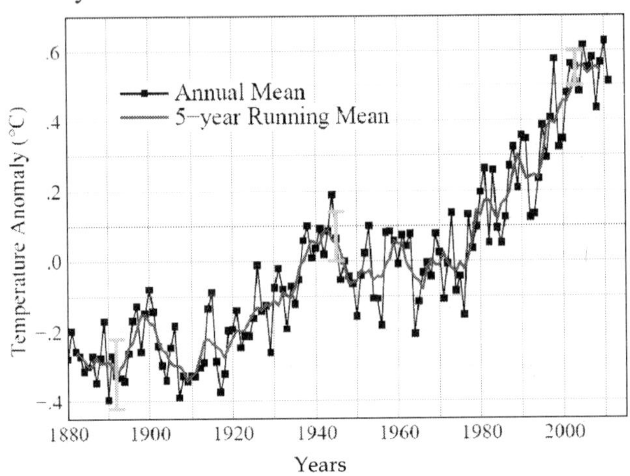

(*Source*: http://data.giss.nasa.gov/gistemp/graphs_v3)

Fig. 3.2: Global Land-Ocean Temperature anomaly

- The updated 100 year linear trend (1906-2005) is 0.74°C, larger than the corresponding trend of 0.6°C for the years 1901 to 2000. It is also reported that the warming trend over the last 15 years is nearly twice that for the last 100 years.

- The analysis of lower and mid-troposphere temperature show warming rates similar to those of surface temperature record.

- The average atmospheric water vapour content increased from 1980 onwards over land and oceans as well as in the upper troposphere. This is in consonance with the extra water vapour that warmer air can hold.

- The average temperatures of the global oceans since 1961 have increased to depths of 3000 meters and the oceans have been absorbing more than 80 percent heat added to the climate system. Such warming causes sea water to expand contributing to sea level rise.

- Mountain glaciers, snow cover have declined on average in both the hemispheres.

3.2.2 National Level

Following are the results with report to climate trend (IMD, 2010) of India

- The monsoon rainfall at All India level did not show any trend and is random in nature over a long period of time.
- Areas of increasing trend in monsoon rainfall are found along the west coast, north Andhra Pradesh and north-west India and those of decreasing trend has been observed over east Madhya Pradesh and adjoining areas, north-east India and parts of Gujarat and Kerala (-6 to -8% of normal over 100 years)
- Surface air temperature for the period 1901-2009 (Figure 3.4) indicated a significant warming of 0.56°C over 100 years.

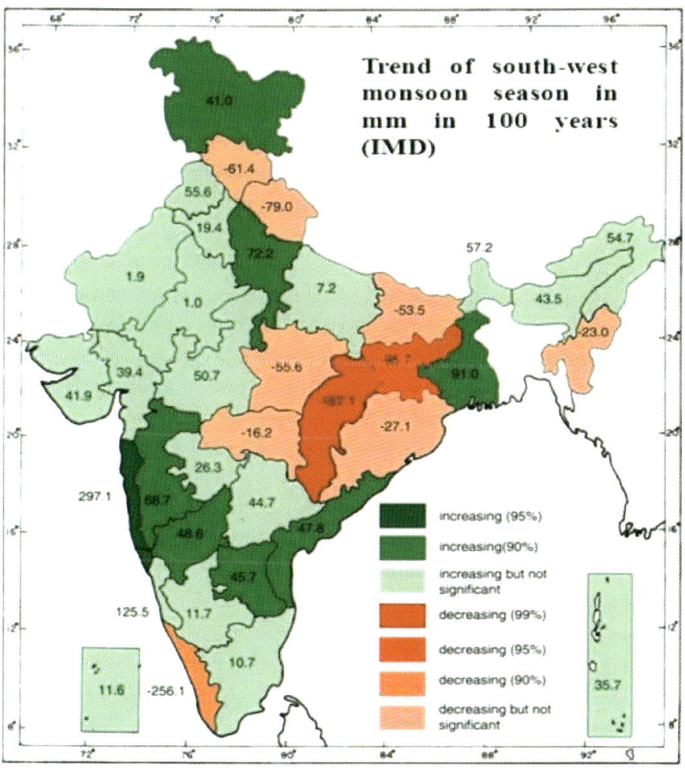

(*Source*: IMD, 2010)

Fig. 3.3: Trend of south-west monsoon season in mm in 100 years

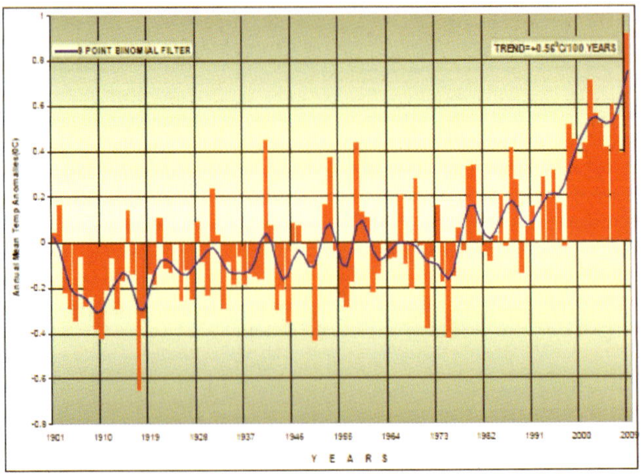

(*Source*: IMD, 2010)

Fig. 3.4: All India annual mean temperature anomalies for the period 1901-2009

- The spatial distribution of temperature changes (Source:IMD,2010) indicated a significant warming trend along the west coast, central India, and interior Peninsula and over north-east India. However, cooling trend has been observed in north-west and some parts in southern India.

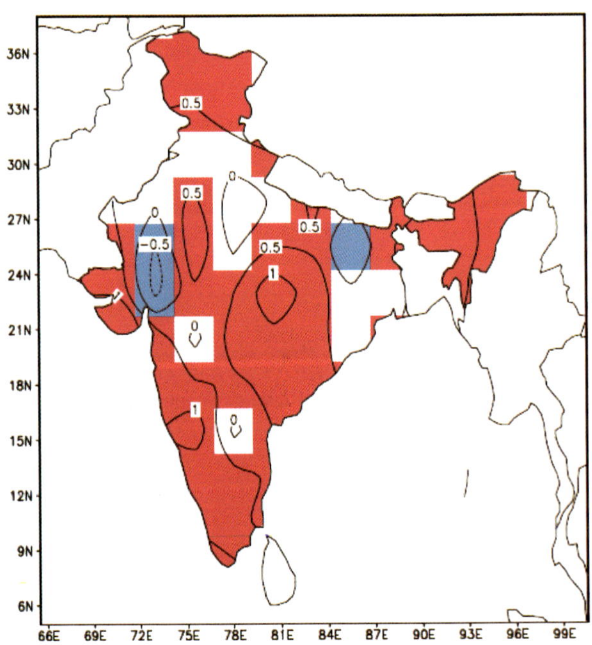

(*Source*: IMD, 2010)

Fig. 3.5: Spatial pattern of mean annual temperature anomalies (1901-2009)

- Instrumental records over the past 130 years have not shown any significant long term trend in the frequencies of large scale droughts or floods in the summer monsoon season.
- The only change is the alternating sequence of multi-decadal periods of more frequent droughts followed by periods of less frequent droughts.
- The total frequency of cyclonic storms that form over Bay Bengal has remained almost constant over the period 1887-1997. A slight decrease in trend in the frequency of cyclonic disturbances was apparent during the monsoon season.

3.3 Climate Projections

3.3.1 Global Level Projections on Future Changes in Climate (IPCC 2001, 2007)

- For the next two decades, a warming of about 0.2°C per decade is projected for a range of (IPCC Special Report on Emission) Scenarios. Even if all the greenhouse gases are kept constant at the year 2000 levels, a warming of about 0.1°C per decade would be expected due to the slow response of the oceans.

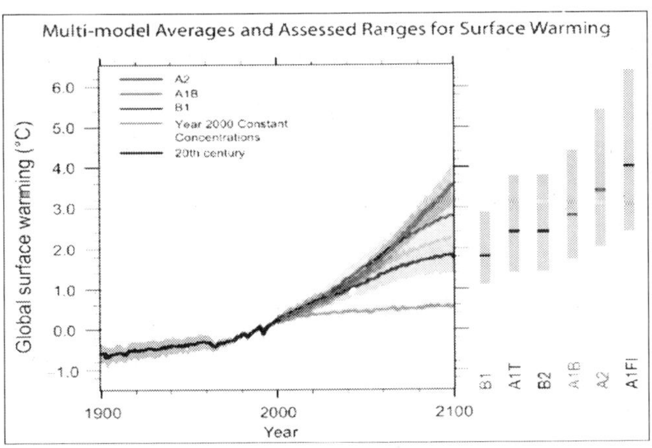

Fig. 3.6: Global Temperature Projections for 2100 (AR4,IPCC 2007)

Solid lines are multi-model global averages of surface warming (relative to 1980-99) for the scenarios A2, A1B and B1, shown as continuations of the 20th century simulations. Shading denotes the plus/minus one standard deviation range of individual model annual averages. The orange line is for the experiment where concentrations were held constant at year 2000 values. The grey bars at right indicate the best estimate (solid line within each bar) and the likely range assessed for the six SRES marker scenarios. The assessment of the best estimate and likely ranges in the grey bars includes the AOGCMs in the left part of the figure, as well as results from a hierarchy of independent models and observational constraints."

- Continued greenhouse gas emissions at or above current rates would cause further warming and induce many changes in the global climate system during the 21st Century.
- Different estimates of temperature change by the end of the Century (2090 – 2099) compared to the temperatures of 1980 – 1999 ranged from 1.8 to 4.0°C under different greenhouse gas emission scenarios.
- Projected warming in 21st Century is expected to be greatest over the land and high over the northern latitudes and least over the southern ocean and parts of North Atlantic Ocean.
- It is very likely that heat waves and precipitation events will continue to come more frequent.
- Sea ice is projected to shrink in both Arctic and Antarctic under all greenhouse gas emission scenarios.
- Extra tropical storm tracks are projected to move pole wards.

3.3.2 Climate Projections at the National Level (Rupakumar et al., 2006)

The increase in rainfall from the baseline period (1961-1990) to the end of the 21st century ranges between 15 to 40 per cent among the models.

- In the case of mean annual temperature, the increase is of the order of 3°C to 6°C.
- Compared to the A2 scenario, the B2 simulations show subdued trends in the future. The temperature, however, shows comparable increasing trends in IS92a and A2, but B2 shows slightly lower trends.
- The magnitudes of projected change differ considerably from one model to the other, when projections of rainfall are considered at state level.
- There is very little or no change noted in the monsoon rainfall over a major part of peninsular India.
- It is important to note here, that the maximum change in rainfall occurs over the climatologically low rainfall region of north-western India.
- Examination of the spatial patterns of annual temperature changes in the two future time slices for different models indicates that the warming is more pronounced over the northern states of India.

- Different models/experiments generally indicate the increase of temperature to be of the order of 2-5°C across the country. The warming is generally higher in the IS92a scenario runs compared to A2 and B2 simulations. Also, the warming is more pronounced during winter and post- monsoon months, compared to the rest of the year.

3.3.3 Climate Projections at the Regional Level

- In the regional climate model, under increasing atmospheric GHG concentrations, the mean surface temperatures are seen to increase everywhere in the region, in all the seasons.
- The warming is more pronounced over land areas, with the maximum increase over northern India.
- The warming is also relatively greater in winter and post-monsoon seasons.
- The summer monsoon season is marked by a relatively lower magnitude of warming.
- However, the spatial patterns of warming during the monsoon season indicate that the maximum warming occurs over northern India, with a secondary maximum over the eastern peninsula.
- Regarding precipitation however, there is a large spatial variation in the relative increase in monsoon precipitation, obviously due to the climatological patterns of rainfall.

3.3.4 Projections of Extremes in Rainfall and Temperature

- Considering the control run of the model as the baseline Climatology representing the present day conditions, the future scenarios representing the 2050s, under the IS92a scenario of GHG emissions, were derived.
- In the IS92a scenario, the model showed an overall decrease in the number of rainy days over a major part of the country.
- This decrease is more in western and central parts of the country (by more than 15 days) while along the foothills of Himalayas (Uttaranchal) and in north-east India the number of rainy days is found to increase by 5-10 days.
- An increase in GHG concentrations may lead to overall increase in the rainy day intensity by 1-4 mm/day, except for small areas in north-western India, where the rainfall intensities decrease by 1 mm/day.

- The model results also indicate that there will be an overall increase in the highest one-day rainfall over a major part of the Indian region.
- This increase may be up to 20 cm/day. However, in some parts of north western India, a decrease in extreme rainfall up to about 10 cm/day has been noticed in the GHG experiment.

3.4 Consequences of Climate Change on Droughts and Agriculture

Climate change is a long term process whereas the variability is a short term fluctuation and it would be high during the period in which particular weather parameter changes. The climate projections up to 2070 for Indian region as given by Watson *et al.* (1998) are given below (Table 3.1).

Table 3.1: Projected climate change over India (Watson *et al.* 1998)

Climatic factors	*Rabi*		*Kharif*	
	2010	2070	2010	2070
CO_2 (ppm)	397-416	605-755	397-416	605-755
Temperature increase (°C)	0.3 to 0.7	1.1 to 4.5	0.1 to 0.3	0.4 to 2.0
Rainfall change (%)	0	-10 to +10	0	0 to +10

It shows that rainfall is likely to increase by 10 percent during *kharif* season and winter rainfall *rabi* shows large variation. It was reported by many authors that frequencies of high intensity rain storms are likely to increase with more gaps in between time events. The climate change projections for India by Lal *et al* (2001) showed that (Table 3.2) the mean temperature may increase by 3.5 to 5.5 by 2080. These projections showed more warming in winter over summer. This is much more than the warming (0.6°C) actually occurred over the country in the past 100 years. The rainfall showed a marginal increase of 7 to 10 per cent of the corresponding period. The future scenario of droughts depends upon how best we could project the monsoon climate over the country. The projections obtained from the models need to be looked into critically and how these changes would impact the characteristics of monsoon and the corresponding agricultural production.

Table 3.2: Climate change projection for India

Year	Season	Temperature change (°C)		Rainfall change (%)	
		Lowest	Highest	Lowest	Highest
2020s	Annual	1.00	1.41	2.16	5.97
	Rabi	1.08	1.54	-1.95	4.36
	Kharif	0.87	1.17	1.81	5.10
2050s	Annual	2.23	2.87	5.36	9.34
	Rabi	2.54	3.18	-9.22	3.82
	Kharif	1.81	2.37	7.18	10.52
2080s	Annual	3.53	5.55	7.48	9.90
	Rabi	4.14	6.31	-24.83	-4.50
	Kharif	2.91	4.62	10.10	15.18

(*Source*: Lal *et al.*, 2001)

Before understanding of the possible changes expected in weather conditions in the future, it would be better to have a quick look how the rainfall during monsoon behaved over the county over a long period. Analysis of 122 years of long term monsoon rainfall data showed that in about 71 per cent of years, the season's rainfall as a whole was normal (±10% of LPA). It was found to be excess (above 110%) in 14% of the years and deficit (below 90%) in 15% of the years (Kulshrestra 1997). The study further state that the monsoon seasonal rainfall has been either normal or in excess in 85% of the years. However, the rainfall is highly variable in time and space on both interannual and intra seasonal scales. Monsoon exhibited an epochal nature of 3 decades either positive or negative anomaly. The monsoon is presently in the positive epoch since 1988. Based on this pattern broadly one can get an idea about the possible nature of monsoon behaviour in the coming years which can help to prepare plans to manage them.

While analysing droughts, Kulshrestha (1997) observed that the distribution of rainfall and the periods of deficits are crucial and important for agricultural production. Though there may be no deficit in the total rainfall and yet the rainfall deficit at crucial stage could lead to drought situation. Have even a small changes in precipitation or snow melt could have adverse consequence if the same scenario repeats in the next season also.

Shifts in major synoptic systems influenced by the general circulation may result in drought. The parameters could be natural in origin (ENSO) or of anthropogenic origin. Identification and understanding of these mechanisms are the main scientific challenges in assessing the climate change on drought. Soil moisture deficit arising due to shortage in rainfall would reduce the vegetation cover that would make the land surface to lose more radiation energy into space that alters global energy balance. Thus, rendering the atmosphere above the ground into stable state that would further reduce rainfall (Kulshrestra 1997).

Climate prediction for monsoon rainfall uses either empirical approach numerical approaches or statistical approaches. In recent years, ELNino has been related to Indian monsoon which is regarded to have negative correlation. The forecasting technique of IMD uses current year's EL Nino as one of the parameters in prediction of monsoon rainfall. The relationship between summer monsoon rainfall and ELNino as given by www.iitm.org.inis shown here.

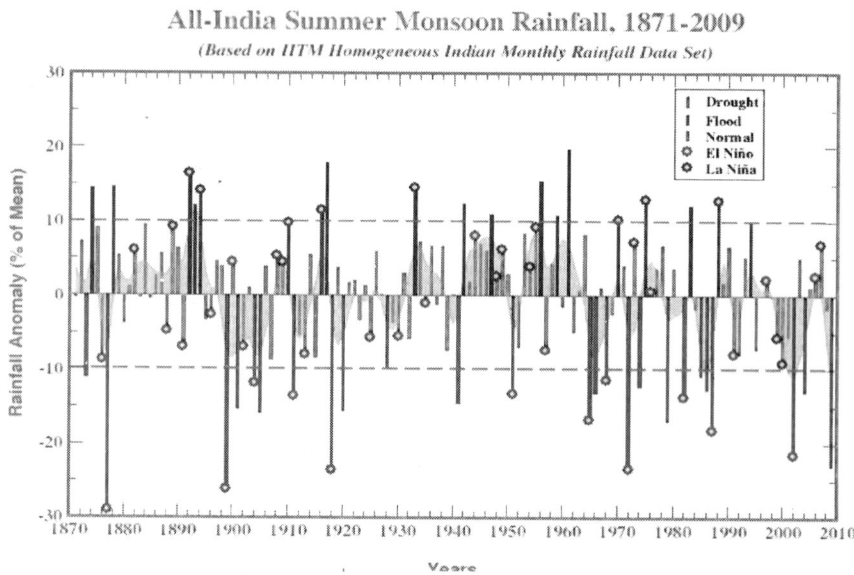

(*Source*: IITM)

Fig. 3.7: Relation between EL Nino and rainfall

It is observed that there is no one-to-one relationship between these parameters. However, it can be observed that all drought years are ELNino years but not all ELNino years are drought years in the country. Many researchers have evaluated the model projected climate on future pattern. American studies have concluded that droughts in thc dry land regions of USA are likely to increase in frequency, intensity, magnitude and duration due to increased levels of GHG in the atmosphere(Rind*et al.* 1990). The study further projects that the frequency of severe droughts in the central dry lands of USA could change from 5 to 50% by the year 2050. The increase in frequency of droughts would alter the dust concentration in the lower atmosphere and reduce the solar radiation thus affecting the energy balance. The impacts of increased levels of dust (Asian Brown Clouds) in the atmosphere and the frequency of drought in the coming years as reported by Ramanathan *et al.*, 2005 are shown below.

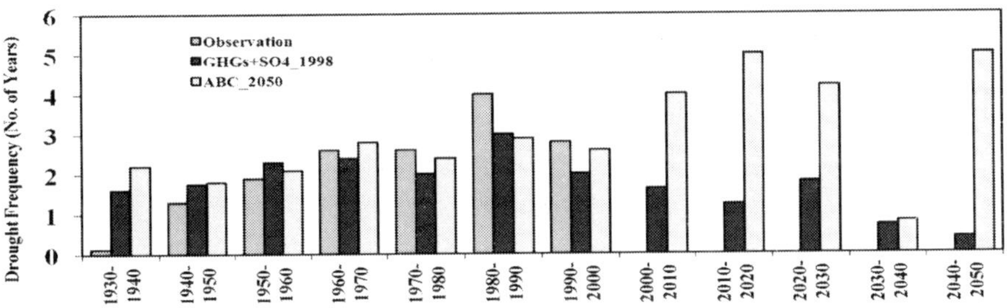

(*Source*: Ramanathan *et al.*, 2005)

Figure 3.8: The frequency of droughts as influenced by climate change

It is seen that frequency of drought due to increased activity of Asian Brown clouds increases up to 2030 and again during 2040-50. However, the influence of GHG on the overall changes in weather parameters is likely to be curtailed in the future due to possible mitigation measures.

3.4.1 Impact of Climate Change on Agriculture in some States of India

Some of the results on the Climate Change on Agriculture over India (2008) reported and published by AICRP on Agro meteorology centre of KAU, Thrissur for few selected states in India are given below.

Punjab & Haryana

Terminal heat stress in rabi season affect the wheat crop productivity. Increase in temperature by 1°C may result in wheat production loss up to 10 percent. *Brassica* (mustard) crop will be less affected due to increased temperature compared with other crops.

Eastern Uttar Pradesh & Bihar

Due to erratic monsoon rainfall both in magnitude and its distribution in east Uttar Pradesh and increased frequency of floods in Bihar affected farming in these regions.

West Bengal

Saline zone facing problems of sea level rise. Due to decrease in pre-monsoon showers – sowing of upland rice and jute was affected. Changing rainfall pattern in the hill zones affected tea production.

Gujarat

Increase in temperature during winter season affected the productivity of wheat crop. Rise in maximum temperature on wheat productivity was found to be more prominent than the minimum temperature.

Odisha

A large part of the state is subjected to seasonal weather aberrations, *viz.*, floods and droughts. Shorter winter season affected the crop productivity.

Chhattisgarh

Climate shifts from sub-humid to semi-arid type was noticed in 16 districts. Rainfall reduced significantly over greater part of the state resulted in decreasing / stagnant rice yields.

Karnataka

Changes in availability of rainwater and shifts in peak rainy season altered the cropped area of major crops, *viz.*, finger millet, chickpea, red gram.

Tamil Nadu

Kharif rice productivity is expected to be reduced by 10-15 percent by 2020, 30-35 by 2050 and upto 80 percent by 2080 over the present yield. Coconut production suffered significantly due to drought conditions.

Kerala

Droughts in 1983 and 2003-04 affected the cardamom production considerably. Drought in 2004 reduced the cocoa yield around 39 percent. Summer season droughts could reduce the coconut yield up to 60 percent.

As these studies are limited, they may further be strengthened. With the Commencement of National Initiative on Climate Resilient Agriculture (NICRA) by GOI during 2010, Climate Change Research programs shall gather momentum further to face challenges posed by climate change issues in the coming years.

References

IMD,2010. Climate profile of India.By S.D.Attri and Ajit Tyagi.Met Monograph. No. Env. Report. -01/2010.122p.

IPCC (Intergovernmental Panel on Climate Change), 2001. Climate Change 2001: Impacts, adaptation, and vulnerability [Contribution of Working Group II to the Third Assessment Report of the Intergovernmental Panel on Climate Change; summary for policymakers] Cambridge: Cambridge University Press. 1032p.

IPCC (Intergovernmental Panel on Climate Change Working Group I) Climate Change 2007: The Physical Science Basis IPPCC Working Group I.

Kulshrestha, S.M., 1997. Drought management in India and potential contribution of climate prediction, Joint COLA/CARE Technical Report No. 1, Centres of the Institute of Global Environment and Society, September 1997, 105p.

Lal, M. 2001. Future climate change: Implications for Indian summer monsoon and its variability, Current Science, Vol 81: No.9. November 10, 1205p.

Rind *et al.* 1990.Potential evapotranspiration and likelyhood of future drought. Journal of Geophysics Res. 95.

Ramanathan, V., C. Chung, D. Kim, T. Bettge, L. Buja, J. T. Kieh, W. M. Washington, Q. Fu, D. R. Sikka ,and M. Wild. 2005.Atmospheric brown clouds: Impacts on South Asian climate and Hydrological Cycle. PNAS. Vol.102:15: 5326–5333.

Rupakumar, K., Sahani, A.K., Krishna Kumar, K., Patwardhan, S.K., Mishra, P.K., Revade Kar, J.V., Kamala, K. and Pant, G.B. 2006. High resolution climate change Scenarios for India for 21st Century, Current Science, Vol 90(3): 10 Feb 2006.

Climate Change and Agriculture over India. 2008. (Eds. PrasadaRao, GSLHV., Rao, GGSN, Rao, VUM., and Ramakrishna,Y.S), AICRP on Agrometeorology Kerala Agricultural University. PP 256.

Watson, R.T., Zinyoera, M.C., and Moss, R.H. 1998. The Regional Impacts of Climate Change: An Assessment of Vulnerability. A Special Report of IPCC Working Group II. Cambridge: Cambridge University Press.

□□□□

Chapter 4

Drought Monitoring

Drought is a temporary phenomenon which must be clearly differentiated from aridity, which is a permanent feature of climate over a region. Seasonal aridity, a well defined dry season also need to be distinguished from drought and the differences in the terminology should be understood properly in drought monitoring studies. It occurs in almost all the countries that are located in dry and humid climates, though it is strongly believed that it is confined mostly to arid and semi-arid and sub-humid climates. Drought in its spatial extent and severity varies on seasonal and on annual time scales (WMO bulletin No-1006, 2006). In many countries viz., Australia, China, India, USA and over Africa, drought occurs over some portion of the country in each year. Drought is not a calamity but it can turn to be a disaster, based on its impact on local people, environment, economies and the time taken to recover from it shoks. Therefore, Drought Risk Management involves improving the capacity of the people that would lead to greater resilience and reduced dependency on relief measures. Drought monitoring and development of early warning systems forms a part of drought management strategy. This chapter deals with the various approaches adopted by selected nations across the globe in monitoring drought using different approaches viz., climatological, hydrological and combination methods.

Some of the challenges in Drought monitoring as identified by WMO (2006) are briefly given below.

1. The data net work systems are inadequate and the data quality is poor due to missing data or inadequate length of record.
2. Data sharing among the Governmental agencies and Research Institutions are negligible. High cost of data, limits their application in drought monitoring, preparedness, mitigation and response.

3. Information delivered through early warning systems is often too technical and detailed, limiting its use by decision makers.
4. Forecasts are often unreliable on the seasonal timescale and lack specificity, reducing their usefulness for agriculture and other sectors.
5. Drought indices are sometimes inadequate for detecting the early onset and end of drought.
6. Drought monitoring systems should be integrated, coupling multiple climate, water and soil parameters and socio-economic indicators to fully characterize drought magnitude, spatial extent and potential impact.
7. Impact assessment methodologies, a critical part of drought monitoring and early warning systems, are not standardized or widely available, hindering impact estimates and the creation of regionally appropriate mitigation and response programmes.
8. Delivery systems for disseminating data to users in a timely manner are not well developed, limiting their usefulness for decision support.

4.1 Approaches for Drought Monitoring

Droughts may be studied with respects to different aspects such as 1) Historical aspects of droughts over a region 2) Real time assessment, its monitoring and status of droughts and 3) Prediction of droughts. For this purpose, a broad scale weather observational network at the regional scale is essential for monitoring droughts at country level. For monitoring multinational continental droughts, the data collection facilities at Regional Meteorological Centre (RMC) and occasionally World Meteorological Centre (WMC) is mostly utilized. Real time monitoring and assessment of drought requires collection of rainfall data and the drought related data like, agricultural crop status, water resources, snow cover, area affected by various intensities of drought, human health and economic growth rates etc. The primary source of data for real time and for research purposes is the synoptic network supported by hydro met and agro meteorological network. To avoid unrepresentative conclusions from sparse raingauge network WMO recommends a good network of raingauge station at 20 to 50 km distance, depending on the local topography. Drought monitoring can be accomplished through weather data collection and its analysis over a large region, periodic measuring of water levels in both surface reservoir and below ground and remotely sensing the crop water stress. In recent years, application of satellites, has gained popularity in monitoring real time crop status over a large area.

Different nations use different methods to monitor droughts in their respective regions. Large international co-operations among various countries in exchange of data and information is working well and several International organizations like WMO, FAO and WHO have been extending support to various drought related networks in developing countries in Africa and Asia through liberal funding. Before discussing the drought monitoring systems adopted by different nations, let us have a look at the basic infrastructure facilities available with selected countries to record weather parameters.

4.2 Weather Network System in Different Countries

A good network of weather observatories is important for monitoring droughts and other calamities. In the past, data collected at other than the surface observatories (Hydro meteorological observatories) often were not processed for long time. With the advancement in communication technologies, real time weather is monitored for applications in various fields *viz.*, Agriculture, Floods, Tourism and Disaster management. The availability of data even from remote areas on real time basis due to improved communication facilities makes it possible to develop Contingency plans and Decision Support System (DSS) for sustainable agricultural production. The National Weather Services (NWS) of United States monitors the surface weather observatory network (first order stations) across the country. Federal network also records daily observations of temperature and precipitation, which are operated by trained volunteers and these stations are considered as of second order. Measurements from Remote Automatic Meteorological Observing Stations (RAMOS) also transmit weather data to NWS through dedicated phone lines, VHF or Satellite links. After the collection and scrutiny of data, it is the climate related information delivery system that assumes significance in transmitting the analysed data and weather based products to various governmental organizations to prepare appropriate actions. The usefulness of the information delivered to clients is inversely related to the period of time that elapses between measurement and delivery (Hubbard, 1987).

With the increasing awareness on climate and its changes over time, importance of collection of weather data has gained momentum around globe. In recent times developing countries in Africa and in Asia have improved their observational network considerably over the years. However they are not adequate and further assistance from international organizations is essential for those nations with limited resources. The status of weather observatory network in the South Asian countries excluding India furnished in the following table.

Table 4.1: Weather station network in south Asian countries

Countries	Surface observatories	Automatic weather stations	Upper air including Pilot balloon	Raingauge stations	Hydromet obs	Agromet obs
Bangladesh	35	NA	14	NA	-	-
Nepal	30	15	6	337	154	22
Pakistan	92	50	NA	500	-	32
Sri Lanka	25	33	6	480	-	-
China	301	220	125	2414	-	-
Korea	52	483	-	-	-	-
Japan	100	1200	47	-	-	-
Vietnam	174	-	13	-	-	27
Cambodia	11	9	-	200	-	-

(*Source* : JMA / WMO Workshop, July 2010, Tokyo)

India has the largest network of weather observatories of various categories (Table 4.2). There are 25 types of atmospheric monitoring networks that are operated and coordinated by India Meteorological Department (IMD). It has 559 surface meteorological observatories, 35 radiosonde and 64 pilot balloon stations, 219 Agrometeorological and 45 radiation observatories (Attri and Tyagi, 2010). IMD in its massive modernization of existing network of observatories to improve the numerical prediction capabilities, it has plans to establish 550 Automatic Weather Stations (AWS) and by the mid of 2012, more than 400 AWS were already established at different locations in the country. Meteorological and Oceanographic Satellite Data Archival Centre (MOSDAC) of Indian Space Research Organization (ISRO) also has established about 1098 AWS across the country to facilitate and promote the exchange of data to the academic and research needs. The distribution of AWS by IMD and ISRO are shown below (Fig.4.1).

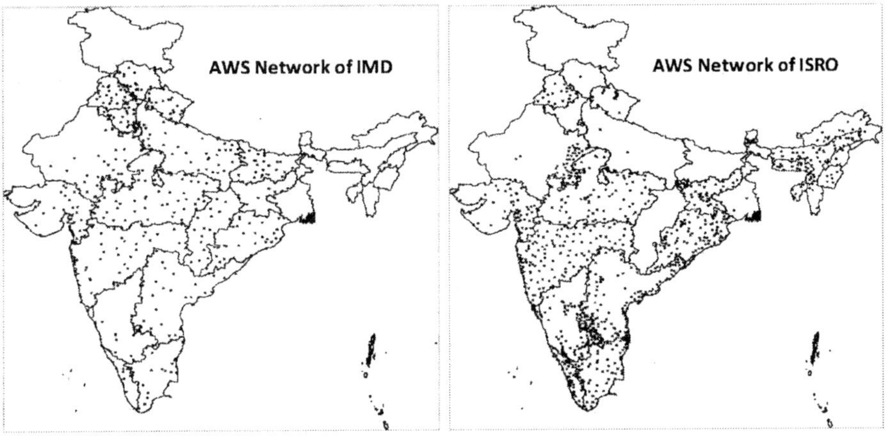

(*Source*: http://www.mosdac.gov.in)

Figure 4.1. Distribution of AWS network of IMD and ISRO across India

Table 4.2: Meteorological observatories network system in India (Tyagi and Attri, 2010)

S.No	Type of observatory	Nos.
1.	Surface observatories	559
2.	Plot balloon observatories	65
3.	RS/RW observatories	35
4.	Aviation current weather observatories	71
6.	Storm detecting radar stations	17
7.	Cyclone detection radar stations	10
8.	High-wind recording stations	4
9.	Stations for receiving cloud pictures from satellites	
	a. Low-resolution cloud pictures	7
	b. High-resolution cloud pictures	1
	c. INSAT-IB cloud pictures (SDUC stations)	20
	d. APT stations in Antarctica	1
	e. AVHRR station	1
10.	Data collection Platforms through INSAT	100
11.	Hydro-meteorological observatories	701
	a. Non-departmental rain gauge stations	
	i. Reporting	3540
	ii. Non-reporting	5039
	b. Non-departmental glaciological observatories (non-reporting)	
	i. Snow gauges	21
	ii. Ordinary rain gauges	10
	iii. Seasonal snow poles	6
12.	Agro-meteorological observatories	219
13.	Evaporation stations	222

(*Contd.*)

S.No	Type of observatory	Nos.
14.	Evapotranspiration stations	39
15.	Seismological observatories	58
16.	Ozone monitoring	
	a. Total ozone and Umkehr observatories	5
	b. Ozone-sonde observatories	3
	c. Surface ozone observatories	6
17.	Radiation observatories	
	a. Surface	45
	b. Upper air	8
18.	Atmospheric electricity observatories	4
	a. Background pollution observatories	10
	b. Urban Climatological Units	2
	c. Urban Climatological Observatories	13
19.	Ships of the Indian voluntary observing fleet	203
20.	Soil moisture recording stations	49
21.	Dew-fall recording stations	80
22.	AWS	550
23.	ARG	1300
24.	GPS	10

Some of the state Governments in India have also started establishing their own network of AWS across the states with the objective to monitor the weather for managing the climate risks and relief measures efficiently. The State Government of Andhra Pradesh took the lead in this regard by establishing 1100 AWS at each Mandal Headquarters. Similarly the State Government of Tamil Nadu also established AWS at Taluka level and the State of Karnataka established a network of Telemetric Raingauge stations at Taluka level. The distribution of network of AWS in Andhra Pradesh is given in the following Figure (Fig 4.2).

Data from AWS, Automatic Raingauge (ARG), Doppler weather Radars (DWR) and Satellite (INSAT-3D) are available on real time basis and is being used by IMD for running number of global and regional NWP models for weather forecast purpose.

The National Agricultural Research System (NARS) in the country consisting of Research Institutes of Indian Council of Agricultural Research (ICAR) and state Agricultural Universities (SAU) and their regional centers have also established Agrometeorological observatories spread across the country and generate data for crop planning and agricultural operations on real time basis. In order to monitor the extreme weather events and to prepare contingency plans, the AICRPAM (All India Coordinated Research Project on Agrometeorology) at CRIDA, Hyderabad had established 100 AWS at different Krishi Vigyan Kendras (KVK) across the country to improve dissemination of Agromet Advisory Services (AAS) to sub district level (CRIDA, 2011).

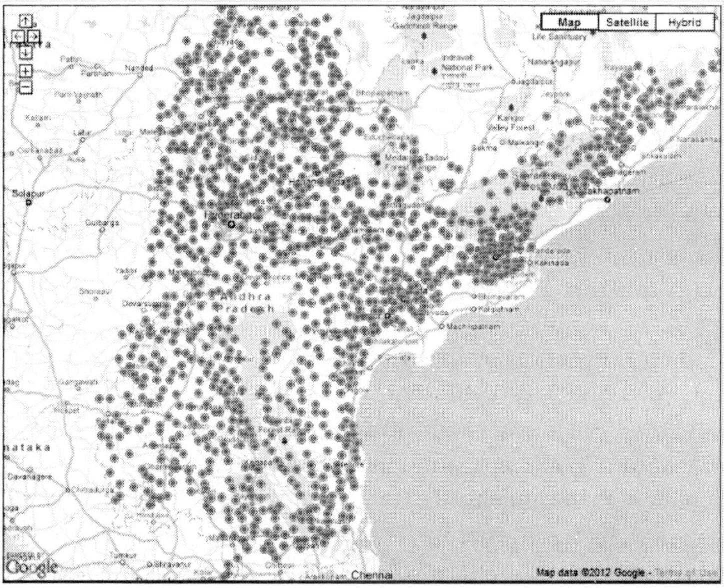

(*Source*: http://apsdms.cgg.gov.in)

Fig. 4.2: Distribution of AWS network in Andhra Pradesh State

4.3 Drought Monitoring Using Climatological, Hydrological and Combination Approaches in Different Countries

Early warning signals of drought helps in reducing the impact if it is delivered to decision makers on a timely and in appropriate format along with measures to be implemented. There are variety of indicators both physical and biological that can be monitored for identification of spatial distribution of drought, its onset and end over a given region. Though rainfall deficiency is the primary factor in computing droughts, other factors such as water levels in the reservoirs, soil moisture, crop conditions etc, need to be considered for assessing the severity of drought. Therefore efficient drought warning systems must integrate rainfall with other parameters. In this endeavor, the United States of America has been the world leader in monitoring droughts scientifically and was successful in integrating several drought indicators such as Palmer Drought Severity index (PDSI) SPI, Stream flows, Vegetation Health, Soil moisture status and impacts on crops and a few ancillary indicators such as reservoir levels, surface water supply index, river basis snow water equivalent and pasture and range conditions for drought monitoring programme on weekly basis. Before 1996, comprehensive drought monitoring system was not in existence in United States. The United States Drought Monitor (USDM) became an operational product on 18 Aug 1999 which is maintained on NDMC

(National Drought Mitigation Center) website (http://www.drought unl.edu/index.htm). The spatial extent of drought in US on a weekly basis is generated by NDMC (Fig. 4.3). This map product has been widely accepted to track drought conditions across the Country by various users.

USDM classifies droughts on a scale from one to four (D1-D4) where D1 denotes Moderate drought, D2- Severe drought D3- Extreme drought and D4-exceptional drcughts. D0- indicates an abnormally dry area which are either heading in to drought or recovering from drought, but still experiencing drought conditions. The map also shows the sectors that are currently experiencing direct or indirect impacts using the symbols viz., A (Agriculture, crops, livestock pasture and grasslands), H (hydrological) and W (water supplies). For example the symbol D2 (A) on the map indicates the severe drought conditions affecting agricultural sector. While preparing maps care was taken not to bring an area in and out of drought too quickly due to slow motion and long recovery processes involved in drought manifestation. Few States in that country are utilizing the Spatial and temporal resolution of information to develop regional based products. For example Oklahoma State has developed a Climate Mesonet that provides detailed climatic information and variety of products that are useful in monitoring drought. At the **gis.usgs.gov link,** users can access to a variety of layers of indices and resolutions for display, which can be straight away used for decision making.

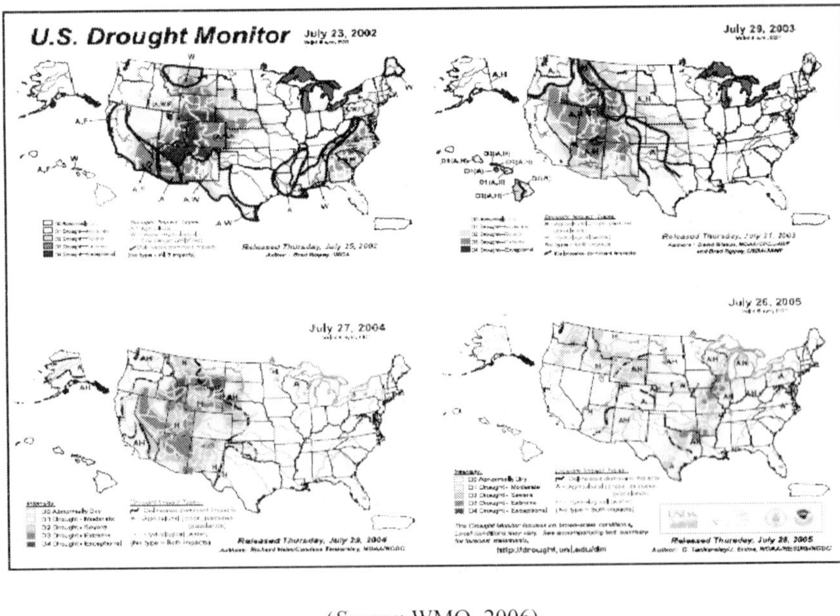

(*Source*: WMO, 2006)

Fig. 4.3: Spatial distribution of drought in U.S. for the years 2002-2005

The same methodology has been applied to the production of North American Drought Monitor (NADM), a collaborative product between US, Mexico and Canada since 2002 onwards. Spatial pattern of drought across North American Continent is given below (Figure 4.4). This product is prepared on monthly basis which is an excellent example of international cooperation on drought monitoring.

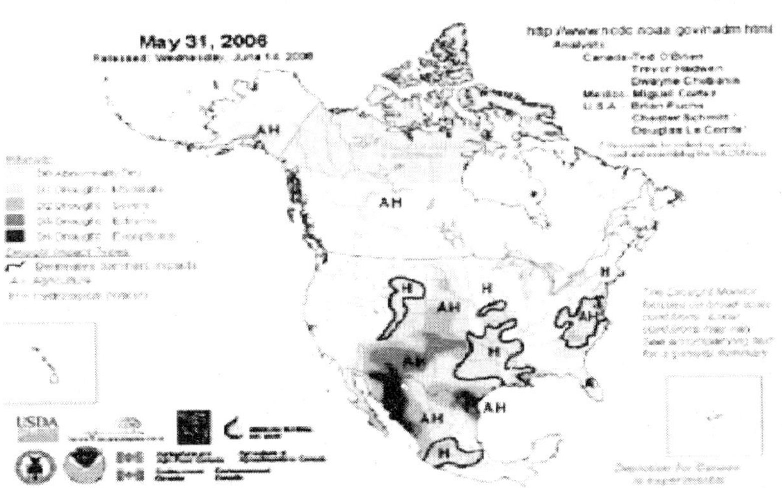

(*Source*: WMO, 2006)

Fig. 4.4: Spatial distribution of drought over North America during May 2006

In addition to US drought monitor, the vegetation stress across the country is being monitored in USA using Remote sensing data and its derived parameter such as Vegetation Drought Response Index (VegDRI) on bi-weekly basis showing the vegetation stress across the country that began initially in seven states during 2006 and now covering all the states from 2009 onwards (Tadesse and Wilhite, 2007). The partners for development of VegDRI are National Drought Mitigation Center (NDMC), the US Geological Survey's National Center for Earth Resources Observations and Science (EROS) and the High plain Regional Climate Center (HPRCC). Remote Sensing data is combined with climate and biophysical data to create vegDRI. Two variables related to general vegetation conditions *viz.,* the Percent Average Seasonal Greenness (PASG) and Start of Season Anomaly (SOSA) are computed from NDVI data acquired by NOAA's Advanced Very High Resolution Radiometer (AVHRR) and incorporated in to vegDRI. The climate related variables incorporated in vegDRI are Palmer Drought Severity Index (PDSI) and the SPI. The bio physical data such as soils, land cover, land use and the ecological settings are incorporated in to the vegDRI. The Spatially temporal pattern of vegDRI maps are qualitatively compared to drought pattern as given in the US-Drought Monitor. The USGS through its Interaction Map Service provides regularly updated geospatial data related to drought conditions including AVHRR VegDRI and eMODIS VegDRI along with others such as US-drought monitor, USGS Stream flows, Precipitation derivatives etc. As an example, the spatial distribution of vegDRI and drought intensity obtained using AVHRR and eMODIS are shown below (Fig 4.5). It is seen that in both the methods, the areas under different intensities of drought are almost the same except at some places where severe and moderate droughts vary.

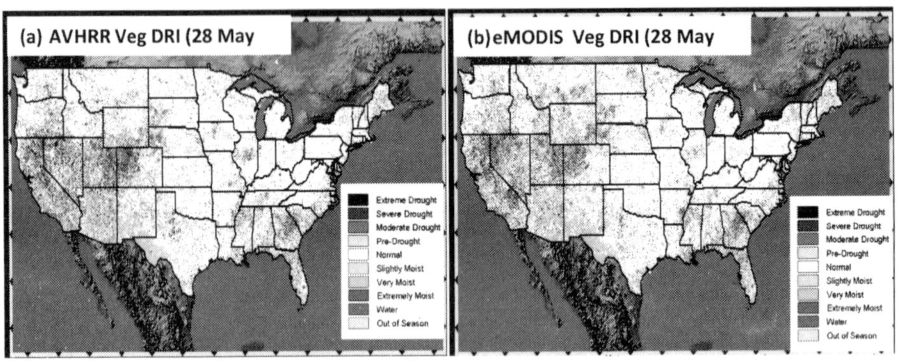

(*Source*: www.gis.usgs.gov)

Fig. 4.5: Spatial distribution of veg DRI in U.S.

Considerable progress has been made in monitoring drought by many countries in the world. The existing cooperation in data sharing and exchange of information related to adaptation and mitigation strategies at global level need to be strengthened further. Efforts in this direction by WMO, FAO and some universities like University of Nebraska, University of Princeton in mobilizing international awareness on management of droughts are commendable. Some of the approaches in monitoring droughts followed by different nations are described below. In a classical article on Monitoring Soil Drought, by Vasselin Alexandrov (2006) reviewed the methods for measurement of soil moisture and drought monitoring adopted by a few countries like Romania, Hungary and Bulgaria in Europe are briefly described below.

Hungary

An interactive internet based irrigation advisory system by Hungarian Meteorological Services calculates daily ET_O (Penman – Monteith method) and uses daily rainfall values to compute the accumulated climatological water shortage (deficit) by balancing between rainfall and the evaporation demand.

From this information, the water requirements of crops are calculated and the irrigation water is regulated. This is a simple and real time basis drought monitoring system (Szalai, 2004) in Hungary.

Romania

In Romania agricultural drought phenomenon at different stages is monitored by assessing the soil moisture. Daily weather data collected over the country is used to compute ETo by either Penman-Monteith method/cropwat model. A 10 day period cumulative soil moisture values for wheat and maize crop at different depths of 0-20, 0-50 and 0-100cm is estimated and soil moisture reserve (m^3/ha) which is the percentage capacity of available soil water to Cumulative Soil Moisture and the soil water deficit (m^3/ha) is computed from simple water Balance model. As an example, the following figure (Fig 4.6) represents the spatial distribution of soil moisture content in Romania at 0-20 cm depth.

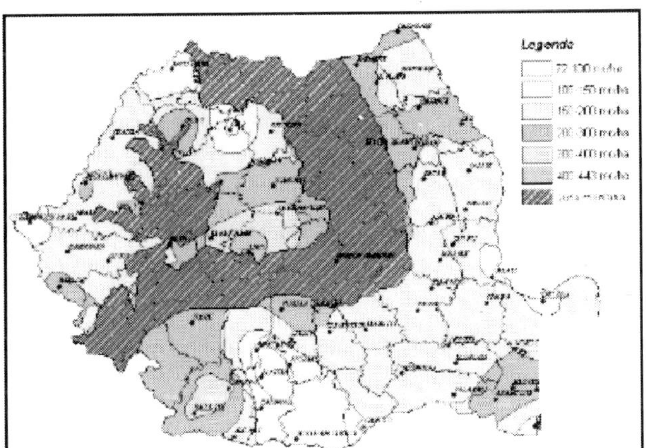

Fig. 4.6: Spatial distribution of the soil moisture content in Romania at 0-20 cm on 22nd September 2004

(*Source* : Vasselin Alexandrov, 2006)

Bulgaria

In Bulgaria, the National Institute of Meteorology and Hydrology (NIMH) is responsible in providing data to research and operational activities in the field of Meteorology, Agrometeorology and Hydrology. The agrometeorological network under NIMH collects and maintains the weather data and implements experiments. The agrometeorological services in the various locations viz. North East, East, South and South western part are being supported by four Regional Branches located at Pleven, Varna, Plovdiv and Kjustendil respectively. The agrometeorological services monitor crop growth and evaluate conditions for carrying out irrigation schedules on weekly basis and disseminate information through National Radio network. Monthly bulletins on crop conditions are also prepared and distributed through mass media communication system. It also monitors soil moisture up to 100 cm depth at a 10 day interval and spatial maps are prepared to identify regions affected by soil droughts (Fig 4.7). Regions with red colours are soil drought affected regions.

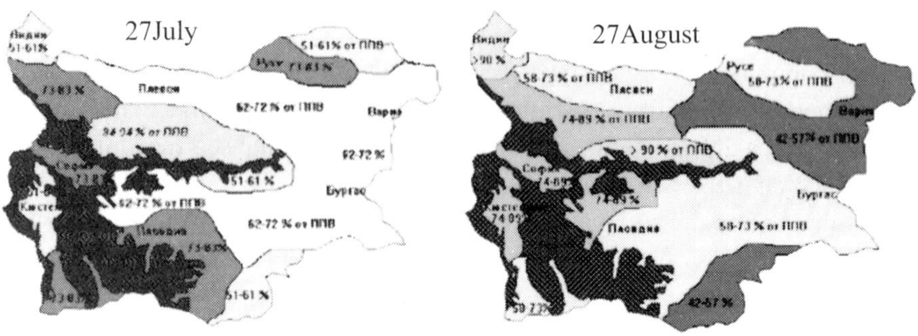

(*Source*: Vasselin Alexandrov, 2006)

Fig. 4.7. Soil moisture (%) under spring crop during 27 July and August 2006 at Bulgaria

Portugal

The Portuguese Meteorological Institute (PMI) monitors the climate and drought conditions in the country and Palmer Drought Severity Index (PDSI) is used to characterize droughts. This index has been adapted and calibrated to climate conditions of mainland of Portugal. A separate unit viz. drought observatory is working in the above institute to look into the problems related to drought. The spatial distribution of drought conditions are presented in monthly PDSI maps which are used to monitor spatial and temporal variations in droughts across mainland. These maps help in taking appropriate measures to minimize the losses associated with droughts.

Poland

The Agricultural droughts in Poland are computed by a special index which is defined as ratio of PE/Pveg. Where PE is potential evapotranspiration and Pveg is the total rainfall during crop season (April-September). In Poland, the index range from 1.6 for areas prone to drought to 1.0 where summer rainfall exceeds PET values (Bochenek Z. *et al.*, 2005). Based on the average values of 60 meteorological stations across Poland the distribution of drought prone areas in the country is shown in Fig 4.8.

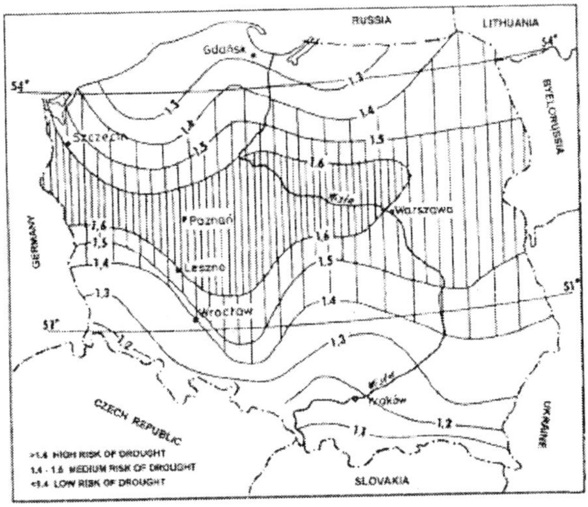

(*Source*: Zbigniew et al., 2005)

Fig. 4.8. Distribution of drought prone areas in Poland

The field assessment of actual crop condition by Central Statistical Office and the drought indices are made available by Ministry of Agriculture in the form of bulletin for taking corrective steps for managing droughts. Remote sensing based monitoring of droughts and crop conditions is also being carried out by the Institute of Geodery and Cartography, Warsaw.

Russia

Till 1917, the Science Committee of Meteorological Agency of the Central Land Organization and Farming Administration monitored droughts in Russia and published maps showing probabilities of dry decades. After this Hydro meteorological Services monitored droughts and published two books *viz.*, Agroclimatic guide book and Agroclimatic Resources for all the region of USSR. These Books provided information on the drought climatology picture for different regions in the country.

Currently drought monitoring in Russia is carried with six popular indices of which three are moisture dependent and the remaining three are temperature dependent, which are widely used by Russian scientists, though they have developed more than 50 indices over the period. They are:

1) Hydrothermal coefficient (HTC) developed by Selyaninov (1928)
2) Moisture index by Shashko (1985)
3) Water supply Index (Protserov, 1949)

4) The number of days with RH < 30% (N_o)

5) The number of days with maximum temperature > 30°C (N_T)

6) Productive moisture reserves in soil layers of 0-20cm (Wo-20), 0-50cm (Wo-50) and 0-100cm (Wo-100) under winter, early spring and late spring grain crops, respectively.

All these indices can be determined using standard hydro meteorological data. Based on the range of values, the intensity of droughts are categorized as extremely severe, severe, moderate, weak or no drought (Table 4.3).

Russian meteorological satellites have also been used for drought monitoring. A Drought Monitoring Center (DMC) is in operation at All Russian Research Institute of Agrometeorology (ARRIAM).

Near East Countries

The Countries in North Africa and West Asia *viz.*, Algeria, Egypt, Gaza strip, Iraq, Israel, Jordan, Lebanon, Libya, Morocco, Syria, Tunisia, Turkey and West Bank are known as Countries in the Near East. This region is characterized by humid, Sub-humid, semi-arid and arid to hyper-arid climates. Large parts of this region are vulnerable to droughts. The common approach to drought in this region has been of crisis management rather than long term planning. When a drought appears over any region, the concerned Government set up drought mitigation programme and the programme terminates once the conditions improve. Due to this approach, Drought Monitoring System that are integrated at National level are not operational (Eddy De Pauw, 2005).

China

Drought monitoring in China is being done since 1995 and the responsibility lies with National Climate Center (NCC) of China Meteorological Administration (CMA) by computing the SPI values at 10 day period. Between 1995 and 1999, a Comprehensive Index (CI) was developed to monitor drought situation and was put it into operation on daily basis. CI is a function of last 30 and 90 day SPI and the corresponding PET values. This index is used to monitor real time drought conditions on daily basis.

$$CI = aZ_{30} + bZ_{90} + cM_{30}$$

Where Z_{30} and Z_{90} are SPI of recent 30 and 90 days which represent abnormal conditions on monthly and seasonal scale.

Table 4.3: Drought categorization based on various indices used in Russia

S.No	Drought assessment index	Category of drought intensity				
		No drought (Class 5)	Weak (Class 4)	Moderate (Class 3)	Severe (Class 2)	Extremely severe (Class 1)
1.	Selyaninov's hydrothermal coefficient (HTC)	>0.76	0.61-0.75	0.40-0.60	0.20-0.89	< 0.19
2.	Shashko's index of moisture (Md)	>0.41	0.31-0.40	0.20-0.30	0.10-0.19	< 0.09
3.	Protserov's index of water (V, %)	>71	61-70	51-60	41-50	< 40
4.	Number of days with relative humidity < 30% (N_o)	0	1-2	3-5	6-7	8-11
5.	Number of days with maximum air temperature >30°C (N_T)	0	1-2	3-4	6-7	8-11
6.	Productive moisture reserves (mm) in the 0-20 cm soil layer (W_{o-20})	>21	16-20	11-15	6-10	< 5
7.	Productive moisture reserves (mm) in the 0-50 cm soil layer (W_{o-50})	>46	36-45	26-35	16-25	< 15
8.	Productive moisture reserves (mm) in the 0-100 cm soil layer (W_{o-100})	>81	61-80	41-60	26-40	< 25

(*Source*: Klenchenko, 2005)

M_{30} is average moisture index (Im) of recent 30 days which represents water deficit

$$M = \frac{P - PE}{PE}$$

Where P is precipitation; PE is Potential evapotranspiration and a,b,c are constants having values of 0.4, 0.4 and 0.8 respectively

Drought intensity is characterized based on the values of CI which is as follows.

Drought intensity	Value of CI
No drought	-0.6 to 0
Weak	-1.2 to -0.6
Moderate	-1.8 to -1.2
Severe	-2.4 to -1.8
Extreme	< -2.4

Daily rainfall data collected from more than 600 stations is used to compute CI value on daily basis and spatial maps of CI are prepared. As an example, Spatial maps of CI for the period October 2010 to February 2011 are shown in (Fig 4.9).

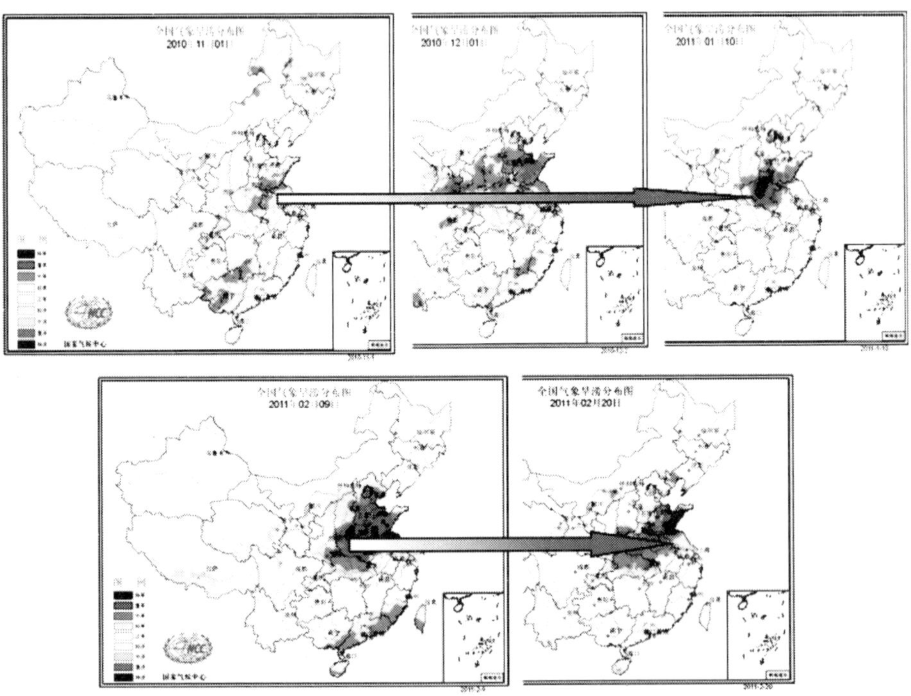

(*Source*: Gao Ge Zhang Peiqun Liao Yaoming, 2011)

Fig. 4.9: Development of drought in China during October 2010 to February 2011

Other drought monitoring products such as soil moisture at top 20 cm soil depth and Remote Sensing based drought monitoring indices are also generated at 10 and 7 days period respectively and put on display through website. Drought watch Bulletins at weekly and monthly intervals are prepared and distributed among different agencies that are directly / indirectly related to drought management issues. Global situation along with country's status is also prepared and made available to public through Information Communication Network System.

Indonesia

Drought in Indonesia is mostly associated with EI-Nino events (Boer and Subbiah, 2005). Rainfall anomalies are closely watched by Bureau of Meteorology and Geophysical (BMG) for monitoring drought in the country. Also Remote Sensing indicators such as NDVI are used for monitoring drought in the country.

Australia

In Australia, nation-wide drought watch on a routine basis is carried by Australian Bureau of Meteorology as Drought Watch Service since 1965 on a simple index of rainfall deficiency. The system utilizes the monthly rainfall datasets of 600 raingauge stations that record and report on daily basis Though rainfall deficiencies over a period and area are essential to declare drought, yet its declarations is a more complex issue as the declaration of drought is the responsibility of local state governments. The Drought Watch Service in Australia uses accumulated rainfall percentiles over succ-essive months to identify regions of rainfall deficiency or excess (WMO 2006). Areas with accumulated rainfall below 10th or 5th percentile for 3 months or more is termed as seriously/ severally deficit. The following figure shows the spatial distribution of rainfall deficiencies during 2002-03, which was recorded as one of the worst drought year (Fig 4.10).

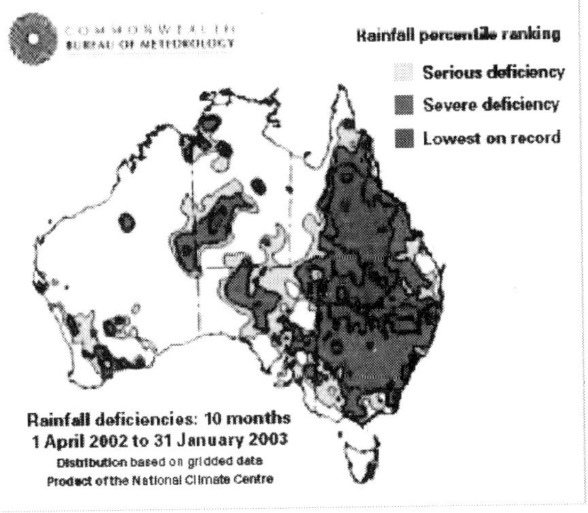

Fig. 4.10: Drought severities during 2002 and 2003 in Australia (WMO, 2006)

African Countries

For monitoring droughts in African countries, the United Nations Development Programme established a regional Drought Monitoring Centre (DMC) at Nairobi and a sub-centre in Harare in 1989 for covering 24 countries in the eastern and southern African sub-region (WMO, 2006). During 2003, The DMC Nairobi became a specialized Institution of the Inter Governmental Authority on Development (IGAD) and was renamed as Institutions for Climate Predictions and Applications Centre (ICPAC). The participating countries are Burundi, Djibouti, Eritrea, Ethiopia, Kenya, Rwanda, Somalia, Sudan, Uganda and United Republic of Tanzania. The centre's responsibility is to monitor climate and predict weather and to develop early warning systems to minimize climate-risks in the greater Horn of Africa. This Institution provides regular weather advisories including 10 day, monthly and seasonal bulletins. The drought products generated for different parts of the region can be viewed through its website **http:// www.Icpac.net.** In addition to monitoring of droughts by ICPAC in the region, a few countries have their own mechanism to monitor drought in their parts.

Ethiopia

In Ethiopia, National Meteorological Services Agency (NMSA) produces 10-day bulletin that provides information on the current period rainfall departures over long term average. The drought intensity is computed based on the range of percentage deviation (D) of rainfall. A region is considered to have more rain if D is above 125 and normal if D lies between 75 to 125% ($75<D<125$), it is termed as insufficient if D lies between 50 and 75 ($50<D<75$) and it is termed as below average if $D<50\%$ of normal value. Also NMSA produces NDVI maps at 10 day interval from the satellites database which is also widely used to monitor vegetation and drought conditions in the country.

South Africa

Droughts are common in South Africa and to monitor drought conditions over the country, South African Weather Services (SAWS) established a Drought Monitoring unit where information on current rainfall and long range forecasts are available at one place. Drought monitoring is carried out by generating SPI spatial maps.

4.4 International Cooperation in Drought Monitoring

There are many organizations operating in Africa and Asia that help to enhance the local government capacities to manage climate-related risks. The role of

ICT in dissemination of information on drought monitoring is well recognized. In order to improve its efficiency, farmers in developing countries may be provided with necessary aid. Following are the few net works that monitor climate and provide necessary advisories.

1. **The African Adaptation programme** (AAP) is a four year programme (2009-12) founded under Japan - UNDP joint frame work for building partnership to address climatic change issues. About 20 countries have taken part in the above programme. The activities of the project can be viewed by browsing their website **www.undp-adaptation.org.**

2. **Economic and social commission for Asia and Pacific (ESCAP)**: Drought monitoring and early warning for the Asia and Pacific region is in operation since September 2010 under Regional Space Application Programme for Sustainable Development. It is intended to provide satellite products for drought monitoring to over 60 member countries of the UN Economic and Social Commission for Asia and the Pacific (ESCAP) and to assist the governments in developing products relevant to local conditions in managing climate risks by providing necessary advisories. Information on satellite data, weather resources collected by respective National and Regional agencies is shared among all the ESCAP countries to help them to achieve effective monitoring and early warnings of drought events that would provide sufficient lead time to the local government in reducing the drought hazards.

The partners to such programme are

i. National Disaster Reduction Center of China

ii. National Remote Sensing Centre, Hyderabad

iii. Korea Aerospace Research Institute

iv. Geo- informatics and Space Technology and Development and Technology Agency of (GISDTA) of Thailand

v. Pakistan Space and Upper Atmosphere Research Commission

vi. Russian Federal Space Agency

vii. South Asian Association for Regional Cooperation

viii. WMO

ix. Asia Pacific Space Cooperation Organization and Typhoonic Committee Secretariat. "More information on drought monitoring and early warning is available at **htpp:// www.unescap.org**"

3. **Agricultural and Hydromet (AGRHYMET) Regional Centre (ARC):** It was established in 1974 in Sahel with a mandate to invest in Research on food security and to combat effects of drought and desertification for ecological balance in Sahel region. It is a specialized institute in developing drought control measures for sustainable food production and food security in the 9 member states *viz.,* Burkina Faso, Capeverde, The Gambia, Guinea Bissau, Mali, Mauritania, Niger, Senegal and Chad. The ARC monitor the disaster trends based on rainfall data, periodic agricultural survey market updates etc, and circulates the early warming information to the partners. A crisis prevention network at sub regional level to improve the coordination among the partners was established and the monitoring system is verified collectively to identify the regions that needed immediate attention / cooperation. More information can be had from their websites **http:// www.agrhymet.ne.**

4. **African Drought monitor:** A modeling frame work that couples available Satellite Remote Sensing data and *in-situ* information which enables to estimate water cycle and drought continuously on the spatial and temporal scale was developed in collaboration with UNESCO and put into operation in Africa. The system contains both data from a historic reconstruction (1950-2008) and real time monitoring of the Hydrological Cycles and drought events. It is updated daily and provides multiple hydrological

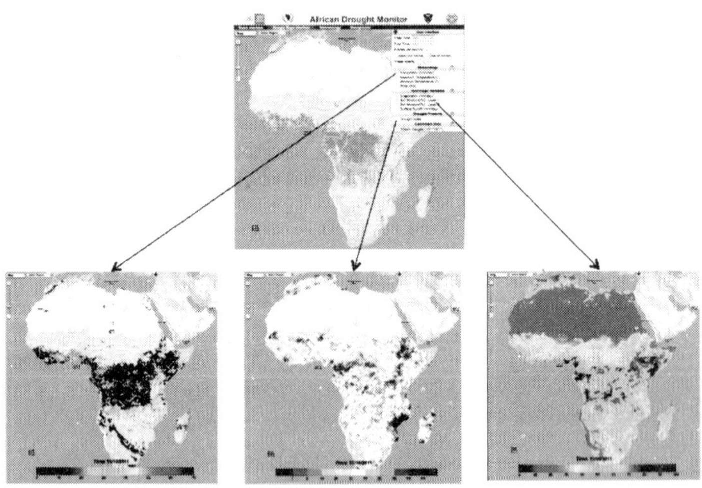

(*Source:* http://hydrology.princeton.edu/~nchaney/Africa_Drought_Monitor_Webpage/GMinterface.php)

Fig. 4.11: Example of the interactive user interface showing the main interface (upper center), precipitation (lower left), drought index (lower center), and the relative soil moisture of the top layer (lower right) on November 10th, 2011

variables at continental and basin scale. The drought monitor data are made available through web based interface, which enables quick accesses to the data and plots of current and past conditions for the whole of Africa and a set of river basins (Fig 4.11).

5. Some of the useful web links that monitor and update about droughts in Africa and Asia are presented and discussed during first Africa-Asia Drought Adaptation Forum (ADAF-2011) held during 14-15 June 2011 at Bangkok, Thailand. The information on drought related website as reported in July 2011 issues of newsletter of UNDP Dry land development center, Nairobi is as follows.

Organization	Website address
Africa	
African Centre of Meteorological Application for Development	http://www.acmad.ne/index.htm
Experimental African Drought Monitor	http://hydrology.princeton.edu/~justin/research/project_global_monitor/
Famine Early Warning Systems Network (FEWSNET) Africa	http://www.fews.net/Pages/default.aspx
Food Security & Nutrition Working Group	fsn-gha-workinggropu@fao.org for monthly FSNWG Updates
IGAD Climate Prediction and Application Centre (ICPAC)	http://www.icpac.net/Forecasts/forecasts.html
Integrated Regional Information Networks (IRIN) Africa	http://www.irinnews.org/IRIN-Africa.aspx
Prevention Web Africa	http://www.preventionweb.net/english/countries/africa/
Relief Web Africa	http://www.reliefweb.int/rw/dbc.nsf/doc115?OpenForm&rc=1
Southern African Development Community (SADC) Climate Service Centre	http://www.sadc.int/dmc/
UNOCHA (Southern & East Africa)	http://ochaonline.un.org/rosea/latestUpdates/tabod/6652/language/en-US/Default.aspx
UNOCHA (West & Central Africa)	http://ochaonline.un.org/Bulletinshumanitaires/tabid/3099/language/fr-FR/Default.aspx
Asia	
Asian Disaster Reduction Centre	http://www.adrc.asia/latest/index.php
East Asian Drought Monitoring System	http://atnis.pknu.ac.kr/~intra2
FEWS NET Central Asia	http://www.fews.net/Pages/default.aspx
IRIN Asia:	http://www.irinnews.org/IRIN-Asia.aspx
Pacifica Disaster Centre/World Natural Hazards Website	http://www.pdc.org/iweb/pdchome.html
Prevention Web Asia:	http://www.preventionweb.net/english/countires/asia/
Relief Web Asia	http://www.reliefweb.int/rw/dbc.nst/doc115? Openform&rc=3
SAARC Disaster Management Centre (SDMC)	http:/saarc-sdmc.nic.in

4.5 Drought Monitoring Mechanism in India

Monitoring of droughts in India is carried out by both State and Central Governments. The rainfall distribution impacts of drought on agricultural production and water bodies collected by the State Governments is used by Central Government to decide the relief amount to be released to the affected regions. Exchange of information on crop conditions at weekly intervals generally takes place between State and Central governments however, with greater frequency during drought years. There are two agencies *viz.,* India Meteorological Department (IMD) and National Remote Sensing Centre (NRSC) of Indian Space Research Organization (ISRO) which are primarily responsible to monitor drought using weather and Remote Sensing data respectively. At the State level, the respective Agricultural Department and the state level Remote Sensing Agency monitor the rainfall distribution and droughts. Few states have their own Drought Monitoring agencies/cell viz., Karnataka is one such example. India Meteorological Department (IMD) under Ministry of Earth Resources, GOI monitor weather across the country on daily basis and prepare rainfall deviation maps at weekly intervals. Similarly the cumulative rainfall departure maps at sub-divisional level for different seasons *viz.,* Mar-May (Spring/summer season), June-September (Monsoon) October-November (Post monsoon) and December-February (Winter) are also prepared and displayed them on its website (www.imd.gov.in) regularly (Fig 4.12).

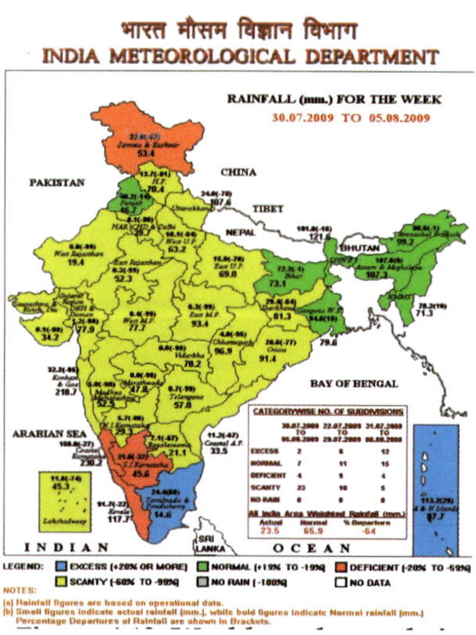

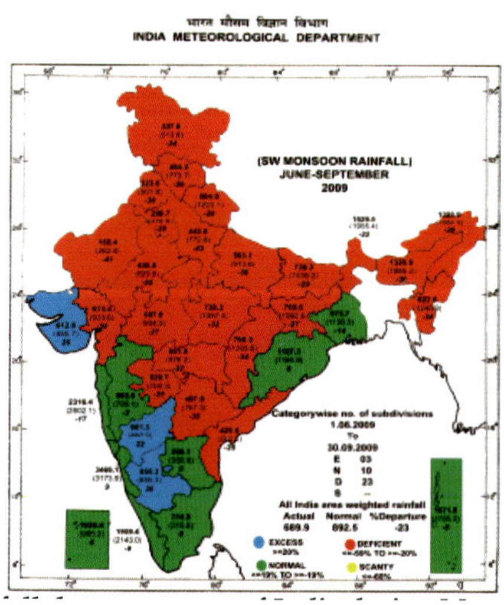

Fig. 4.12: Weekly and cumulative rainfall departure map of India during Monsoon 2009

IMD monitors the incidence, spread, intensification and cessation of drought (near real time basis) on a weekly time scale, based on aridity anomaly index (Fig 4.13a) and publishes an outlook for drought in the subsequent weeks. These maps shall help to assess moisture stress experienced by crop plants in different regions. Based on the above, Aridity anomaly report for southwest monsoon season for the whole country and for five meteorological sub-division *viz.,* Coastal Andhra Pradesh, Rayalaseema, South Interior Karnataka, Tamil Nadu, Pondicherry and Kerala for northeast monsoon season are prepared by IMD H.Q and sent to various Agricultural authorities of State and Central Government and Research Institutes on operational basis. Also IMD computes SPI at monthly and seasonal time scale which are uploaded in the IMD website regularly (Fig 4.13b).

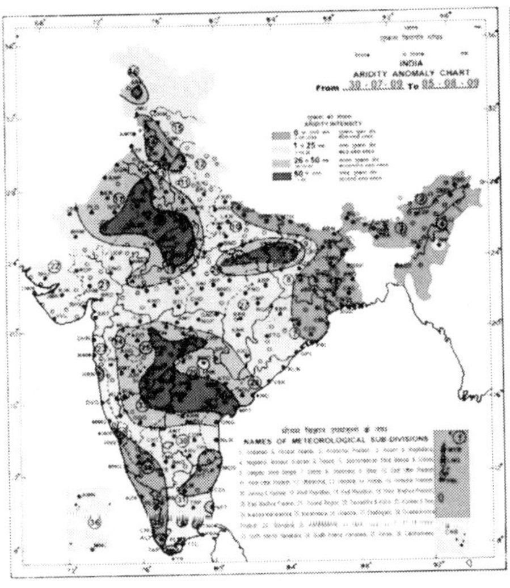

Fig. 4.13a: Aridity anomaly index for 30th SMW 2009 in India

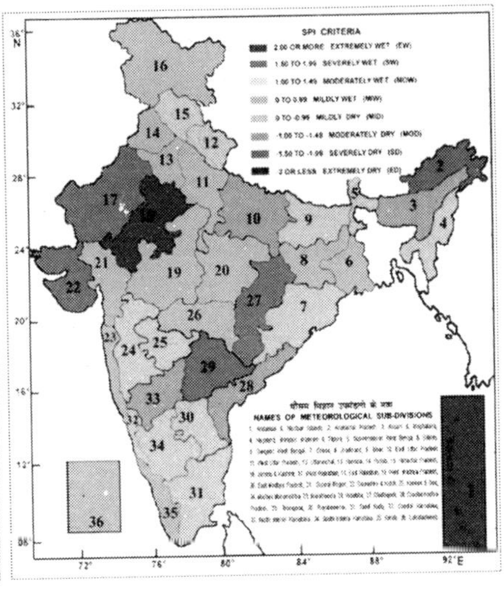

Fig. 4.13b: SPI map for September 2011 in India

National Agricultural Drought Assessment and Monitoring Systems (NADAMS) operated by National Remote Sensing Centre (NRSC) uses both moderate and coarse resolution data for drought monitoring by computing NDVI values at different spatial scales. It monitors agricultural droughts using space technology inputs since 1989 and provides near real time information on prevalence, severity and persistence of agricultural drought at sub-district level. Along with agricultural crop conditions, it also generates products on drought related parameters such as rainfall, cropped area etc at fortnightly/monthly intervals. At present it covers 13 states in India which are predominantly agricultural based and prone to droughts. The information is communicated through surface, e-mails to the users' *viz.,* Ministry of Agriculture, Government of India, and Directors of Agriculture of different states, relief/revenue Commission of States, IMD and State Remote Sensing Application centres. The products related to agricultural droughts generated by NRSC for the states of Andhra Pradesh, Bihar, Chhattisgarh, Gujarat, Haryana, Jharkhand, Karnataka, Madhya Pradesh, Maharashtra, Odisha, Rajasthan, Tamil Nadu are put on its website is shown below (Fig 4.14).

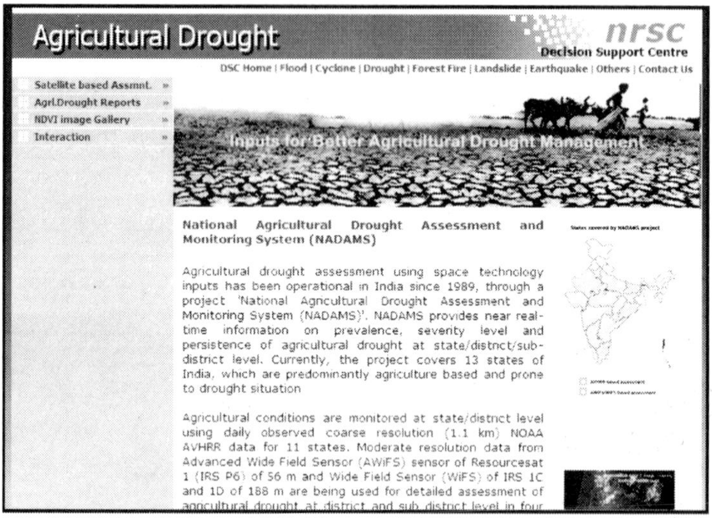

Fig. 4.14. Agricultural drought situation in different states shown in NRSC webpage

As an example, the Agricultural drought situation through NDVI values at country and as well at state level at monthly interval is generated and posted in web (NRSC-DSC) regularly. At the end of season (October) Agricultural drought report for each state is summarized from the rainfall deviation and NDVI maps and the number of tehsils / Blocks/ mandals affected by drought are identified. As an example the number of mandals affected by different intensities of Agricultural drought in each district in Andhra Pradesh during the year 2011 kharif season is given in Fig 4.15.

At the National level, Crop Weather Watch Group (CWWG) an inter ministerial mechanism headed by Ministry of Agriculture, GOI has been working as a nodal agency since 1979 in all matters related to droughts which includes, monitoring, assessing and management. This group evaluates information and data furnished by IMD and other scientific and technical organizations during the rainy season (June-September) on the likely impacts of deviations in weather parameter on the agricultural crop conditions and its productivity.

District	Moderate-Severe	Mild	Normal
Adilabad	17	1	34
Anantpur	60	3	0
Chittoor	26	36	6
Kadapa	45	5	4
Eastgodavari	1	5	51
Guntur	33	8	16
Karimnagar	33	9	14
Khammam	32	1	13
Krishna	9	11	28
Kurnool	53	0	2
Mahabubnagar	45	8	12
Medak	32	7	6
Nalgonda	47	6	7
Nellore	0	0	46
Nizamabad	3	9	24
Prakasam	56	0	4
Ranga Reddy	17	12	10
Srikakulam	17	0	20
Visakhapatnam	1	12	29
Vizianagaram	0	13	21
Warangal	35	12	3
West Godavari	0	6	40
Total	562	164	391

Fig 4.15: Number of mandals affected by drought in each district of Andhra Pradesh during *Kharif* 2011 (NRSC, 2011)

Drought monitoring at State government level is carried out mostly by Disaster management cells operating in the respective states. A better coordination among various agencies shall help in taking corrective and timely actions for effecive management of droughts. Some states like Karnataka monitor drought by its Drought Monitoring Cell from rainfall departure and SPI maps prepared at weekly intervals and storage levels in the surface reservoirs during the season. It also gives the weekly and seasonal soil moisture adequacy index maps (Figure 4.16).

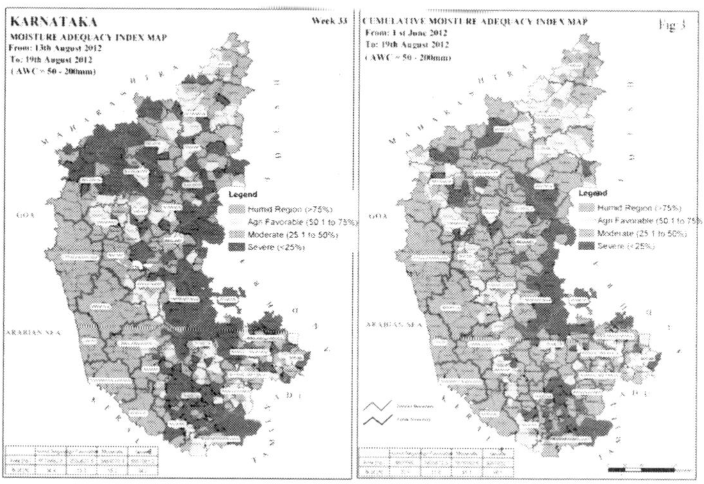

(*Source*: Drought Monitoring Cell, Karnataka)

Fig. 4.16: Weekly and seasonal moisture adequacy index maps of Karnataka

4.6 Early Warning and Decision Support System

Drought being a slow motion disaster, its early detection and warning allows for activation of drought management plans which include both mitigation and emergency purposes. This type of information to the administrators, policy makers etc, would give ample time to react and minimize the losses associated with droughts. Early warning systems along with the actions to be carried out identified from Decision Support System shall act as drought proofing mechanism against all odds. It requires uninterrupted flow of information on climate, soil moisture, stream flow, ground water, lake and reservoir levels from various agencies and integrating them with early warning system. The resource inventory of historical data on the above parameters plus the social and economic data helps in developing location specific Decisions Support System (DSS) for taking appropriate measures to reduce the losses. The system functions at three levels *viz.,* 1) Receiving forecasts and advisories from scientific organisations. 2) Monitoring key drought indices at National and Regional level and 3) Provide necessary advisories, region wise through effective dissemination mechanism. There are many institutions and research workers across the world that have developed methodology for early detection of drought and DSS for various region.

In India, several organization *viz.,* IMD, Agromet Division of IMD, ICAR Research Institution like CRIDA, CAZRI and ISRO Research centre like NRSC and State Agricultural Universities and several private organizations like skymet *etc.,* prepare early warming cum advisories to the farming communities. A number of private firms *viz.,* Reuter Market Light, Virtti solutions, IFFCO Kisan Sanchar limited etc are providing the IMD forecast and advisory to the contact farmers through SMS (cell phone technology). India Meteorological Department provides long range forecast of seasonal and total rainfall for the entire country and also rainfall summaries at weekly intervals. The Agromet unit of IMD in coordination with SAU's issues weekly /bi-weekly Agromet advisory bulletins from 127 Agromet field units (AMF). With the expansion of Agro-advisory network in the recent years, currently Agromet advisories are being issued at district level and likely to expand this facility at sub-district level shortly.

The State Agricultural Universities prepare crop planning bulletins on annual basis considering the long term weather and other data such as on soil, land use, water resources, the changing cropping pattern and etc. Based on the crop conditions and weather forecast during the current season, agro-advisory bulletins are prepared and disseminated through effective mass communication media systems. Each SAU has its own website and the advisories, crop plans and contingency plans etc, prepared in local languages are made available to the public. As an example, comprehensive weather advisory bulletins prepared by different SAUs are given below (Fig 4.17).

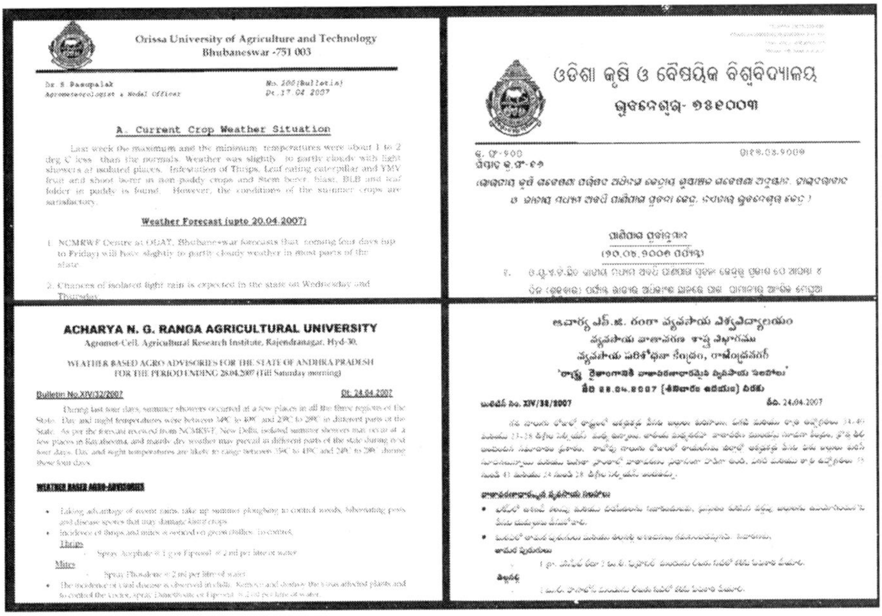

Fig. 4.17: Weather based Agro advisory bulletins by different SAUs

The Central Research Institute for Dryland Agriculture (CRIDA), Hyderabad, a Research Institute under Indian Council of Agricultural Research (ICAR), monitors the weather conditions daily and closely watches the moisture deficit zones during rainy season (June-September) and provides latest crop status in the country based on the input provided by the 25 centers of AICRP on Agrometeorology and other 25 centers of AICRP on Dryland Agriculture. State wise agro-advisories prepared by the respective centre on weekly basis during the crop growing season (Jun-Sep) were sent to ICAR H.Q, New Delhi for preparation of crop contingency plans at National level and the same is also updated in the *"webpage of www.cropweatheroutlook.in"* regularly. Success of such programs depends on accuracy of forecast, which should be about 90%.

In an another study at CRIDA during 2001-2003 under National Agriculture Technology Programme (NATP) a "Near real time monitoring of weather parameters in Andhra Pradesh for managing adverse weather condition" was conducted in collaboration with the local Agricultural University and

Department of Agriculture, Government of Andhra Pradesh (Rao, *et.al.* 2003). From the average weekly rainfall received in 23 districts of AP and from the water balance computations using Thornthwaite & Mather (1955) technique, the soil moisture Index (SMI) which is defined as the ratio of actual soil moisture to the maximum available soil moisture expressed in percentages for different regions spread across the State were computed and spatial map at weekly intervals of SMI were prepared and presented as an example for the year 2002 which happen to a severe drought year in Fig 4.18. Monitoring of the weekly changes of SMI for the weeks 24 to 40 is a good indicator of region wise water deficit or water surplus for better agricultural operations and for preparation of contingency crop plans. Based on the spatial Soil Moisture Index (SMI) maps and the forecasted weather, the Collaborative Agricultural University (ANGRAU) prepared Agro-Advisories and communicated to the respective state Department for dissemination and preparation contingency measures (Fig 4.19).

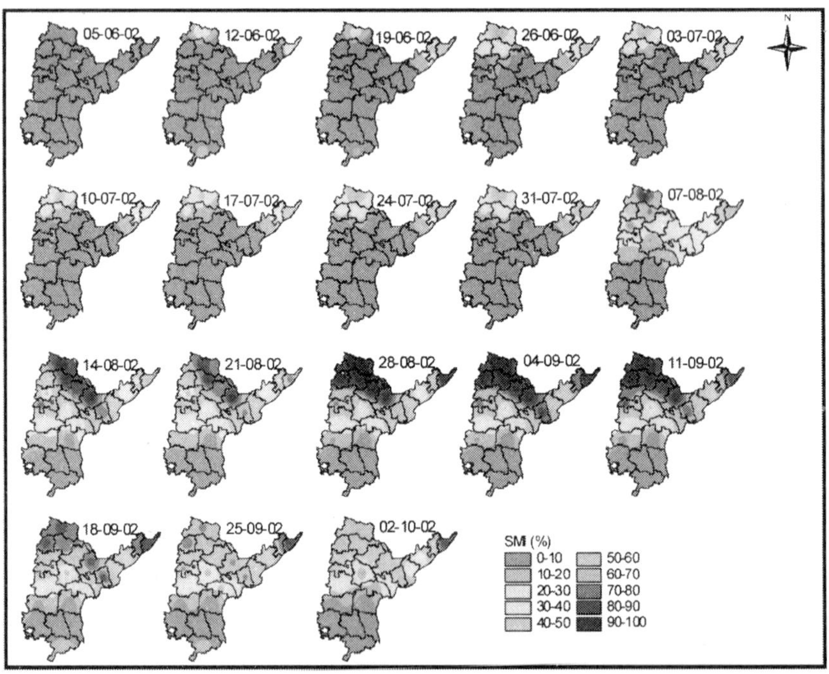

Fig. 4.18. Spatial distribution weekly soil moisture index in Andhra Pradesh during Monsoon 2002

A methodology to monitor the incidence and spread of drought in the western Rajasthan using departure of aridity index (Ia) from its normal values and expressed as percent was developed by Ramakrishna *et.al* (1984). The study was intended to identify the pattern and behaviour of drought occurrence and its spread in the subsequent months. As a case study, two severe drought years *viz.,* 1918 and 1968 were analysed. The monthly departures of Ia values during the crop season June – September for the 11 district of western Rajasthan were computed and monthly departure maps of Ia were prepared which are shown in Fig 4.20.

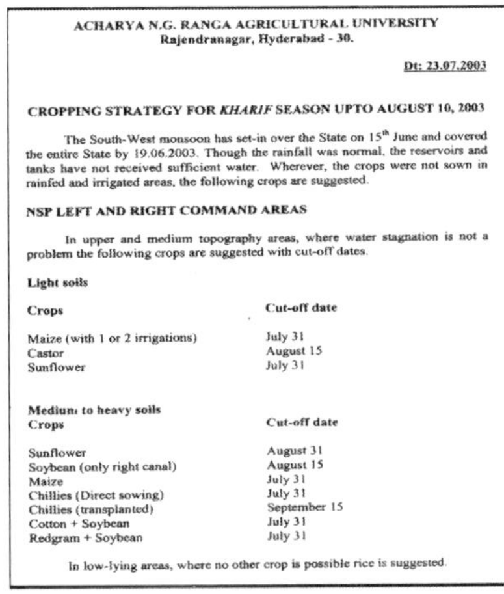

ACHARYA N.G. RANGA AGRICULTURAL UNIVERSITY
Rajendranagar, Hyderabad - 30.

Dt: 23.07.2003

CROPPING STRATEGY FOR *KHARIF* SEASON UPTO AUGUST 10, 2003

The South-West monsoon has set-in over the State on 15th June and covered the entire State by 19.06.2003. Though the rainfall was normal, the reservoirs and tanks have not received sufficient water. Wherever, the crops were not sown in rainfed and irrigated areas, the following crops are suggested.

NSP LEFT AND RIGHT COMMAND AREAS

In upper and medium topography areas, where water stagnation is not a problem the following crops are suggested with cut-off dates.

Light soils

Crops	Cut-off date
Maize (with 1 or 2 irrigations)	July 31
Castor	August 15
Sunflower	July 31

Medium to heavy soils

Crops	Cut-off date
Sunflower	August 31
Soybean (only right canal)	August 15
Maize	July 31
Chillies (Direct sowing)	July 31
Chillies (transplanted)	September 15
Cotton + Soybean	July 31
Redgram + Soybean	July 31

In low-lying areas, where no other crop is possible rice is suggested.

Fig. 4.19: Weather based advisory for crop contingency planning for Andhra Pradesh during 2002

A perusal of the above diagram indicates that the zone of maximum drought intensity changed both in its intensity and location indicating a sequential pattern of drought spread and decay over the region. It can be further observed that the droughts initially originate in northeast region around July (in Churu and Jhunjhunu), spread further in a south-westerly direction during August (Barmer region) and dissipate with an easterly movement (Pali region) during September. Also drought occurs early in regions receiving comparatively higher rainfall than in region of lower rainfall. Similar pattern of drought were observed for the drought years 1972, 1974 and 1979 (Rao *et al.,* 1984).

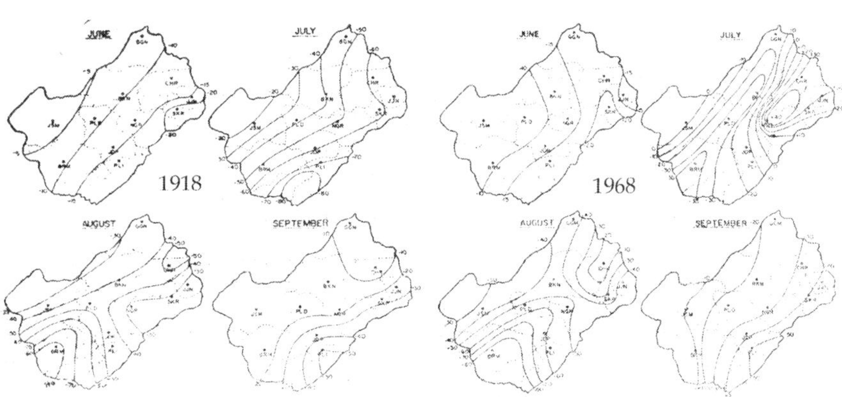

Fig. 4.20. Monthly departures of aridity index during 1918 and 1968 over western Rajasthan

A simple weather based decision system for growing *Kharif* sorghum in Telangana Region of Andhra Pradesh developed by AICRP on Agrometeorology, CRIDA (Rao *et al.*, 2009), A ready reckoner was developed for suggested action in the event of deficit or excess rainfall during the crop growing season. It provides first-hand information to plan various operations well in advance. As the sowing window for the crop spread over 3 weeks after the commencement of rainy season, the crop at any place may not be at the same growth stage. For example based on resource availability with the farmers and the receipt of rainfall, commencement of agricultural operations varies due to differential growth stages. Moisture regime is most important for crop growth during *kharif* season under tropical conditions. Therefore only the distribution of rainfall with different combinations has been considered. As an example, crop management options for anticipated weather during 26 Std. week in Telangana region is given in Table 4.4.

Such bulletins provide necessary input base for developing Decision Support System for better management crops. Similar type of bulletins were developed for the other crops *viz.*, Cotton, Rice, Sorghum, Pearl Millet at Kovilpatti, Mohanpur, Solapur and Hisar respectively under the All India Coordinated Research Project on Agrometeorology. Such studies are useful in agro advisories if they are extended to all crops grown under different agroclimatic regions of the country.

4.6.1 Software for Drought Detection, Monitoring and Early Warning

A software for drought detection and monitoring and early warning was developed by the Scientists of CRIDA as a part of consultancy programme with Government of Andhra Pradesh (Vijayakumar *et al.*, 2011). This was intended to use in each district for early detection of drought. In this software, moisture adequacy index and water requirement satisfaction index were used to identify drought. A crop-specific water balance model was used for obtaining these indices. Other State Governments in India may also develop such soft ware for detecting early signals of droughts. The input and output for the model is given below:

Input

Rainfall, Maximum and minimum temperature, Crop coefficients, Yield response factor, Available water holding capacity (AWC) of the soil, Latitude and longitude, maximum crop yield and crop duration

Output

Runoff, Deep drainage, Drought signals, and Relative yield

Table 4.4: Crop management options for anticipated weather during 26 Std. week in Telangana region of Andhra Pradesh

Existing crop condition	Anticipated weather likely to be observed during the week	Likely effect of anticipated change in weather on the crop and management decision to be taken
Vegetative stage (22 Std. week sown crop).	• Normal conditions prevail	• Inter cultivation should be done in between the rows with country plough to remove weeds. Weeding within the row can also be done. Monitor for stem borer attack. • The second dose of nitrogen can be applied to crop as top dressing @ 16kg/acre or 40 kg/ha should be applied.
	• No rain, cloudy sky, low RH	• Monitor for stem borer attack.
	• No rain cloudy sky, High humidity • Scanty rainfall/ intermittent cloudy weather /moderate humidity	• Inter cultivation should be done as above. • The second dose of Nitrogen can be applied to crop as top dressing @ 16kg/acre or 40kg/ha should be applied. The first dose of Nitrogen can be applied as above. Monitor for pest incidence.
	• Wide spread rains/cloudy sky/high humidity	• Monitor for rust incidence. If severity is more than 3 on 109 scales spray Mancozedb 75WP foliar application @ 2g/litre a.i. • Monitor for anthracnose and zonate leafspot and leaf blight.
	• Heavy rain followed by clear weather	• After a dry period of 6-12 hrs above operations (Inter cultivation and top dressing) can be done. In black soil areas, drain out the excess water from fields. • Identify the incidence of striga weeds in the field. Striga weed is the most important noxious weed, which causes heavy yield loss. Spray of Ammonium sulphate @ 50 g or Urea @ 200 g/ litre of water can control the striga infestation.

Source: Rao, *et al* 2009

The input and output frames are given in Fig 4.21. The software that was developed, not only provides drought signals during the crop-growing period, but also estimates yield at the end of the season. The software was so designed that even a technical officer at sub district level can use it. Provision was made in the software to estimate the sowing date based on daily rainfall. From the actual onset date, the appropriate crops and crop varieties that are to be grown will be displayed for different soil types of the State.

The software would assess the water requirement of a crop from the library in which weekly water requirement of all the major crops is stored. During the crop-growing period, whenever drought signal appears, the software estimate the water requirement from that date till the end of crop. It also estimates the amount of rainfall likely to occur at 50% probability till the end of crop (stored in the library). The amount by which the likely occurrence of rainfall going to fall short of the water requirement of the crop would also be displayed. The software had the capability to record and updates at weekly intervals the information on the total area sown in different sub district levels. A library of the normal cropped area (total of all crops) in each mandal is stored in the software and percentage of area sown is displayed at weekly intervals for declaring drought or no drought situation. Mandal may be declared, as drought affected, if the total cropped area does not exceed 50% of the normal by July 31 as this is considered as one of the norms followed by Government of India.

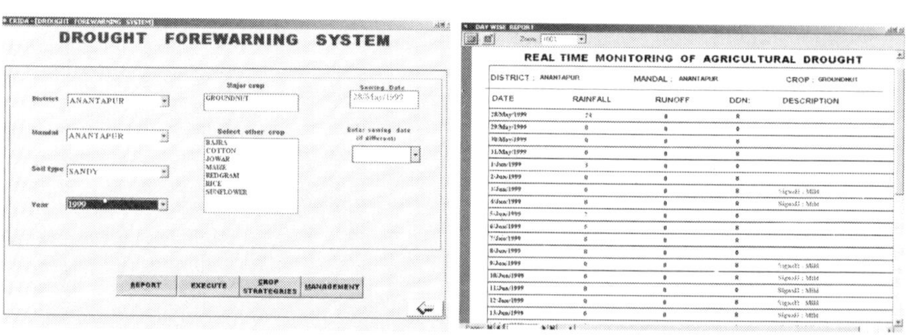

Fig. 4.21. Drought forewarning system (sample input)

4.6.2 Decision Support System for Droughts Developed Outside India

Drought Management consists of judicious usage of Water Resources and improving the soil moisture storage by adopting viable soil and water conservation methods. Many countries have evolved methods to augment the limited water resources for sustainable production even in the deficit rainfall years. In the USA, the congress has enacted the Agricultural Risk Protection Act of 2000 to encourage USDA and Risk Management Agency (RMA) and farmers to be more proactive in managing drought risks. The National Science Foundation (NSF), RMA have been working along with University of Nebraska, Drought Mitigation Center (DMC) and the High plains Regional Climate Centre to develop Geospatial Decision Support tools to address the Agricultural drought hazards and identify regions vulnerability in management of drought risks. Accordingly a National Agricultural Decision support system (NADSS) was developed by Kozal Dustin *et.al.,* 2004, of University of Nebraska to assess drought risk. The NADSS is a collection of Decision support tools that can help the farmers to assess variety of risks. The architecture of the DSS has four layers viz; knowledge, information, data and presentation. The data layer contains unmodified data meant for access. The information layer uses the data from data layer and develops it into information layer such as data interpolation and map servers which are domain independent (Spline method) or domain Specific (Terrain regression). The knowledge layer utilises the information layer to develop domain specific knowledge and it incorporates sequential data mining techniques, simulation modeling and other knowledge analysis algorithms. Knowledge layer tools allow the expert to evolve meaningful interpretation for the complex problems in drought risk management. The decision makers will interact with this layer and gather domain specific knowledge. The web based DSS is an effective drought risk management tool for farmers, experts and general end users.

A Decision support system for drought forecasting and Reservoir Management in Northeast Brazil was developed by Marcos Airton Desousa Freitas (1998) from university of Fortaleza, Brazil. Identification of meteorological parameters for prediction of drought have been done by analyzing global circulation systems, especially the ENSO (El Nino – Southern Oscillation) and the patterns of atmospheric and ocean conditions of tropical Atlantic. Forecasting of rainfall for the Northeast Brazil was attempted by developing correlation between Rainfall and sea surface temperature (SST) anomaly patterns in Pacific and Atlantic Oceans and air Pressure gradient in the Pacific. Drought prediction was carried by modeling the time series of rainfall data by the use of neural network for reference stations and neuro-fuzzy system analysis for the pattern recognition of the SST data of both pacific and the tropical Atlantic.

The use of SST data from Atlantic Ocean improved the Rainfall predictability in NE Brazil. Similarly based on drought prediction, the actual reservoir management was developed by Stochastic simulation technique. The results of drought management approach by different methods gave satisfactory results in development of appropriate tools for managing reservoirs in this zone.

Andrew *et. al* (2009) described how the drought can be managed using DSS, a tool that can provide information for drought risk estimation and assessment in Jucar River basin (42989 sq.km) including several adjacent river basins in eastern Spain. Special Drought awareness and Mitigation plans (SDP) have been formulated according to proactive approach drought preparedness. These include long term (planning), medium term (alert) and short term (emergency and mitigation) measures that were prepared using Standardized Drought Monitoring Indicator (SDMI) which is a combination of data viz., Rainfall, Storage in reservoirs, ground water levels and flows in Rivers. Besides the regular use of drought indicators for assessment of drought condition, DSS are also put to use for real time management of droughts in River basins.

For example the DSS was developed using "Aquatool" a generalized tool or DSS shell (DSSS) is exclusively used for integrated water resources planning and management. It has friendly graphical design and database management capability and provides control on the execution of models, analysis and reporting of results. After the system is defined completely, the user can perform simulation runs for the management of multiple alternatives. Therefore, it was useful for the evaluation of alternatives to analyse planning decisions and to assess tradeoff between alternatives. This system was effectively used for management of drought in Jucar River basin during the period 2005-2008. Some of the measures taken by the respective organizations to manage drought situation were as follows.

- Old wells that were not in use were recovered and new wells were identified and drilled
- lowering the application rates of irrigation water
- Conjunctive use of ground and surface water

A Decision Support System for water management in Thomson Reservoir, Victoria, Australia was developed by Mills *et. al* (2009) to manage drinking water supply to the Melbourne residents and for agricultural purposes in the Victoria region during drought period. The Aquatic Real time Management System (ARMS) was used to optimise field sampling and run scenarios forecasting as a part of a comprehensive liminological study of Thomson Reservoir. The 3D Estuary, Lake and Costal Ocean model (ELCOM) was coupled with ARMS to develop operational and planning decision over short and long range time scales.

The System addressed the following key issues,

- Melbourne water's drought management policy
- Environmental flows
- Management of the overall supply system

The ARMS allows the end-user to view the reservoir data in real time and evaluate management strategies through scenario forecasting of alternative operating regimes under changing reservoir conditions during drought.

4.7 GIS and Drought Monitoring

The Geographical Information system (GIS) is an organized collection of spatial data and to efficiently capture, store, update and manipulate, analyze and display all forms of geographically referenced information. The final output of GIS is creation of digitized spatial data bases from which spatial maps, images are prepared. These maps are quite useful in monitoring various parameters over different periods for taking right decision

In recent years GIS and Remote sensing played key role in many operational sectors especially in Agriculture for evaluation of regional drought conditions. Many agencies have been using this tool to study and monitor the drought situation on a regional scale. Wilhelmi and Wilhite (2002) employed GIS for assessment of agricultural drought vulnerability in Nebraska. The key biophysical and social factors that contribute to drought vulnerability of a region are climate, soil, land use and access to irrigation (Table 4.5). The agricultural drought vulnerability map was created through development of numerical weighting scheme to evaluate drought potential of the classes within each factor (Fig 4.22).

Table 4.5: Weighing scheme for assessing agricultural drought vulnerability in Nebraska

Agricultural drought vulnerability	Vulnerability class	Drought vulnerability factor class's score (weight)
Land use types	Rangeland	1
	Cropland	2
	Water, wetland, urban, forest	20 (maksing)
Soil root zone available	More than 200	1
Water holding capacity (mm)	150-200	2
	100-150	3
	Less than 100	4
Probability of seasonal	Less than 30 (low)	2
Crop moisture deficiency (%)	30-50 (moderate)	3
	50-70 (high)	4
	More than 70 (very high)	5
Irrigation support	Available	100 (masking)
	Not available	4

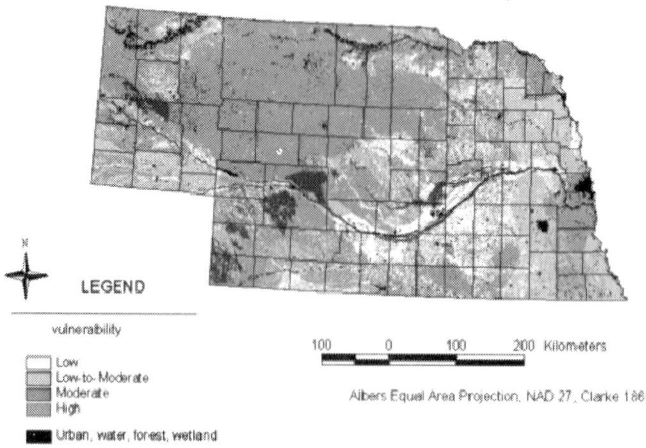

Source: Wilhelmi and Wilhite (2002)

Fig. 4.22: Agricultural drought vulnerability of Nebraska

The results showed that non-irrigated croplands, rangelands on sandy soils were most vulnerable to agricultural droughts which are located in regions where seasonal crop moisture deficiency is very high.

The Colorado state experienced series of droughts from 2000 to 2003 that resulted in reduced water levels in the reservoirs which were unable to meet the water requirements for outdoor irrigation. The city council of Greeley city in this state set a goal in reducing the water consumption for outdoor irrigation by 15 per cent in order to ensure water availability in the coming years. To ensure proper water pressure system in the event of new regulations, the water resources division of the city used GIS to address the situation that has arisen due to drought conditions. GIS provided the necessary data to engineers and managers to take sound decision. (Crorin, 2003).

Drought Risk areas in Lopburi province in Thailand were examined by Prathumchai *et al.* (2001) by analyzing the meteorological and physical factors based on the criteria adopted by Ministry of Science, Technology and Environment (MOSTE) of Thailand. NDVI changes between normal year (1995) and a drought year (1997) was analyzed for each drought risk regions and observed that NDVI is lower in high drought risk areas.

A GIS based Water Resources and Agricultural Permitting and Planning System (GWRAPPS) to estimate the water demand due to drought in five water management districts in Florida State was developed by Satti and Jacobs (2003). This model integrates the Agricultural Field Scale Irrigation Requirements Simulation (AFSIRS) Crop model, a Geographical Information System (GIS) and Database management system within Arc GIS framework. The model quantifies the irrigation water for regional planning and farm scale operations under statistically averaged to drought conditions using spatially distributed soil, land use and long term daily climate data. It also estimates daily water withdrawals that are essential for input into Conjunctive surface/ ground water models. This system can function as a practical tool that is applicable to wide range of water resources management and developmental problems under different drought conditions.

A land based information system for drought analysis using GIS for Jhodnadi river basin a tributary to River Godavari in Nanded district of Maharashtra, was reported by Nagarajan and Mahapatra (2000). The study was aimed to store information of the past and compare it with the current data to assess the vulnerability of an area to droughts and to workout mitigation plans. In addition to biophysical parameters like climate, soil, crop data, surface and groundwater, social factors data infrastructure and satellite data were used to identify the regions vulnerable to drought conditions. Various spatial maps for different parameters were prepared and over laid one over the other to arrive at vulnerability maps for drought in the Jhodonadi basin. The information generated helped the administrators and planners for developing operational and strategic planning.

Drought risk assessment using RS and GIS for Gujarat State was reported by Parul Chopra (2006). The study was aimed to delineate drought risk regions facing drought using NOAA-AVHRR based NDVI and SPI values for the year 1981-2000. Correlation and regression analysis was carried out between NDVI, SPI , Rainfall anomaly and food grain anomaly. SPI values were interpolated to obtain spatial pattern of meteorological droughts. Food grain yield trend was plotted and an equivalent NDVI threshold values were identified to get agricultural drought risks. Rainfall anomaly and NDVI were correlated and a threshold defined by IMD for Meteorological drought was used to derive meteorological drought risk. A risk map of drought showing the areas prone to hazards for the study area was prepared by integrating agricultural and meteorological risk maps. The results indicated that central and northern parts of Gujarat are more prone to droughts and some mitigation measures have been suggested.

A GIS based modeling for drought assessment in Gubbi Taluka of Tumkur district in Karnataka was developed for assessing the damages and helped in planning relief measures (Shiva Prakash *et.al.*, 2006). In this work, 17 thematic data layers were integrated and a map showing regions of drought severity was obtained. The approach followed by the authors includes analyzing the remotely sensed data to derive land use information, creation of spatial database and its integration in GIS environment by developing suitable rating scheme for generation of drought severity map. The methodology used in preparation of GIS map is as follows (Fig 4.23).

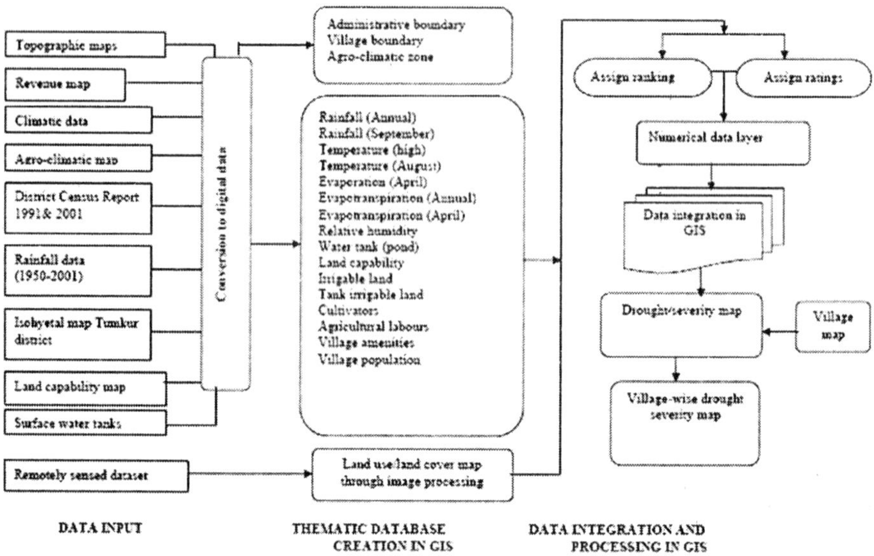

(*Source*: Shiva Prakash et al. 2006)

Fig. 4.23. Methodology used in preparation of GIS map

Remote sensing based Drought information for Palar and Thamiravaruni Basins located in NE and SW parts of regions in Tamil Nadu was reported by Ravikumar *et al.* (2003). The study integrated the meteorological, hydrological and agricultural drought areas using GIS and prepared maps showing drought risk areas. Drought Information System (DIS) was developed comprising of model viz. water related data base, meteorological droughts, hydrological and agricultural drought and drought risk area analysis.

References

Andreu, J. Ferrer-Polo, J. Perez, M.A. and Solera, A. 2009. Decision support system for drought planning and management in the Jucar River Basin, Spain. 18th World IMACS/MODSIM Congress, Cairns, 13-17 July 2009, 3223-3229.

Attri, S.D and Tyagi, A. 2010. Climate Profiel of India. Met Monograph, No. Environment Meteorology-01/2010. India Meteorological Department. Ministry of Earth Sciences, New Delhi.129p.

Bochenek, Z., Dabrowska-Zielinska, K., Clokosz, A., Drupka, S. and Boken, V.K. 2005. Monitoring Agricultural Drought in Poland. In : Monitoring and predicting agricultural drought : A global study. (Eds. Boken, V.K., Cracknell, A.P. and Heathcote, R.L.). Oxford University Press Inc., New York.171-195.

Boer, R. and Subbiah, A.R. 2005. Agricultural Drought in Indonesia. In : Monitoring and predicting agricultural drought : A global study. (Eds. Boken, V.K., Cracknell, A.P. and Heathcote, R.L.). Oxford University Press Inc., New York.330-344.

CRIDA. 2011. Annual Report 2010-11. Central Research Institute for Dryland Agriculture. Hyderabad. 162p.

Cronin, S.T. 2003. Using GIS to deal with drought. Arc User. April-June Redlands, C.A.22-25.

Eddy De Pauw. 2005. Monitoring Agricultural Drought in Near East. In : Monitoring and predicting agricultural drought: A global study. (Eds. Boken, V.K., Cracknell, A.P. and Heathcote, R.L.). Oxford University Press Inc., New York.208-224.

Gao Ge Zhang Peiqun Liao Yaoming. 2011. Drought monitoring operational system and early warning in China Meteorological Administration. Presentation made at WCRP Workshop on Drought Predictability and Prediction in a Changing Climate, held at Barcelona, Spain during 2-4 March 2011.

Hubbard. K.G. 1987. Surface weather monitoring and the development of drought and other climate information delivery systems. In : Planning for drought : Toward a reduction of societal vulnerability. (Eds. Wilhite, D.A., Easterling, W.E. and Wood, D.A.), Westview Press Inc., Colorado.97-112 .

JMA / WMO. 2010. JMA / WMO workshop on quality management in surface, climate and upper air observations in RA II (Asia). 27-30 July 2010, Tokyo, Japan. 197p.

Klenchenko, A.D., Zoidze, E.K. and Boken, V.K. 2005. Monitoring Agricultural Drought in Russia. In : Monitoring and predicting agricultural drought : A global study. (Eds. Boken, V.K., Cracknell, A.P. and Heathcote, R.L.). Oxford University Press Inc., New York. 196-207.

Kozal, D., Culver, M. and Harms, S. 2004. A knowledge-based geo-spatial decision support system for drought assessment. Proceedings of the annual national conference on Digital Government Research. Digital Government Society of North America., USA.

Marcos Airton desousa Freitas. 1998. A Decision support system for drought forecasting and reservoirs management in Northeast-Brazil. In: VIII Latin American and Iberian Congress of Meteorology, Brasilia.

Mills, R. Marti, C. Yeates, P. Haydon, S. and Imberger, J. 2009. A decision support system for water management in Thomson Reservoir, Victoria. 18th World IMACS/MODSIM Congress, Cairns, 13-17 July 2009.3259-3265.

Nagarajan R. and Mahapatra Subrata. 2000. Land based information system for drought analysis. Proceedings of Map India 2000, April 10-11, 2000, New Delhi, India.

NRSC, 2011 Agricultural Droughts. www.nrsc.gov.in

Parul Chopra. 2006. Drought risk assessment using remote sensing and GIS. A case study of Gujarat. M.Sc. Thesis. Submitted to International Institute for Geo-Information Science and Earth Observation, Enschede, The Netherlands.

Prathumchai, K., Honda, Kiyoshi and Nualchawee, K. 2001. Drought risk evaluation using remote sensing and GIS: A case study in Lop Buri Province. Proceedings of 22nd Asian Conference on remote sensing, 5-9 November 2001. Centre for Remote Imaging, sensing and Processing (CRISP), National University of Singapore, Singapore.

Rama Krishna, Y.S., Sastri, A.S.R.A.S., Rao, G.G.S.N and Ramana Rao, B.V. 1984. On prediction of droughts in the Indian arid region. Mausam 35 (3): 349-354.

Rao, G.G.S.N., Rama Krishna, Y.S. Ramana Rao, B.V. and Purohit, R.S. 1984. Incidence of droughts and its impact on food production. Annals of Arid Zone. 23(4): 327-334.

Rao, G.G.S.N., Kesavsa Rao, AVR, Ramakrishna, Y.S. Vijaya Kumar, P. and Raji Reddy, D. 2003. Final report of "Near real-time monitoring of agrometeorological conditions for contingency planning in Andhra Pradesh". RRPS-32. NATP-Project 2000-2004. CRIDA, Hyderabad.

Rao, G.G.S.N., Ramana Rao, B.V., Srinivasa Rao, M., Ravindra Chary, G., Rao, V.U.M., Desai, S., Rao, A.V.M.S. and Venkateswarlu, B. 2009. Weather based decisions for growing *kharif* sorghum in Telangana region of Andhra Pradesh – A Practical Manual. AICRPAM Tech. Bul.No.2/2009, CRIDA, Hyderabad. 42p.

Rao, G.G.S.N. Rao, V.U.M, Vijaya Kumar, P and Rao, A.V.M.S. 2010. 25 years of Research on Agrometeorolgy, AICRP on Agromteorology, CRIDA, Hyderabad 111p.

Ravikumar, G. Prakashvel and M. Kaarmegan, M. 2003. Remote Sensing based drought information system for Palar and Thamirvaruni Basins using GIS. Proceedings of Map Asia Conference 2003. October 13-15 2003, Kuala Lumpur, Malaysia.

Samra, J.S. 2004. Review and analysis of drought monitoring declaration and management in India. Working paper 84. Drought series. Paper 2. International Water Management Institute, Colombo. 25p.

Satti, S.R. and Jacobs, J.M. 2004. A GIS based model to estimate the regionally distributed drought water demand. Agricultural Water Management. 66 (1):1-13.

Shiva Prakash, P.H. Garg , P.K. and Ghosh. S.K. 2006. GIS based modeling for drought assessment. Proceedings of ESRI Users Group Conference 2006.

Szalai, S. 2004. Drought monitoring in the early warning system. Hungarian Meteorological Service. Booklet and Presentation.

Tadesse, T. and Wilhite, D.A. 2007. Modern drought monitoring tool for decision support system. In: Encyclopedia of Digital Government, IGI Global.1212-1218.

Thornthwaite, C.W. and Mather, J.R. 1955. The water balance. Drexel Institute of Technology, Publications in Climatology. 8(1): 104p.

Vasselin Alexandrov. 2006. Monitoring soil drought (a review). National Institute of Meteorology and Hydrology, Sofia, Bulgeria, 34p.

Vijaya Kumar, P. 2011. Characterization of droughts and development of forewarning system. In: Techniques of water conservation and rainwater harvesting for drought management, SAARC Training Program. (Eds. P.K. Mishra, M.S. Osman, Satendra and B. Venkateswarlu). Central Research Institute for Dryland Agriculture, Hyderabad, 428-440.

Wilhelmi, O.V. and Wilhite, D.A. 2002. Assessing Vulnerability to agricultural drought : A Nebraska Case Study. Natural Hazards. 25: 37-28.

WMO. 2006. Drought monitoring and early warning: Concepts, progress and future challenges. World Meteorological Organization Technical Report No.1006. Geneva. 24 p.

□□□□

CHAPTER 5

Impacts on Agricultural Production

Agricultural crops are sensitive to climatic parameters. Its variability and their production is often affected by climate extremes such as droughts, floods, heat and cold waves etc. Among all, the impact of droughts on the fall in agricultural production is significant as droughts influence large areas at regional or at national scale that threatens the food security, while other events are confined to smaller scale for a very short period and have limited impact on national agricultural production. In addition to weather parameters other key elements that influence agricultural production are population growth, land use pattern, water availability, improved agricultural practices inclusive of fertilizer and pest/disease management and development of high yielding varieties. Among the above, rapid population growth especially in the developing countries is of major concern along with the changing climate in the recent past. It is estimated that out of the current global population of 6.9 billion, 16 per cent of them live in developed countries and remaining 84 per cent live in developing world (Hisas, 2011). It is further projected that the total human population over the globe would reach by 7.8 billion by 2020, which is about 13 percent increase over the current population. Though there is good growth in respect of food grain production from 1960 onwards, yet the population growth rate especially after 1985 in developing countries surpassed the production rate thus the availability of food grain per person has been falling (Pinstrup-Anderson and Pandaya–Lorch, 1996) steadily. If adequate measures are not taken soon, it will have adverse repercussions on additional needs of food grain. To throw more focus on this issue, food grain production with respect to global and national level is discussed in this chapter. Rearing of livestock and fisheries are also part of human food chain, therefore their variability with respect to weather parameters are also briefly discussed.

5.1 Global Food Production

5.1.1 Area under cultivation

The area under cultivation, water availability and the climate are directly related to food grain production. The arable land available for cultivation in the world totals to 1,411.1 M ha out of which 41 per cent (576.2 M ha) lies in the developed countries and the remaining 59 per cent (834.9 M ha) is distributed in the developing countries (Hisas, 2011). The percentage distribution of arable land in different regions Fig 5.1 showed that highest land under cultivation is in Asia (36%) followed by Europe and Africa.

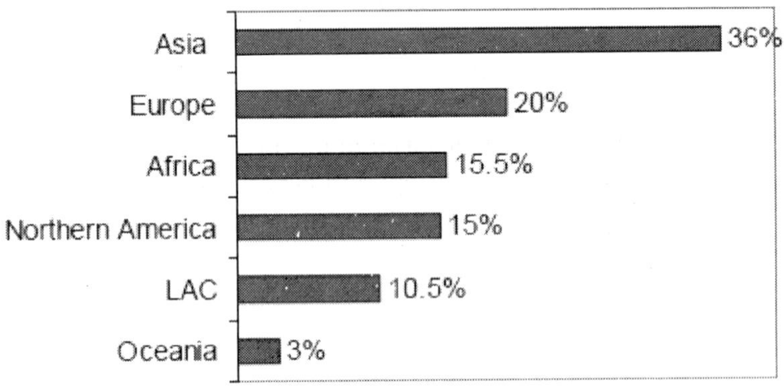

Source: (Hisas, 2011).

Fig. 5.1: Arable land: Share of world's total, by region

It was reported that during the past 15 years, the arable land in the world expanded by 7.9 M ha. However, intensive cultivation and increased use of fertilizers over the years degraded 40 per cent of arable land. About 80 per cent of the total agricultural land over the globe is rainfed and the rest is irrigated for which about 70 per cent of fresh water is being used (Hisas, 2011). The arable land is distributed in the tropical and temperate zone where 60 and 38 per cent of world's population lives respectively.

5.1.2 World food production

World's food production consists of production from major crops, meat and milk from livestock and fish from marine and fresh water catch and captive fishes under aquaculture. The major cereal crops that account for food grain are wheat, rice, coarse grains (Maize, Sorghum, Millets and Soybean). Wheat is grown both under rainfed conditions in temperate climates and irrigated in tropical regions. Hisas (2011) reported following statistics on the impacts of climate change on food production. He had observed that about 86 per cent of wheat is produced by 20 countries in the world and 82 per cent of the production is consumed where it is grown. He further observed next to wheat, rice is grown in tropical warm climate and 95 per cent of it grown in 20 countries in this climate and about 94 per cent of the rice produced is used where it is grown. More than 56 per cent of the worlds' coarse grain viz. maize, sorghum, barley, rye, oats, millets, and soybean is used for animal food while 17 per cent is used for human consumption. Among the coarse grain two crops *viz.,* maize and soybean are important as 12 per cent of global production of maize was utilized to produce ethanol, 60 per cent for animal feed and remaining 28 per cent for food consumption, whereas 10 percent of world's production of soybean was used for food and 75 per cent to produce edible oil. In respect of meat and milk production, around 48 per cent of beef production and 53 per cent of milk comes from rainfed mixed farming system. Considerable difference in consumption of meat and milk between developing and developed countries was noticed. On average, the meat consumption was about 31.5 kg/year in developing countries while it was 81.9 kg/year in developed countries. Similarly, milk consumption among developing and developed were 65.5 and 246.9 kg/ year respectively. Out of 141.6 Mt of fish produced over the globe, 80 per cent of it was used as food. (Hisas, 2011).

5.2 Variability in Global and Indian Food Grain Production

The world food grain production for the years 1960-2010 as compiled by FAO and reported by USDA (2010) is reproduced in the figure 5.2. It is observed that the global food grain production which was 824 Million Tons during 1960 increased to 2213 Million Tons in 2010 at a rate of 27.86 Million Tons/year.

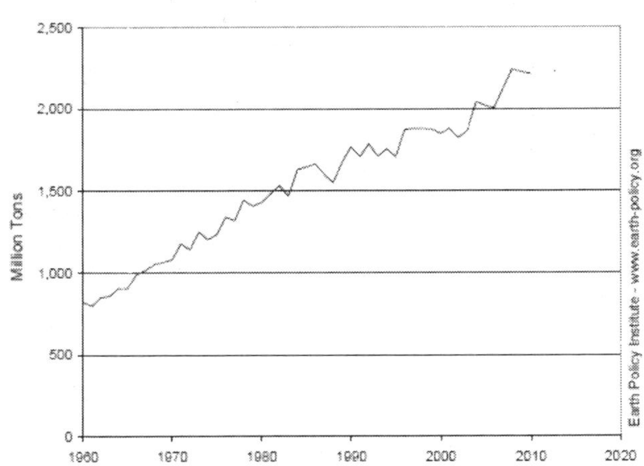

(*Source*: USDA, 2010)

Fig. 5.2: World Food Grain Production, 1960-2010

Though the production curve shows near linear trend over the period, variability in total annual production is noticed, this can be mostly attributed to prevailing weather conditions in different parts of the globe during those years.

The food grain production in India over the years 1950 to 2010 along with the monsoon season rainfall is given in Fig.5.3. It is observed that the food grain production increased from 50.8 Mt to 241.6 Mt over the period 1950-2010 at a rate of 3.08 Mt/year with fluctuations directly related to monsoon rainfall. A close look at the figure shows a good relation between monsoon rainfall and all India food grain production. Whenever monsoon rainfall decreased below its mean (890 mm), correspondingly the food grain production also declined. During the period under study (1950-2010) the years 1965, 1966, 1972, 1979, 1987, 2002 and 2009 were considered as drought years. Among these years 1965,1966,1987,2002 and 2009 were considered as phenomenal drought years. The fall in food grain production in these drought years over its previous year's production were 19, 17, 7, 17, 6.6, 17.8 and 7 per cent respectively. Droughts are frequent in the arid zone of India located in the northwest part of the country and their influence on agricultural production was studied extensively by the scientists from Central Arid Zone Research Institute, Jodhpur.

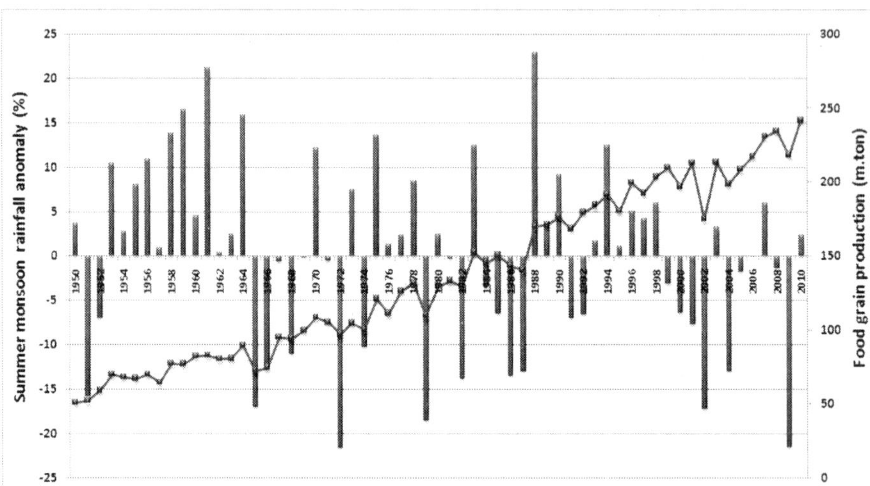

Fig. 5.3: All India Summer monsoon rainfall anomaly and food grain production over the period (1950-2010)

Sastry *et al.* (1982) have studied the impact of various intensities of droughts on pearlmillet and pulses and found that the productivity of the pearlmillet crop fell by 37, 71, 80 per cent and that of pulses by 6, 20, 76 per cent respectively from no drought conditions to mild, moderate and severe drought conditions. Similarly Rao *et.al* (1986) have found out the impact of droughts on forage productivity of natural grasslands that were grouped under 3 rainfall zones *viz.*, < 200 mm, 200-400 and > 400 mm/year in western Rajasthan and reported that about 41 to 48 per cent fall in forage productivity due to mild drought conditions in all rainfall zones. Under severe drought conditions, significant fall in forage productivity to an extent of 74 to 80 per cent over no drought in low and medium rainfall zone was noticed. Rao *et.al.* (1984) reported the influence of droughts during 1972, 1974 and 1979 on the crop production of *viz.*, pearl millet, pulses, jowar and maize in Rajasthan when compared with normal rainfall year 1970. About 68, 84 and 86% fall in pearl millet, 71, 75 and 89 per cent fall in pulses. 37, 35 and 69 per cent fall in jowar and 11, 45 and 37 per cent fall in maize productivity was observed in the State when compared with normal year during 1970. Inter annual monsoon rainfall variability culminated into drought and floods affected the food grain production in India (Parthasarathy and Pant 1985, Selvaraj 2003, Krishna Kumar *et.al* 2004). Rainfall received at the end of monsoon season provides an excellent stored soil moisture conditions for better establishment of *rabi* season crop and subsequent growth.

Selvaraj (2003) analysed the relation between Indian Summer Monsoon Rainfall (SMR) and Food Grain Production (FGP) and observed close relation between inter annual variations in SMR and food grain production. The magnitude of change in food grain production is smaller than the rainfall (Fig 5.4).

The correlation between SMR and total Food Grain Production (FGP) was 0.71, significant at 1% level. SMR shows high correlation (r=0.80) with *kharif* food grain production and moderate correlation with (0.41) with *rabi* season food grains.

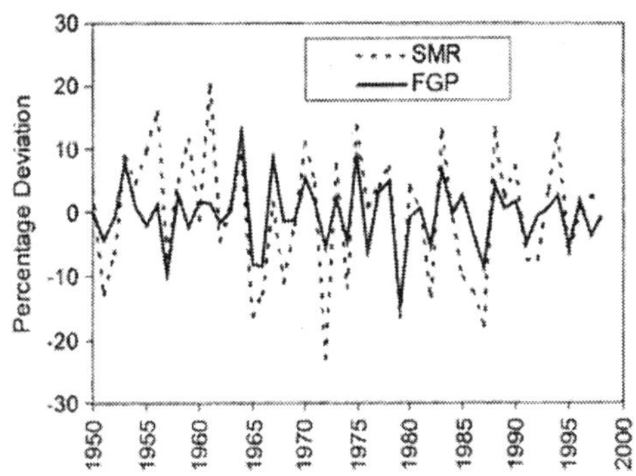

(*Source*: Selvaraj, 2003)

Fig. 5.4. Relationship between SMR and total food grain production (FGP) over India

Monsoon failure is common in some parts of the country every year even during good monsoon rainfall years and the regional production gets affected due to local rainfall conditions. Almost in each year, there were some regions suffered from severe fall in production of food grains and severity was much larger than that was recorded for the country as a whole. This can be seen from the fluctuations in output of *kharif* food grains over the years

1981-82 to 2005-2006 given in the Table 5.1. The table presents the estimates of deviation in output at the country level and the negative deviations from the top three states in terms of fall in production for the years 1981-82 to 2005-06 (Ramesh Chand and Raju, 2009). A perusal of the table reveal that though the fall in production was of 2.1% at all India level during 1981-82, the food grain production in the states of Uttar Pradesh, Bihar and Orissa, the production fell by 12 to 23 per cent over its mean which is lower than the trend values. Similarly, food grain production in some of the states was less than two thirds of normal level during 1985-86 when the all India production was only 1 per cent less than the normal. During the year 2002-03 drought year, the fall in production varied between 38 to 63 per cent in Tamil Nadu, Orissa and Rajasthan when all India food grain production was lower by 18.5 per cent. Krishna Kumar *et.al* (2004) studied the climate impacts on Indian agriculture by analyzing the crop climate relationships using historical production statistics for major crops. The analyses has brought out the influence of monsoon rainfall and identified some of the potential predictors' viz., Pacific and Indian Ocean sea-surface temperatures, and Darwin Sea-level pressure on crop production. He further observed that state level crop production statistics and sub divisional rainfall were generally consistent with all India results but demonstrated some spatial variations. It was reported that sub-divisional rainfall of western and central peninsula is strongly related to ENSO and Indian Ocean sea surface temperature.

Table 5.1: Deviation (percent) in *Kharif* food grain production from the trend at All-India level and in three worst affected states (*Source:* Ramesh Chand and Raju, 2009)

Year	All India (%)	Decline in production in three Top Ranking states in (%)		
1981-82	-2.1	Uttar Pradesh (-22.9)	Bihar (-15.4)	Orissa (-12.4)
1982-83	-15.1	Uttar Pradesh (-33.9)	Bihar (-33.2)	Orissa (-30.4)
1983-84	6.8	Uttar Pradesh (-20.2)	West Bengal (-12.4)	J&K (-10.6)
1984-85	-0.3	Andhra Pradesh (-16.2)	Maharashtra (-12.2)	Orissa (-11)
1985-86	-0.9	Gujarat (-41.1)	Rajasthan (-36)	Maharashtra (-14.6)
1986-87	-8.1	Rajasthan (-33.1)	Maharashtra (-30)	Gujarat (-22)
1987-88	-15.8	Rajasthan (-67.3)	Haryana (-42.1)	Himachal Pradesh (-27.2)
1988-89	13.0	Punjab (-17.2)	Himachal Pradesh (14.7)	Assam (-12.6)
1989-90	11.0	Haryana (-4.6)	Karnataka (-3.1)	Assam (-1.7)
1990-91	7.8	Karnataka (-13.2)	Tamil Nadu (-2.6)	-
1991-92	-2.0	Rajasthan (-37.6)	Gujarat (-21.3)	Maharashtra (-19)
1992-93	7.0	Madhya Pradesh (-35.9)	Bihar (-12.3)	Orissa (-0.3)
1993-94	4.2	Rajasthan (-35.1)	Gujarat (-10)	Haryana (-8.6)
1994-95	4.0	Karnataka (-7.4)	Andhra Pradesh (7.5)	-
1995-96	-3.4	Rajasthan (-38.7)	Bihar (-38)	Haryana (-19.5)

(*Contd.*)

Year	All India	Three Top Ranking states in Decline		
1996-97	4.2	Orissa (-25.6)	J&K (-10.2)	Punjab (-6.6)
1997-98	0.6	Andhra Pradesh (-20)	Madhya Pradesh (-18)	Maharashtra (-11.3)
1998-99	0.7	Rajastha n(-22.8)	Bihar (-14.2)	Kerala (-13)
1999-2000	2.0	Rajasthan (-38.5)	Orissa (-18.7)	Gujarat (-14.7)
2000-01	-2.6	Gujarat (-47)	Madhya Pradesh (-33.3)	Rajasthan (-22.8)
2001-02	5.7	Karnataka (-11)	Maharashtra (-7.4)	J&K (-2.6)
2002-03	-18.5	Rajasthan (-62.7)	Orissa (-46.6)	Tamil Nadu (-37.7)
2003-04	7.9	Tamil Nadu (-37.9)	Karnataka (-27.4)	Kerala (-11.7)
2004-05	-5.8	Uttar Pradesh (-19.3)	Maharashtra (-15.5)	Assam (-12.1)
2005-06	-0.9	Bihar (-38.2)	Himachal Pradesh (-19.5)	Rajasthan (-13.6)

Note: Figures in paren this is an the diviations with respect to its average
Source of Basic data: DES, MOA, GOI

5.3 Impact of Phenomenal Drought During 2009 on Food Grain Production in Different Countries

Droughts occur somewhere or the other over the world every year and there are very few drought years in which majority of the countries in the world are affected and global grain production gets badly affected. Several studies were conducted at various International Institutes to record the impact of droughts on agricultural crops in different countries for each drought year. As an example the impact of drought on agricultural production during the recent drought year 2009 is given below.

The year 2009 happened to be a severe drought year and considered to be a humanitarian disaster around the world in recent times and many nations including China, India, and Argentina etc were subjected to its severity resulting in steep fall in production. The Catastrophic fall in 2009 global food production was reported in Global E-news letter by Eric de Carbonnel (2009). The news items on fall in production at many locations are briefly summarized below.

The status of different intensities of drought experienced by various countries during 2009 over the world as reportd by (Eric de Carbonnel, 2009) is given in Figure 5.5. Except over the greater part of Europe, Canada, Central and Western parts of Africa, Southeast Asian parts, Northwest parts of South America, all other countries in the world were affected by droughts of different intensities. In China it was reported that 4.37 million people and 2.1 million livestock were affected due to scarcity in drinking water. Wheat crop in eight provinces was badly affected. The Chinese Government allocated about $12.69 billion as drought relief to the drought hit regions. Also as a part of improving the soil moisture status, cloud seeding with 2392 Rockets and 409 cannon shells loaded with chemicals was conducted.

Australia's water resources were badly hit by the worst drought recorded in the last 117 years. The Murray River stopped flowing at its terminal point and its mouth has closed up. Many of the lower lakes are a meter below sea level due to high evaporation from lake surfaces. Argentina's food production was reduced approximately by 50 per cent. The wheat yield during 2009 was 8.7 Mt. whereas it was 16.3 Mt in 2008. Soybean and corn crops were badly affected due to drought conditions experienced in various regions. More than 8.0 Mt of fall in soybean production was recorded. In Paraguay the government declared agricultural emergency. Crops that produce fodder for cattle were damaged badly and soybean crop in some areas was totally ruined. Uruguay also declared agricultural emergency and agricultural crops were badly hit. Fifty rural districts in Chile were badly affected and agricultural emergency declared and the water levels in the reservoirs were at all time low.

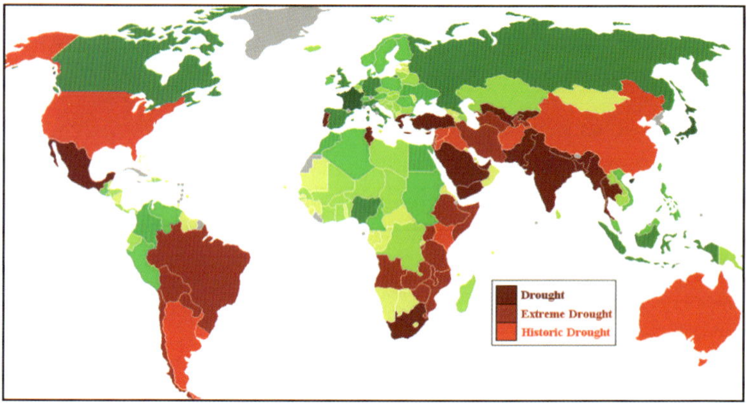

(*Source:* Eric de Carbonnel, 2009)

Fig. 5.5: Different intensities of drought experienced by different countries during 2009

In US, the California State witnessed worst drought in recorded history. Thousands of acres of row crops were fallowed. Most of the important reservoirs were only at 49 per cent of their storage capacity. The 2009 year drought in Texas states was comparable only with the droughts during 1917-18. Eighty eight per cent of the state was affected by droughts. Lack of rainfall left pastures barren and cattle production was hit. Florida was also hit by winter droughts damaging crops as about half of state experienced some level of drought conditions.

The countries in Africa too suffered food shortages due to insufficient rainfall. Kenya was the worst hit nation and the food was imported to bridge the shortfall. Tanzania recorded poor harvest and banned export of food. Burundi, a tiny

East African country faced severe food shortage as the crops failed in the northern part of the country. Severe food shortage was noticed in Uganda due to acute drought conditions that prevailed in north-eastern Uganda. South Africa reported lowest wheat production in the last 30 years due to low rainfall conditions. Other African countries like Malawi, Zambia, Swaziland, Somalia, Zimbabwe, Mozambique, Tunisia, Angola and Ethiopia also suffered from drought conditions during the year 2009.

The countries in Middle East and central Asia suffered from worst drought during 2009. The total wheat production in the drought affected region declined by 22 per cent. Major reservoirs in Turkey, Iran, Iraq and Syria were all at low level. In Iraq, Syria and Afghanistan wheat production was down by 45, 50 and 60 per cent respectively in 2009. Other countries that suffered from drought in this region are Lebanon, Israel, Turkey, Bangladesh, Myanmar, Tajikistan, Turkmenistan, Thailand, Nepal, Pakistan, Kyrgyzstan, Uzbekistan, Cyprus and Iran.

India's total food grain production during 2009 fell by about 7 per cent over 2008, though the country's average rainfall during southwest monsoon season was less by 23 per cent and more than 50 per cent geographical area was affected due to severe drought, which is one of the worst droughts witnessed over the last 125 years. *Kharif* season production was badly hit as only 63, 65 and 73 per cent of the normal sown area of Rice, Jowar, Bajra respectively could be sown by the mid August (NRAA, 2009). However, the country's food grain production was adequately compensated by good *rabi* crop production due to the receipt of post monsoon season rainfall and favourable winter rains. Rice output alone was less by 12 Mt over the 2008 production. Similarly coarse cereal and pulse production also fell by 5.7 and 0.47 Mt respectively as crops were grown under rainfed conditions (Review of Economy, 2009-10)

5.4 Impacts on Livestock and Fisheries

5.4.1. Livestock

Livestock is considered as a valuable asset among the farming households and any loss to the animals makes them poorer and more vulnerable to climate induced adversities such as droughts and heat waves etc. Drought losses in the pastoral sector affect the farmers by way of changes in the availability of fodder and the increased rates. The impacts of droughts on farming sector with respect to crops and livestock as visualised reported by FAO is shown in Fig 5.6.

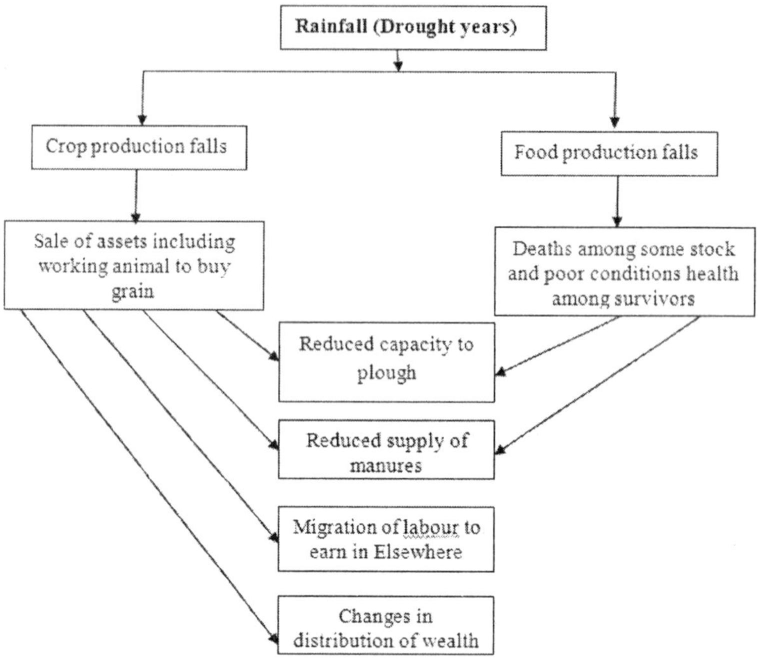

(*Source*: FAO)

Fig. 5.6: Effects of drought on farming activities

Low rainfall results in poor pasture growth and insufficient fodder around the villages lead to weight loss and increased number of deaths among the stock. The situation becomes worse with immigrant herd which exert further pressure on the local pastures. In general, the nomadic type group tends to migrate in search of green pastures. For the farmers who own less number of livestock there is no option as they are less familiar with migration problem. However, a few farmers whose economic conditions are relatively better can hire labor to take their animals for grazing areas and continue with farming activity. Thus the sedentary herds are badly hit during drought years. The immediate affect of drought is fall in production of both crop and fodder. To fulfill the social commitments during drought years, sale of livestock act as a buffer in times of hardship. Initially, the small animals like sheep and goats, which make least contribution to farm production are sold and if drought induced food deficit continues, the farmers will start sell their transport and draft animals. Wood (1976) reported that in Ethiopia, stocks are usually disposed off in the following order *viz.,* sheep and goats, then younger cattle with horses, donkeys and work oxen sold as a last resort. The loss of livestock during the drought year reduces the availability of dung among the farming community, an excellent fuel where firewood is scarce. Dung is estimated to be the second most important product obtained from livestock after draft power and providing 80% household fuel needs in many areas (Gryseels and Anderson, 1983). In many parts of West Africa, dung is a major element in the exchanges between pastoral and farming community. In south Shelian region, drought conditions force many herd-owners to move further south in search of fodder and water. The farmers in their region rely on the visit of herds for their dung supplies and this greatly gets reduced during a drought year. However dung supplies will increase for those farming communities in Savannah areas, which are hosts to immigrant pastoral herds in a drought year (Toulmin, 1986). A survey of deaths among the draft animals in Niño areas of northwest Mali was between 50-70 per cent over 1983-84 (FAO, 1984). The distribution of work oxen losses between deaths and distress sale vary according to the circumstances in which drought has occurred. A report by the Relief and Rehabilitation commission of Ethiopia for the province of Wollo in 1974 shows that all the losses incurred by the farmers were due to death of livestock rather than distress sales. Seventy one per cent of the animals died due to starvation as against 19 per cent sold, 2 per cent disposed of by other means and 8 per cent remaining (Wolde Mariam, 1984). On the other hand, Wood (1976) observed that livestock losses in Ethiopia were due to distress sale in order to raise cash.

In India, the impact of droughts on livestock are only available for Rajasthan state as it is located in the arid regions of Northwest India where water resources are negligible and drought frequency is high. Livestock censuses conducted from the year 1951 to 2003 at different intervals are given in the Table 5.2.

Table 5.2: Growth and composition of livestock in Rajasthan (Figures in millions) (*Source*: Rathore, 2005)

Year	Cattle	Buffalo	Sheep	Goat	Camel	Others	Total
1951	10.787	3.045	5.387	5.562	0.341	0.399	25.5
	(42.26)	(11.93)	(21.11)	(21.80)	(1.34)	(1.56)	(100)
1961	13.136	4.019	7.361	8.052	0.570	0.371	33.5
	(39.20)	(11.99)	(21.97)	(24.03)	(1.70)	(1.11)	(100)
1972	12.470	4.593	8.556	12.162	0.745	0.353	38.9
	(32.07)	(11.81)	(22.01)	(31.28)	(1.92)	(0.90)	(100)
1977	12.896	5.072	9.938	12.307	0.752	0.394	41.3
	(31.18)	(12.26)	(24.03)	(29.76)	(1.82)	(0.95)	(100)
1983	13.504	6.043	13.431	15.480	0.756	0.436	49.6
	(27.20)	(12.17)	(27.05)	(31.18)	(1.52)	(0.88)	(100)
1988	10.916	6.340	9.913	12.593	0.721	0.418	40.9
	(26.69)	(15.60)	(24.24)	(30.79)	(1.76)	(1.02)	(100)
1992	11.639	7.765	12.497	15.352	0.744	0.454	48.5
	(24.0)	(6.0)	(25.08)	(31.77)	(1.5)	(0.9)	(100)
1997	12.159	9.756	14.312	16.937	0.668	0.490	54.3
	(22.4)	(18.0)	(26.3)	(31.2)	(1.2)	(0.9)	(100)

(Contd.)

Year	Cattle	Buffalo	Sheep	Goat	Camel	Others	Total
2003	10.871	10.511	9.759	16.774	0.497	0.480	48.9
	(22.2)	(21.5)	(19.9)	(34.3)	(1.0)	(0.10)	(100)
% change w.r.t 1997	-10.59	7.74	-31.81	-0.96	-25.60		-9.99

Data Source: Government of Rajasthan, Department of Animal Husbandry, Jaipur
*Figures in parenthesis are percentages to the total livestock population

It provides information on the changes in the composition of livestock in Rajasthan (Rathore, 2005) over the above period. The impacts of droughts could not be captured exactly during the period, though there were drought years in between the livestock census years. Only the impacts could be captured for the census year 2003, as the state was reeling under drought continuously for two years 2002 and 2003. About 10 to 32 per cent livestock population perished in Rajasthan. The primary survey during the drought years showed that (Rathore, 2004) losses in cow population were much higher than the state level census data. About 25 to 50 per cent of cows were abandoned due to acute shortage of fodder during the above years and there were no buyers from slaughter house as there is a ban on cow slaughter in the state. Small animals were relatively less affected compared to large animals as they graze from common property resources *viz.,* grazing lands, waste lands, forests *etc.* out of the total of 296.46 million about 150 million livestock in the country were affected during 2002 drought year (Karunakaran, 2009), consequently the milk production fell significantly in a few locations. Also migration of small animals to the adjoining areas is a traditional practice followed in the state of Rajasthan. It was reported that about 32 per cent sheep population perished due to widespread drought during 2002-03.

It is difficult to obtain the impacts of droughts on livestock as droughts do not occur simultaneously in all the countries. The recent drought year 2009 has been considered as one of worst drought year over the last 100 years. The impact on food grain production during the period has already been discussed earlier in this chapter. The summary of worldwide impacts of 2009 drought on cattle sector as reported by Charlotte Johnston (2011) available in the web **www.thebeefsite.com/articles/2011/** is as follows. In Kenya's eastern province, Isiolo, cattle were lifted in trucks to the slaughter houses as they could not walk longer distances in search of food. The Ugandan news vision reported a steep fall in cattle prices due to lack of fodder and water. The cattle population in the Nairobi National park recorded all time high due to movement of livestock across the boundaries of neighbouring countries that has brought out several environmental issues like increased grazing pressure and health hazard due to non disposal of carcasses of the dead animals.

The lake Mburo National park in Uganda has reported similar issues where pastoralists moved thousands of cattle into the park land to access the only constant supply of water. Concerns were expressed about the increased risks of zoonotic diseases between cattle and wildlife.

Beef producing southern states of US were badly hit by drought during 2009 summer. Markets and abattoirs have seen large influx of cattle as farmers were affected with shortage of feed and water. Texas Agrilife extension service economist predicted the loss at $3.6 billion. Mexico was also affected by drought and about 1000 cattle died due to lack of rainfall and fodder availability. Argentina, one of the top five beef exporting nation in the world market experienced increased cattle slaughter during July and even the breeding stock were slaughtered which could affect the meat supply in the long run. Farmers in New Southwales in Australia were forced to sell the cattle for slaughter as feed and water supplies dwindled. The Meat and Livestock Australia organization predicted the impact of drought on the cattle to continue for five more years. The worldwide drought also has a severe affect on the livestock population in India, China and Middle East.

5.4.2. Milk Production

Next to animal meat, milk production from the livestock assumes significance in the human food chain. Its production varies with availability of fodder, water, the animal health and nearness to market etc. The world's milk production during 2010 was 713.6 Mt and an increase in production by 2 per cent i.e 728 Mt was projected for the year 2011 (FAO, 2011). The report further states that much of the anticipated expansion in the milk production is likely to occur in Asia, where the world's largest milk producing country like India is expected to increase its production by 5 Mt. Similarly, a substantial increase in production was also expected in China and Turkey during 2011. Despite rising cost, China recorded 10 per cent growth during 2010, while in 2011 it was around 7 per cent. In Africa, the milk output fell during 2010 due to drought conditions in the Horn of Africa. The high cattle mortality during 2009 affected the milk output in Kenya, Ethiopia and Somalia. The rest of the countries like Egypt produced 6 Mt. New Zealand is one of the top 10 countries that produced 17.0 Mt of cow's milk during 2010 which corresponds to 2.84 per cent of total world production (599.62 Mt) (World Milk Production database source: **http://www.dairyco.org.uk/resources-library/market-information/market-information/world-milk-production/**).

Occurrence of droughts increased the demand for water due to expansion in crop irrigated area since 1900 onwards in New Zealand. The average milk-solids production per cow in the New Zealand during the drought years between the period 1992 and 2008 (histograms) is presented in Fig 5.7.

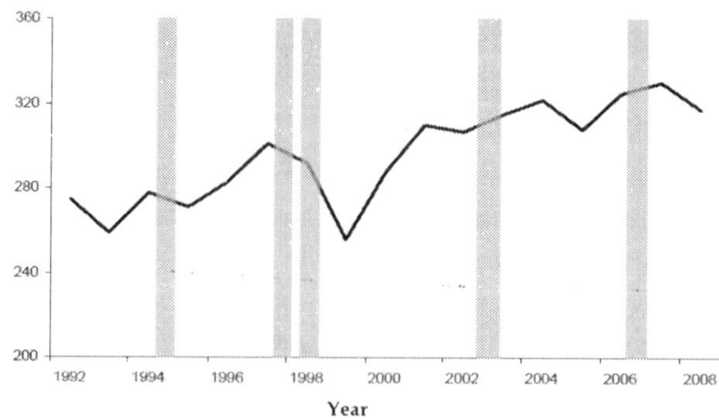

Fig. 5.7: Drought affect on milk-solid production (Average kg of milk-solids produced per cow in New Zealand; Shaded areas = periods of drought)
(*Source*: NZIER, 2008)

It is observed that the national milk-solid production either slowed or decreased during three out of four years of severe droughts after 1992 (NZIER, 2008). The effects of droughts on national milk-solid production are largely region specific and time of occurrence etc. For example, the impact of drought during 2006-07 is not apparent at national level data. This was because dairying is not common in east coast region where the drought was experienced. However it was reported that milk output during 2006-07 in the east coast was less by 31.9% of the regional average. It was also observed that the occurrence of dry summer in New Zealand has less effect on the milk output than dry spring, which is the maximum milking period.

In India, 2002 drought affected the livestock population in over 18 states and among them Rajasthan was badly hit and in this state alone 226 lakh cattle were affected by severe drought conditions followed by Madhya Pradesh (192 lakh), Maharashtra (189 lakh) Orissa (117 lakh), due to which milk production fell by 22 per cent in Rajasthan, 8 and 7% in Madhya Pradesh and Tamil Nadu (Karunakaran, 2009). Drought in 2002 left behind a large deficit of 70,000 lts/ day of milk supply from Tamil Nadu and Karnataka to Kerala (Anonymous, 2003). On the contrary, drought like conditions in Gujarat helped the dairy industry to register higher milk production. A study conducted by International Water Management Institute (IWMI) found that some states in the country inclusive of Gujarat, the dairy economy emerges as an adaptive strategy for the farmers during drought like situation. During 2002, the value of paddy and oilseeds output fell over 20 per cent and wheat by 10 per cent, but the value of livestock and milk production increased by 3 per cent on All India basis. In Andhra Pradesh, the value of milk production increased by 12 per cent, which is almost close to the value of crop production. In Gujarat, the milk production value increased by 3.6 per cent. (Anonymous, 2009). Milk production in drought prone regions of Maharashtra viz., western Maharashtra and Marathwada recorded a surplus stock of 28 lakh liters during summer 2012 equivalent to half of Mumbai's daily requirements as farmers focused more on supplementary business when income from agriculture is not stable.

Thus rearing or livestock act as a protective shield against drought for the marginal and small farmers especially under rainfed farming situation. So far, the policies of any government have not yet given any thought to rebuild the animal assets in reducing the poverty and vulnerability of society to climate extremes. Therefore there is a need to frame future animal husbandry policies of Government at National and International level to reduce the poverty in the drought prone regions.

5.4.3. Fisheries

Fisheries assume significance next to crop production and livestock. Annually world's fisheries production which includes both capture and aquaculture in the recent year 2008 is estimated to be as 142.3 Mt, out of which 49.9 Mt was from aquaculture. (FAO, 2010). It was reported that 81 per cent (115 Mt) of worlds fish production in 2008 was used for human consumption, while the rest (27 Mt) was utilized for non-food purposes such as fish meal and fish oil (20.8 Mt), culture, bait and pharmaceutical uses and direct feeding in aquaculture and for fur animals (FAO, 2010). Among the countries China remains the largest fish producer with a production of 47.5 Mt (32.7 from aquaculture and 14.8 Mt from capture fish respectively). World's capture fisheries production has been relatively stable (Fig 5.8) with exception of fluctuations driven by the catches of *anchoveta*- a species extremely sensitive to oceanographic conditions influenced by ENSO conditions in the South Pacific.

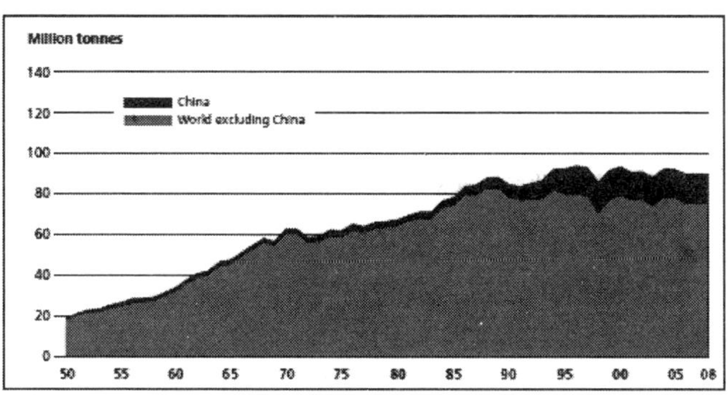

Years

Fig. 5.8: World capture fisheries production (*Source*: FAO, 2010)

The aquaculture fish production is driven by the water resources and their fluctuations due to climate conditions prevailed over the region. As droughts are manifestation of short/long terms climate variability, its impacts has both direct and indirect influence on fish stock. Direct effects of climate occur due to changing patterns in rainfall, snow melt and rising sea level which affect the hydrology and water quality that will act on physiological behavior, growth rates, reproductive capacity, mortality its distribution and migration. Indirect effects of climate resulting from alteration in food chain and increased soil erosion may alter the productivity, structure, and composition of the ecosystem on which fish depend for food and shelter.

Droughts are mostly associated with low precipitation and greater evaporation with increased temperature. The threats to aquaculture due to droughts arise from 1) stress due to increased temperature and oxygen demand and decreased pH, 2) uncertain future water supply, 3) increased frequency of diseases and toxic elements and 4) uncertain future supply of fish meal (IPCC, 2001). Fresh water aquaculture of many countries in the Central and Eastern Europe is likely to be affected by flooding, droughts and high temperature due to projected changes in climate (Tomaszmoth Poulsen, 2009). The impacts of climate related drivers viz., temperature, salinity, wind fields, oxygen, pH and density structure of water column were studied on individual fish species through a combination of experimental studies and modelling. Brander (2007) reviewed various studies on impact of climate on fisheries taken up at various locations. The increase of temperature by 2°C produced positive effects on the appetite, growth, protein synthesis and oxygen consumption in water during winter in respect of Rainbow tront (*Oncorhyncus mykins*), but the same temperature increase in summer producced negative effects (Morgan *et.al*, 2001). The mortality rate of *Salmon* fish in the Fraser River, Canada increased when summer temperature exceeded the 60 year mean during summer (Portner and Knust, 2007). Climate aberration within a short period *viz.,* floods, droughts, heat waves etc will have strong influence on the regional fish productions. Experiments on the changes in Salinity on the reproductive nature of Atlantic cod (*Gadus morhua*), principal demersal fish species of Baltic Sea showed that at salinities less than 11, the cod spermatozoa cease to be active and eggs sink to deeper layers and cannot survive due to anoxic condition (Nissling *et.al*, 1998).

The ENSO (El Niño-Southern Oscillation) events associated with warm waters in the coastal and equatorial ocean of Peru and Equador for 6-18 months have adverse effect on *peruvian anchovy* production in the eastern pacific (Jacabson *et al.* 2001). The impact of droughts, flow regime and fishing on the fish assemblage in southern Australian largest temperate estuary was reported by Ferguson *et.al* (2010). Using a 25 year time series data of fish catch for the four large–bodied native fish species, it was reported that fish assemblages generally differed at 5 yearly intervals. Species richness deeply declined over 25 years in fresh water and the estuarine habitats and the species diversity also declined after 2001 in the estuarine habitats. The implications of overfishing results in loss of older / large species suggests reduced capacity to withstand or recover from the deteriorating environmental conditions due to extreme drought conditions experienced in lower Murry River system.

The fish production in India both marine and Inland for the years 1991-92 to 2009-10 is given in the Table 5.3.

Table 5.3: Fish production in India (Lakh tones)

Year	Marine	Inland	Total
1991-1992	24.47	17.10	41.57
1992-1993	25.76	17.89	43.65
1993-1994	26.49	19.95	46.44
1994-1995	26.92	20.97	47.89
1995-1996	27.07	22.42	49.49
1996-1997	29.67	23.81	53.48
1997-1998	29.50	24.38	53.88
1998-1999	26.96	26.02	52.98
1999-2000	28.52	28.23	56.75
2000-2001	28.11	28.45	56.56
2001-2002	28.30	31.20	59.56
2002-2003	29.90	32.10	62.00
2003-2004	29.41	34.58	63.99
2004-2005	27.80	35.20	63.04
2005-2006	28.16	37.55	65.71
2006-2007	30.24	38.45	68.69
2007-2008	29.20	42.07	71.27
2008-2009	29.78	46.39	76.17
2009-2010p	26.89	48.62	78.51

P: Provisional

(*Source*: Annual Report, MoA 2009-10)

It is observed that the fish production increased from 41.5 lakh tons in 1991-92 to 76.08 lakh tons in 2008-09 @ 1.92 lakh tons/year. The increase in total fish production is due to increased inland aquaculture activity, whereas, the marine fish production showed a small increment of 5.25 lakh tons over the period 1991-2009. The influence of droughts on country's total fish production seems to be nil. However, the rate of fish production declined during the drought year compared to the previous year's production. For example, the year 2002-03 was a phenomenal drought year, during which the rate of inland fish production was only 2.8 % over its previous year's (2001-02) production.

However in a normal year *viz.*, 2007-08, the rate of increase was 9.4% over its previous year (2006-07).

A survey was conducted by Central Inland Fisheries Research Institute, Barrackpore in two major fish seed producing districts *viz.*, North 24 Paraganas and Bankura in West Bengal on fish seed production with respect to anomalies in air and water temperature and rainfall during fish breading season (March-September) during all India drought year 2009 (Das *et.al*, 2011). It was reported that due to increase in air minimum and water temperature and deficit rainfall (24% in North 24 Paraganas and 15% in Bankura) during optimum fish breeding month (June-September), majority of hatcheries investigated (68%) were affected by high temperature and water deficit. In North 24 paraganas the major factor for low spawn production was due to high larval mortality (45%) while in Bankura, it was due to inadequate water area for stocking restricted the purchase of fish seeds from hatcheries. The economic analysis of spawn production during 2009 in North 24 Paraganas and Bankura indicated 61 and 73 per cent loss in income compared to a normal year 2006. The impact of climate on inland fisheries in River Ganges and its plains in India (Vass *et.al* 2009) indicated that the river Ganges at Haridwar has become more favourable habitat for warm water fishes due to increase in mean annual minimum water temperature by 1.5°C during 1987-2003 compared to period 1970-86. In recent years, the breading phenomenon of Indian Major carps *viz* "*Catla Catla, Labco rohita and cirrlinus mrigala* was noticed as early as in March under captive conditions. Rise in temperatures influenced the reproductive cycle in fishes. This parameter along with rainfall and photoperiod stimulate endocrine glands that help in advancement of maturation. It was also observed that aquaculture farm hatcheries of Gangetic plains of west Bengal advanced the breading programme of Indian major carps from last week of May (24-31 May,1980) to mid April during 2005 and attribute this to rise in water temperature Survey work conducted by Dey *et. al* (2007) on fifty fish seed forms in four districts of West Bengal *viz.* North 24 Paraganas, Bankura, Burdwan and Hoogly showed an extended breeding period 45-60 days.

A Comparative study on various ecological factors of the Pulicat lake, one of the largest salt water lagoons on the East coast of India in relation to fishery during drought season 1975 and a normal season in 1976 was reported by Raman *et.al* (1977). High salinity and temperature conditions during drought year affected the plankton production and consequently the fish production. Also it was noticed that the closure of mouth of the lake during June-July has profound influence on the environmental features and fisheries in estuaries and lagoons.

5.5 Estimation of Food Grain Production in India Using Rainfall Indices

Climate determines the distribution of plant species and annual weather conditions control the production from plants, both perennial and annual. The seasonal weather in the country is greatly influenced by the monsoon system, which brings in life over the entire country. Several Research scientists and organization tried to link agricultural production with monsoon rainfall and developed relationships to understand the influence of rainfall and to forecast the food grain productions well in advance (Mooley *et.al* 1981a, 1981b; Mooley and Parthasarathy, 1982; Garnett and Khandekar, 1992; Parthasarathy *et.al* 1988 and Gadgil, 1996). Food grain estimation by the National Crop Forecasting Centre (NCFC) set up by Ministry of Agriculture uses the sampling data collected from 80-120 crop cutting experiments for crops in a major district where the cropped area in the district exceeds 80,000 ha or lies between 40,000 and 80,000 ha but exceeds the average area per district in the state. Otherwise, the district is considered as minor for a given crop and about 44 or 46 experiments are planned in a minor district (Anil Rai, 2007).

Generally crop cutting is done in a plot of 5m x 5m size for most of the crops in most of the states. However in Uttar Pradesh the shape of plot is of an equilateral triangle of 10m size and in West Bengal a circular plot of radius 1.745m is taken for crop cut. The average yield is obtained after harvesting, threshing, weighing and recording the weight of the product from selected plots. In sub-sample experiments, further processing of the harvested produce is carried out to determine the percentage recovery of dried grains or the marketable grains of the produce depending on the nature of crop. This procedure is used to estimate crop yields in three non-land record states *viz.* Kerala, Orissa and West Bengal (Anil Rai, 2007).

The advanced estimates of crop area and production are released through four forecast bulletins for principal food and non food crops that cover nearly 87 percent of agricultural output. The first forecast is issued in middle of September, the second in January, third by end of March and the fourth by end of May. The advance estimates released in September are related to kharif crops, which is mostly based on reports submitted by states (Anil rai, 2007). He further reported that the second forecast bulletins covers both *Kharif* and *Rabi* and is released in January taking into account additional information obtained from various sources including agricultural inputs, incidence of pests and diseases, weekly reports from state government regarding area coverage, conditions of standing crops etc. Currently the estimates obtained through remote sensing are also used at this stage. The third forecast, issued in March, revises the estimates of *Kharif* and *Rabi* seasons based on information received from sources such as Market Intelligence Units, Meteorological Department and the Crop Weather

Watch Group (CWWG). The forecast made by the end of May is based on actual figures supplied by state Agricultural Statistics Authorities (SASAs) using yield estimates obtained through General Crop Estimation Survey (GCES).

Attempts are being made to estimate the crop acreage and production using remote sensing data under the nationwide projects, via Crop Acreage and Production Estimation (CAPE) and Forecasting Agricultural output using Space, Agrometeorology and Land based Observations (FASAL) funded by the Ministry of Agriculture. Most of the short-term climate variability in the tropics is related to the El Nino/Southern Oscillation (ENSO) phenomenon, which influences/determines the location of tropical- convection and ultimately changes the global atmospheric circulation. The Indian summer monsoon rainfall, which is the ultimate water resource for the agricultural production in India, is known to have significant relationship with the Southern Oscillation Index (SOI). Nageswara Rao, (1999) have evaluated the sub divisional composite mean rainfall during the positive and negative phases of SOI. Besides rainfall, other weather parameters that affect the crop yields are cloud cover which reduces solar radiation, temperature and evaporation but increases humidity. There have been several studies linking the agricultural production with SOI (e.g., Nicholls, 1985; Rimmington and Nicholls, 1993; Kuhnel, 1994; Phillips *et al.*,1996; Meinke and Hammer, 1997; Hansen *et al.*,1998 and Mjelde and Keplinger, 1998).

Any deviation in the quantum and spread of monsoon rains can lead to substantial variability in the agriculture production in India. Though the current increasing trend in the national foodgrain production is considered satisfactory, yet the per capita availability of foodgrains per annum is still below the nutritional norm of 182.5 kg. Therefore, advance estimates of foodgrain production are essential for evolution of mitigation measures to ensure greater food security in the country, as well as for planning the country's economy. Parthasarathy *et al.* (1988) have attempted to estimate the annual food grain production using all-India summer monsoon rainfall. Similarly, Arif (1988) and Parthasarathy *et al.* (1996) have made efforts to estimate the *kharif* foodgrains production from all-India monsoon rainfall. All these studies have attempted to relate the all-India annual food grain production with all-India summer monsoon rainfall amount. Rao G.N. *et al.* (2003) have obtained more accurate estimates of all-India annual food grain production well in advance of harvesting from monthly (June-September) rainfall indices (RI) derived for the purpose based on the sub-divisional rainfall and the Southern Oscillation Index (SOI), as parameters. The data and methodology used and development of model and its verification in the study as given by Rao, G.N. et al (2003) are as follows.

Monthly (June-September) rainfall series for all-India and 29 meteorological subdivisions in India during 1953-2000 were extracted from the 'IITM Indian regional/sub divisional Monthly Rainfall data set', and sub-divisional rainfall for the year 2001 were collected from NCMRWF, New Delhi. The monthly SO Indices (Tahiti minus Darwin sea level pressures) for the study period (1953-2000) were extracted from the Bureau of Meteorology, Australia website. From these monthly SO Indices, the seasonal indices for winter (DJF), spring (MAM) and the winter to spring seasonal pressure tendency (MAM-DJF) were computed. The total annual food grain production, for the period from 1953-54 to 2000-01 are extracted from the Directorate of Economics and Statistics, Ministry of Agriculture, website Government of India. The entire study period has been divided into two parts, 1953-1992 for establishment of relationships and development of regression models and 1993-2000 for verification of the model estimates with actual data

The all-India annual food grain production series has two distinct components namely, (a) the trend component which can be attributed to non-meteorological factors such as, increased gross sown area, Green Revolution, improved technology, application of fertilizers, crop pest and disease control *etc*., and (b) the detrended production whose interannual variations are assumed primarily to be due to weather.

5.5.1 Trend in the annual food grain production

To find out the impact of climate variations on the food grain production, the trend component was removed from the original series by fitting various trend equations (e.g., Thompson 1969a, 1969b and 1970; Mooley *et al*., 1981 a and 1981b; Parthasarathy *et al.*, 1988 and 1996). The following exponential equation, fitted to the annual food grain production series which was used to remove the trend component.

$$T = EXP\,(0.0256264*YR)*1.18E\text{-}020 \quad (1)$$

Where, T is the trend component and YR is the year. The trend equation with a multiple correlation coefficient (MCC) of 0.94 is highly significant (at 0.001% level). The mean trend in the all-India annual food grain production was found to be to be 2.5% per year.

5.5.2 Detrended Production Index

For estimating detrended production, the actual food grain production (P) in each year is expressed as a percentage of the corresponding trend value (T). The detrended Production Index (DPI) is as follows:

$$DPI = (P/T)*100 \quad (2)$$

These detrended PIs are expected to be free form the technological trend and their year-to-year variations are assumed to be primarily due to weather.

5.5.3 Construction of Rainfall Indices (RI)

Earlier studies attempted to relate the detrended PI with all-India monsoon seasonal rainfall alone (e.g., Parthasarathy *et al.,* 1988 and 1996). Since, the Indian summer monsoon rainfall is known to have significant variations in space and time and the crop production depends on the behaviour of monsoon, the monthly sub-divisional rainfall (June-September) and the Southern Oscillation Index were considered for correlating with DPI.

Results showed that the detrended PI was significantly correlated with the rainfall of some specific subdivisions in each month. Surprisingly, the meagre rainfall in July in Subdivision No.30 (Tamil Nadu), which is dominated by the winter monsoon also showed significant correlations with all-India annual food grain production index. Such subdivisions having highly significant (1%) correlations with annual DPI during 1953-92 (model development period) were identified. Monthly rainfall indices (RI) were then computed from the area-weighted rainfall of these subdivisions with which the annual food grain productions (PI) has significant correlations

$$RI = \Sigma A_i R_i / \Sigma A \quad (3)$$

Where A_i is the area and R_i is the rainfall of the ith subdivisions that have shown significant relationship with the annual DPI.

5.5.4 Development of model for estimation of production index

The detrended all-India annual food grain production indices (DPI) evaluated were found to have better correlation with the monthly sub divisional area-weighted rainfall *indices* (RI) than with the corresponding all-India monthly rainfall *amounts,* except in September. Also DPI showed a correlation of 0.48 (significant at 1 % level) with SOI during 1953-92. In view of the significant relationship of DPI with both RIs and SOI, a multiple regression equation was developed for estimation of all-India annual food grain production index (PI) from monthly rainfall indices (RI) and SOI and is given below.

$$PI = 72.88+0.039*RI\ Jun+0.031*RI\ Jul+0.027*RI\ Aug+0.034*RI\ Sep+0.141*SOI \quad R^2 = 0.68 \quad (4)$$

5.5.5 Verification of Model Performance

Annual food grain PIs have been estimated for the periods, 1953- 1992 (development period) and 1993-2000 (verification period). The trend component of the food grain production, in each year during the above periods is also estimated from the fitted trend equation (Eq.1). The estimated production indices (PI) are then converted into annual productions (P) using the corresponding trend values (T) with Eq.2. The actual (compiled by the Directorate of Economics and Statistics, Government of India) and estimated production of all-India annual food grains thus evaluated show that the modelled production is in close agreement with the actual values during both the development and verification periods. The percentage deviations of the production estimates from their actual values during the verification period varied from -2% to +2%, except in one year (1996-97).

Using the above model, food grain production for 2001-2002 was worked out as 210.9 million tons in October 2001, compared to the official final estimate of 211.3 million tons released in June 2002. Thus, this model facilitates estimation of the all-India annual food grain production by October itself, nearly eight months before the official final estimates generally become available.

5.5.6 Estimation of *Kharif* Price Production

Similar attempts were made to estimate rice production using above procedures by the following multiple Regression equation that were developed to estimate annual Rice PI from the Monthly RI and SOI.

Where PI = Production Index, RI = Rainfall Index and SOI = Southern Oscillation Index.

$$PI= 58.293+0.0396* RI\ Jun+0.088* RI\ Jul+0.0381 * RI\ Aug+0.0253*RI\ Aug+0.0226*SOI\ (R^2= 0.70) \quad (5)$$

This could explain the variability upto 70 percent of the variance in PI. Annual Rice production was evaluated from the PI estimated from the above equation and the trend value (T) from equation.

$$T= Exp\ (0.0272931*YEAR)* 1.79309E\text{-}22 \qquad (R^2= 0.96) \qquad (6)$$

And production (P) = PI*T/100.

The estimates of annual rice production can be obtained by October (well before the crop harvest) and was found to be in close agreement with the actual value (Fig 5.9).

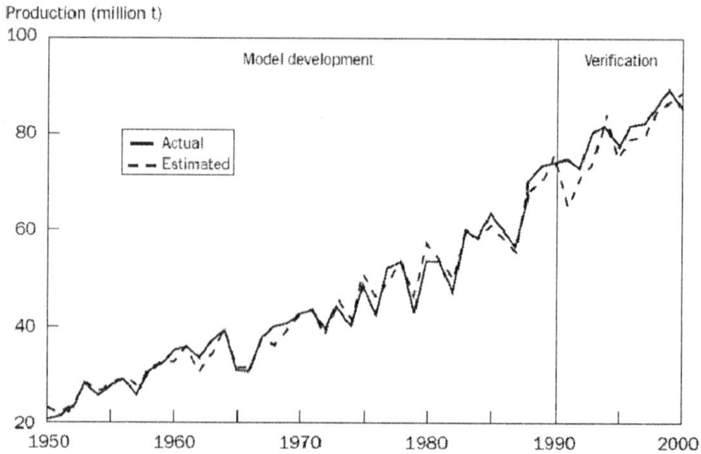

(*Source*: Rao, G.N. *et al.* (2003))

Fig. 5.9: Actual and estimated values of annual rice production in India during model development (1950-89) and verification (1990-2000) periods

The all India annual rice production for 2001-02 was predicted by October with in ± 2% deviation with the final estimates available in June 2002 (8 months in advance). Aggarwal (2009) reviewed various statistical methods *viz.*, regression models, time series models, probabilistic models that are widely used in estimating agricultural production.

5.6 Forecasting of Agricultural Output using Space, Agrometeorology and Land Based Observation (FASAL)

To enhance the existing systems capabilities to forecast agricultural output, adoption of technological advancements such as Remote Sensing (RS) and GIS was initiated by Government of India in 1987 under the project Crop Acreage and Production Estimates (CAPE). This project was implemented through Space Application Centre (SAC), Ahmedabad and provided platform for development and standardization of procedures, models, and software packages for the crop area and production estimates (Dalip Singh, 2009). In addition to Remote Sensing data, the other inputs such as weather and land based observations and economic parameters influencing farmers' decisions have been included in this model. The concept of integrating inputs from the diverse sources such as Remote Sensing, econometric and crop simulation models using previous years crop statistics, market prices and current season weather data was attempted in FASAL scheme to make forecast of desired accuracy and coverage. This scheme is in operation since 2006 and the crop yield forecasts at different growth stages as reported by Dalip Singh (2009) as follows:

- First yield forecast (Fo) is given by Indian Economic Group (IEG) using econometric model before the start of the crop season using variables such as prices of last year, availability of irrigated area, price of fertilizer (sowing season), availability of fertilizer, rainfall during sowing month and season and other relevant variables in area and yield functions etc.
- IMD provides mid season yield forecast (F2) using agromet models developed by the research centre of AICRP on Agrometeorology, located at various agricultural universities and at ICAR Research Institutes and the crop area estimation provided by SAC.
- Pre-harvest yield forecast (F3) is provided by IMD and communicated to Department of Agriculture and Cooperation (DAC).

References

Agrawal, R. 2009. Forecasting techniques in crops. E-book of Indian Agricultural Statistical Research Institute, New Delhi. 13 p.

Anil Rai. 2007. Crop area and production surveys. In : E-book on Statistical methods for agricultural research. Indian Agricultural Statistical Research Institute, New Delhi. 9p.

Anonymous. 2003. Drought hits milk supply in Kerala. Financial Express. 24 March 2003.

Anonymous. 2009. Drought -like conditions may lead to higher milk production in state: study: The Indian Express. 11 September 2009.

Arif, R.R. 1988. *Kharif* production and monsoon rainfall: An empirical study, Reserve Bank of India Occasional Papers, 9(4): 375-393.

Brander, K.M. 2007. Global fish production and climate change. Proceedings of the National Academy of Sciences. 104 (50) : 19709-19714.

Charolette Johnston. 2011. Worldwide impact of drought on the cattle sector. Available in http://www.thebeefsite.com

Dalip Singh. 2009. Forecasting techniques under FASAL scheme. In : E-book on Techniques of Estimation and Forecasting of Crop Production in India. (Eds.Sud, S.D., Hukum Chandra and Kaustav Aditya). Indian Agricultural Statistical Research Institute, New Delhi. 186-191.

Das, M.K., Srivastava, P.K., Dey, S., Liakat Mandal, MD. And Sahu, S.K. 2011. Impact of drought on hatchery fish seed production in West Bengal – A case study. J. Inland Fish. Soc. India. 43 (1):25-32.

De, U.S. and Joshi, K.S. 1996. A climataological-cum-statistical analysis of monthly rainfall in the meteorological subdivisions of India. In: Advances in Tropical Meteorology, Meteorology and National Development (Ed.R.K.Dutta), Concept Publishing Co., New Delhi. 384-393.

Dey, S., Srivastava, P.K., Maji, S., Das,K., Mukhopadhyaya, K. and Saha, P.K. 2007. Impact of climate change on the breeding of Indian major carps in West Bengal. J.Inland Fish. Soc. India, 39(1):26-34.

Eric deCarbonnel. 2009. Catastrophic fall in 2009 global food production. Global Research E-Newsletter. 11p.

FAO. 1984. Rapport de la mission FAO/PAM Multi-donateurs. Republique du Mali. Evaluation de la situation de l'alimentation de l'agriculture et de l'elevage. OSRO Rapport no 15/84/F. Rome.

FAO.2010. State of world fisheries and aquaculture 2010-Part 1. Food and Agriculture Organization, Rome, 218p.

FAO.2011. Food Outlook : Global Market Analysis - June 2011. Report of FAO Trade and Markets Division. FAO. Rome. 116p.

Ferguson, G.J. , Ward, T.M., Ye.Q., Geddes, M.C. and Gillanders B.M. 2010. Impacts of drought, flow regime and fishing on the fish assemble in southern Australia's largest temperature estuary. Fishery Stock Assessment Report for PIRSA Fisheries. South Australian Research and Development Institute (Aquatic Scienc es), Adelaide. SARDI Publication No.F2010/ 000709-1. SARDI Research Report Series No.498. 32p.

Gadgil, S. 1996. Climate change and agriculture - An Indian Perspective. In: Climate variability and Agriculture (Eds. Y.P. Abrol, S. Gadgil and G.B. Pant), Narosa Publishing House, New Delhi, 1-18.

Garnett, E.R. and Khandekar, M.L. 1992. The impact of large-scale atmospheric circulations and anomalies on Indian monsoon droughts and floods and on world grain yields- a statistical analysis: Agrl. and Forest Meteorology, 61(1-2):113-128.

Gryseels G. and Anderson F. 1983. Research on farm and livestock production in the central Ethiopian Highlands: Initial results 1977-80. ILCA Research Report No.4, ILCA, Addis Ababa.

Hansen, J.W., Hodges, .W. and Jones, J.W. 1998. ENSO influences on agriculture in the southeastern U.S., Journal of Climate, 11:404-411.

Hisas, L. 2011. The Food Gap: The impacts of climate change on food production: A 2020 perspective. Technical Rport. Universal Ecological Fund (FEU-US), Alexandria, USA.45p.

IPCC. 2001. Climate Change 2001: Impacts, Adaptation, and Vulnerability. Intergovernmental Panel on Climate Change. 1005p.

Jacobson, L.D., De Oliveira, J.A.A., Barange, M., Cisneros-Mata, M.A., Fe´lix-Uraga, R., Hunter, J.R., Kim, J.Y., Matsuura, Y., Niquen, M. and Porteiro, C. Rothschild, B., Sanchez, R.P., Serra, R., Uriarte, A. and Wada, T. 2001. Surplus production, variability, and climate change in the great sardine and anchovy fisheries. Can J Fish Aquat Sci 58 (9) :1891–1903.

Joshi, K.S., Vaidya, D.V. and De, U.S. 1981. An objective aid to forecasting weekly rainfall over India. Vayu Mandal, 11:19-22.

Karunakaran,N. 2009. Drought in India : Challenges & initiatives. Poorest Areas Civil Society, New Delhi. 50p.

Krishna Kumar, K., Rupa Kumar, K., Ashrit, R. G., Deshpande, N. R. and Hansen, J. W. 2004. Climate impacts on Indian agriculture. Int. J. Climatol. 24: 1375–1393.

Kuhnel, I.1994. Relationship between the Southern Oscillaiton Index and Australian sugarcane yields. Aust.J.Agrl.REs., 45:1557-1568.

Meinke, H. and Hammer, G.L. 1997. Forecasting regional crop production using SOI phases: a case study for the Australian peanut industry. Aust.J.Agrl.Res., 48:789-793.

Mjelde, J.W. and Keplinger, K.1998. Using the Southern Oscillation to forecast Texas winter wheat and sorghum crop yields. Journal of Climate, 11:54-60.

Mooley, D.A., and B. Parthasarathy. 1982. Fluctuations in deficiency of summer monsoon over India and their effect on economy. Arch. Meteorol. Geophys. Bioklimatol., Ser. B., 30:383-398.

Mooley, D.A., Parathasarathy, B. and Munot, A.A. 1981a. Large scale droughts over India and their impact on economy during the period 1871-1978. Ann. Natl. Assoc. Geogr. Ind., 2: 17-26.

Mooley, D.A., Parathasarathy,B., Sontakke, N.A. and Munot, A.A. 1981b. Annual rainwater over India, its variability and impact on the economy. Journal of Climatol., 1:167-186.

Morgan, I., McDonald, D.G. and Wood,C.M. 2001. The cost of living for freshwater fish in a warmer, more polluted world. Global Change Bioloiy. 7(4):345–355.

Nageswara Rao, G. 1999. Variations of the SO relationship with summer and winter monsoon rainfall over India. 1872-1993. Journal of Climate, 12:3486-3495.

Nicholls, N. 1985. Impact of the Southern Oscillation on Australian Crops. Journal of Climatol., 5:553-560.

Nissling A, Larsson R, Vallin L, Frohlund K. 1998. Assessment of egg and larval viability in cod, Gadus morhua: methods and results from an expeiimental study. Fisheries Research. 38(2):169–186.

NRAA. 2009. Drought management strategies-2009. National Rainfed Area Authority, New Delhi. 71p.

NZIER. 2008. Climate and production : The impact of climate on agricultural production in New Zealand. NZ Institute of Economic Research Inc, Wellington. 28p.

Parathasarathy B, Pant G.B. 1985. Seasonal relationship between Indian summer monsoon rainfall and southern oscillation. Journal of Climate. 5:369-378.

Parathasarathy, B., Munot, A.A. and Kothawale, D.R. 1988. Regression model for estimation of Indian foodgrain production from summer monsoon rainfall. Agricultural and Forest Meteorol., 42:167-182.

Parathasarathy, B., Rupa Kumar, K.and Munot, A.A.1996 Relationship between India summer monsoon rainfall and kharif foodgrain production. In: Advances in Tropical Meteorology, Meteorology and National Development (Ed.R.K.Dutta), Concept Publishing Co., New Delhi, 291-297.

Phillips, J., Cane, M.A. and Rosenzweig,C. 1996. ENSO, seasonal rainfall patterns and maize yield variability in Zimbabwe. Agrl. and Forest Meteorol., 90:39-50.

Pinstrup-Andreson, Per., Pandya-Lorch, Rajul. 1996. The world food situation and a 2020 vision. Agroalimentaria. 2(2):24-33.

Pörtner, H. O. and Knust, R. 2007. Climate change affects marine fishes through the oxygen limitation of thermal tolerance. Science. 315: 95-97.

Raman, K., Kaliyamurthy, M. and Joseph K. 1977. Observations on the Ecology and fisheries of the Pulicat lake during drought and normal periods. J. Mar. Biol. Ass. India, 19(1):16-20.

Ramesh Chand and Raju, S.S. 2009. Dealing with effects of monsoon failures. Economic and Political Weekly. 64(41):29-34.

Rao, G. N., Ramakrishna, Y. S., Rao, A. V. R. K., Rao, G. G. S. N. 2003. Advance estimation of rice production in India using weather indices. International Rice Research Notes 28(2): 56-58.

Rao, G.G.S.N., Rama Krishna, Y.S. Ramana Rao, B.V and Purohit R.S. 1986. Impact of droughts on productivity of pearlmillet in Rajasthan. Mausam. 37:2 197-202.

Rao, G.G.S.N., Rama Krishna, Y.S. Ramana Rao, B.V and Purohit R.S. 1984. Incidence of droughts and its impact on food production in Rajasthan. Annals of Arid Zone. 23(4): 327-334.

Rathore, M.S. 2004. Adaptive strategies to droughts in Rajasthan. Jaipur: Institute of Development Studies.

Rathore, M.S. 2005. State level analysis of drought policies and impacts in Rajasthan, India. Working paper 93: Drought Series Paper No.6. International Water Management Institute, Colombo, Sri Lanka: 40p.

Review of the economy. 2009/10. Sectoral – Agriculture, Industry and Services. Chapter 4. Economic Advisory Council to the Prime Minister, Vigyan Bhawan, New Delhi. 43p.

Rimmmington, G.M. and N. Nicholls.1993. Forecasting wheat yields in Australia with the Southern Oscillation Index, Aust.J.Agrl. Res., 44:625-632.

Sastri, A.S.R.A.S., Ramana Rao, B.V., Rama Krishna, Y.S. and Rao, G.G.S.N. 1982. Agricultural Droughts and Crop Production in the Indian Arid Zone. Archives for Meteorology, Geophysics and Bioclimatology. 31:405-411.

Selvaraju.R. 2003. Impact of El Niño-southern oscillaiton on Indian foodgrain production. International Journal of Climatology. 23:187-206.

Thompson, L.M. 1969a. Weather and technology in the production of corn in the U.S. corn belt. Agron.J., 61:453-456.

Thompson, L.M. 1969b. Weather and technology in the production of wheat in the United States. J.Soil Water Conserv., 24:219-224.

Thompson, L.M. 1970. Weather and technology of soybeans in central United States. Agron.J., 62:232-236.

Tomasz Moth-Poulsen. 2009. Effect of climate change on fisheries and aquaculture with emphasis on central and eastern Europe. Food and Agriculture Organization, Rome. 5p.

Toulmin, C. 1986. Drought and the farming sector: loss of farm animals and post-drought rehabilitation. Network Paper No. 10. International Livestock Centre for Africa, Addis Ababa, Ethiopia. 27p.

USDA. 2010. Production, Supply and Distribution: Electronic database of United States Department of Agriculture.

Vass, K.K. Das, M.K., Srivastava, P.K. and Dey.S. 2009. Assessing the impact of climate change on inland fisheries in RiverGanga and its plains in India. Aquatic Ecosystem Health & Management. 12(2):138-151.

Wolde Mariam. 1984. Rural vulnerability to famine in Ethiopia: 1958-1977. Vikas Publishing House, India.

Wood A. 1976. Farmer's responses to drought in Ethiopia. In: Hussein A.M, Rehabilitation: Drought and Famine in Ethiopia. African Environment Special Report No.2. London.

□□□□

CHAPTER 6

Impacts on Water Resources

Of all the natural resources of the earth, none is as important to life as water. The properties of water in its three physical states make it by far the most useful compound for all the living organisms on this planet. The high specific heat of water, its ability to exist in all forms *viz.,* gaseous, liquid and solid forms under natural conditions and its capacity to store and release latent heat with changing states plays greater role in atmospheric processes. The constant movement of atmosphere helps in redistribution of water on earth's surface though the atmosphere holds a minute fraction (1 percent) of earth's water. If all the atmospheric water vapour were precipitated, a layer of water of 2.5 cm depth would develop over the entire globe (Critchfield, 1983).

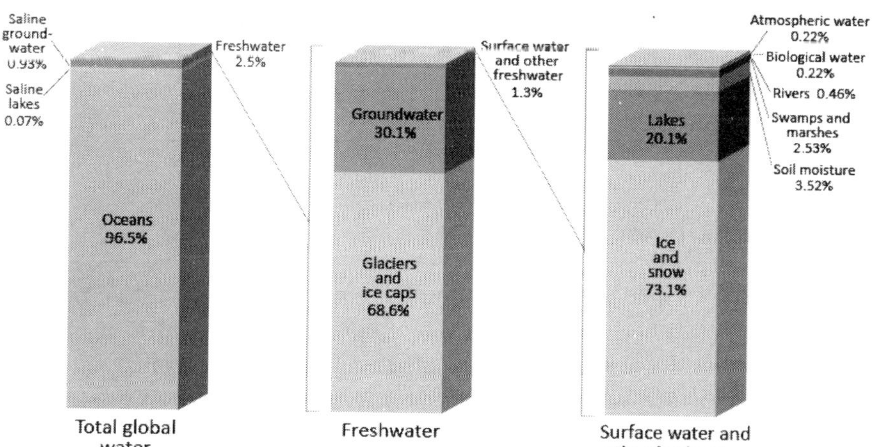

(*Source*: Igor Shiklomanov's chapter "World fresh water resources" in Peter H. Gleick (editor), 1993)

Fig. 6.1: Distribution of earth water resources

The distribution of earths water as given by United States Geological Survey (USGS) water Science School **(http://ga.water.usgs.gov/edu/watercycle.html)** indicate that out of the total global water of 1386 million km^3, 96 per cent is saline (Fig 6.1). And, of the total fresh water, over 68 percent is locked up in ice and glaciers; another 30 percent is in the ground. Fresh surface water sources like rivers, lakes are about 93,100 km^3 which is approximately 1 percent of total water. Various agencies estimate the quantum of water resources at different storages. A slight difference in estimates can be noticed.

6.1 Hydrological Cycle

Constant movement of water from oceans to air and back again to oceans in the form of precipitation over Oceans and subsurface flows from land surfaces broadly generates the hydrological cycle or the water cycle over the globe. It has no beginning or end. The Hydrological cycle used to model the storage and movement of water between the biosphere, atmosphere, lithosphere and hydrosphere (Fig.6.2).

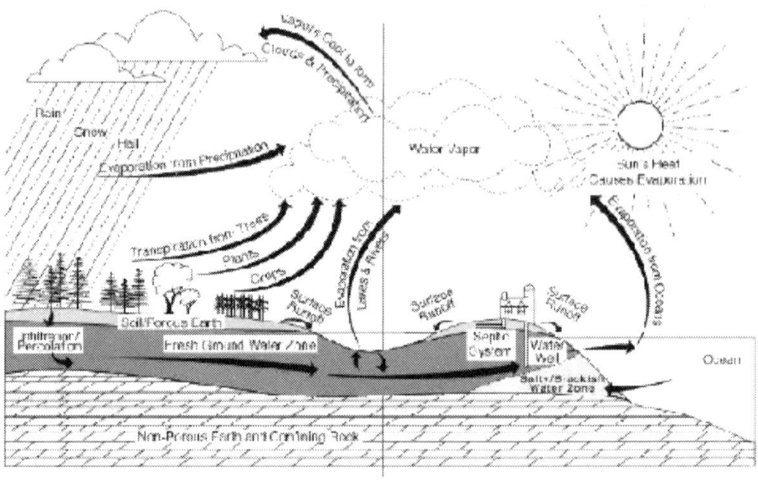

Fig. 6.2: Hydrological cycle (*www.b-u.ac.in/sde_book/msc_water.pdf*)

The figure shows it as an intricate combination of evaporation, transpiration, air mass movements, condensation, precipitation, runoff and ground water movement and water is continuously cycled between the reservoirs. The greater part of the moisture which eventually turns into precipitation comes from Oceans. However some water enters the air directly through evaporation and transpiration from soil and vegetation. Evaporation from oceans, evapotranspiration from plants combinely falls as rain over land or over sea. Rain falls directly on land and converts into runoff and flows into oceans through surface drainage network systems comprising of rivers and streams. Part of it enters into the soil and adds to ground water table. Lateral flows into the streams, lakes, river system reaches the oceans through surface stream flows and evaporation into the atmosphere. Also some part of it joins the sea directly. Some of the rain that falls on vegetation is intercepted and evaporates from its surface into the atmosphere. A portion of it travels from leaves, branches and flows down to the ground and joins the surface flow or increases the soil water content through infiltration. The typical resident times of water in the major reservoirs are given in the Table 6.1.

Table 6.1: Typical residence times of water found in various reservoirs

Reservoir	Average Residence Time
Glaciers	20 to 100 Years
Seasonal Snow Cover	2 to 6 Months
Soil Moisture	1 to 2 Months
Groundwater: Shallow	100 to 200 Years
Groundwater: Deep	10,000 Years
Lakes	50 to 100 Years
Rivers	2 to 6 months

(*Source*: *www.b-u.ac.in/sde_book/msc_water.pdf*)

The annual exchanges of water between the storages and the volumes of water contained in the land as reported by Department of Atmospheric Sciences of University of Illinois at Urban-Champaign **(http:// ww2010.unic.edu)** are as follows (Fig 6.3).

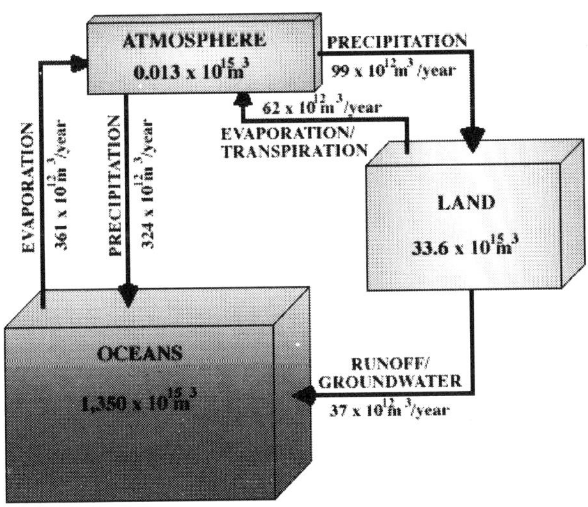

Fig. 6.3: The earth's water budget: Storage and fluxes (http:// ww2010.unic.edu)

The annual precipitation over the earth is more than 30 times the atmospheric total capacity to hold water, which indicates the rapid recycling of water between the earth's surface and the atmosphere. Also the amount of precipitation falling on oceans is much more, approximately 324 times more than the precipitation falling on ground.

6.2 Global Surface Waters

The world's average annual country-wise fresh water resources as compiled by Peter Gleick *et. al* (2011) in world's water volume 7 collected from various sources is given in following Table 6.2.

Table 6.2: Total renewable Freshwater Supply, in different Countries

Region	Country	Annual Renewable Water Resources (km^3/yr)	Year of Estimate	Source of Estimate
Africa	Algeria	11.6	2005	c
	Angola	184.0	1987	b
	Benin	25.8	2001	l
	Botswana	14.7	2001	l
	Burkina Faso	17.5	2001	l
	Burundi	3.6	1987	b
	Cameroon	285.5	2003	m
	Cape Verde	0.3	2005	c
	Central African Republic	144.4	2005	c
	Chad	43.0	1987	b
	Comoros	1.2	2005	c
	Congo	832.0	1987	b
	Congo, Democratic Republic (formerly Zaire)	1283	2001	l
	Core D'Ivoire	81	2001	l
	Djibouti	0.3	2005	c
	Egypt	58.3	2005	c
	Equatorial Guinea	26	2001	l
	Eritrea	6.3	2001	l
	Ethiopia	110.0	1987	b
	Gabon	164.0	1987	b
	Gambia	8.0	2005	c
	Ghana	53.2	2001	l
	Guinea	226.0	1987	b
	Guinea-Bissau	31.0	2005	c
	Kenya	30.7	2005	c
	Lesotho	5.2	1987	b
	Liberia	232.0	1987	b
	Libya	0.6	2005	c
	Madagascar	337.0	2005	c
	Malawi	17.3	2001	l

(Contd.)

Region	Country	Annual Renewable Water Resources (km³/yr)	Year of Estimate	Source of Estimate
	Mali	100.0	2005	c
	Mauritania	11.4	2005	c
	Mauritius	2.8	2005	c
	Morocco	29.0	2005	c
	Mozambique	217.1	2005	c
	Namibia	17.7	2005	c
	Niger	33.7	2005	c
	Nigeria	286.2	2005	c
	Reunion	5.0	1988	m
	Rwanda	9.5	2005	c
	Senegal	39.4	1987	b
	Sierra Leone	160.0	1987	b
	Somalia	14.2	2005	c
	South Africa	50.0	2005	c
	Sudan	64.5	2005	c
	Swaziland	4.5	1987	b
	Tanzania	91	2001	l
	Togo	14.7	2001	l
	Tunisia	4.6	2005	c
	Uganda	66.0	2005	c
	Zambia	105.2	2001	l
	Zimbabwe	20.0	1987	b
North and Central America	Antigua and Barbuda	0.1	2000	j
	Bahamas	nd	nd	j
	Barbados	0.1	nd	m
	Belize	18.6	2003	j
	Canada	3300.0	2000	t
	Costa Rica	112.4	1985	j
	Cuba	38.1	2000	j
	Domnica	nd	nd	j
	Dominican Republic	21.0	2000	j
	El Salvador	25.2	2001	j
	Grenada	nd	nd	j
	Guatemala	111.3	nd	j
	Haiti	14.0	2000	j
	Honduras	95.9	2000	j
	Jamaica	9.4	2000	j
	Mexico	457.2	2000	j
	Nicaragua	196.7	2000	j
	Panama	148.0	2000	j
	Saint Kitts and Nevis	0.02	2000	j
	Trinidad and Tobago	3.8	2000	j
	United States of America	3069.0	1985	n
South America	Argentina	814.0	2000	j
	Bolivia	622.5	2000	j

(Contd.)

Region	Country	Annual Renewable Water Resources (km^3/yr)	Year of Estimate	Source of Estimate
	Brazil	8233.0	2000	j
	Chile	922.0	2000	j
	Colombia	2132.0	2000	j
	Ecuador	432.0	2000	j
	Guyana	241.0	2000	j
	Paraguay	336.0	2000	j
	Peru	1913.0	2000	j
	Suriname	122.0	2003	m
	Uruguay	139.0	2000	j
	Venezuela	1233.2	2000	j
Asia	Afghanistan	65.0	1997	f
	Bahrain	0.1	2008	x
	Bangladesh	1210.6	1999	h
	Bhutan	95.0	1987	b
	Brunei	8.5	1999	h
	Cambodia	476.1	1999	h
	China	2738.8	2008	u
	India	1907.8	1999	h
	Indonesia	2838.0	1999	h
	Iran	137.5	2008	x
	Iraq	75.6	2008	x
	Israel	1.8	2008	x
	Japan	430.0	1999	h
	Jordan	0.9	2008	x
	Korea Democratic People's Republic	77.1	1999	h
	Korea Republic	69.7	1999	h
	Kuwait	0.02	2008	x
	Laos	333.6	2003	m
	Lebanon	4.5	2008	x
	Malaysia	580.0	1999	h
	Maldives	0.03	1999	h
	Mongolia	34.8	1999	h
	Myanmar	1045.6	1999	h
	Nepal	210.2	1999	h
	Oman	1.4	2008	x
	Pakistan	233.8	2003	k
	Philippines	479.0	1999	h
	Qatar	0.1	2008	x
	Saudi Arabia	2.4	2008	x
	Singapore	0.6	1975	d
	Sri Lanka	50.0	1999	h
	Syria	16.8	2008	x
	Taiwan	67.0	2000	r
	Thailand	409.9	1999	h
	Turkey	213.6	2008	x

(Contd.)

Region	Country	Annual RenewableWater Resources (km³/yr)	Year of Estimate	Source of Estimate
	United Arab Emirates	0.2	2008	x
	Vietnam	891.2	1999	h
	Yemen	2.1	2008	x
Europe	Albania	41.7	2001	p
	Austria	84.0	2007	v
	Belgium	20.0	2007	v
	Bosnia and Herzegovina	37.5	2003	m
	Bulgaria	107.2	2010	v
	Croatia	105.5	1998	o,q
	Cyprus	0.3	2007	v
	Czech Republic	16.0	2007	v
	Denmark	16.3	2007	v
	Estonia	12.3	2007	v
	Finland	110.0	2007	v
	France	186.3	2007	v
	Germany	188.0	2007	v
	Greece	72.0	2007	v
	Hungary	116.4	2007	v
	Iceland	170.0	2007	v
	Ireland	46.8	2003	o
	Italy	175.0	2007	v
	Luxembourg	1.6	2007	v
	Macedonia	6.4	2001	p
	Malta	0.07	2005	s
	Netherlands	89.7	2007	v
	Norway	389.4	2007	v
	Poland	63.1	2007	v
	Portugal	73.6	2007	v
	Romania	25.7	2007	v
	Slovakia	50.1	2007	v
	Slovenia	32.1	2007	v
	Spain	111.1	2007	v
	Sweden	183.4	2007	v
	Switzerland	53.5	2007	v
	United Kingdam	175.3	2007	v
	Serbia-Montenegro	208.5	2003	m
Former Soviet Union	Russia	4498.0	1997	e.g
	Armenia	7.8	2008	x
	Azerbaijan	34.7	2008	x
	Belarus	58.0	1997	g
	Estonia	12.8	1997	g
	Georgia	63.3	2008	x
	Kazakhstan	109.6	1997	g
	Kyrgyzstan	46.5	1997	m
	Latvia	337.3	2007	v

(*Contd.*)

Region	Country	Annual Renewable Water Resources (km^3/yr)	Year of Estimate	Source of Estimate
	Lithuania	24.5	2007	v
	Moldova	11.7	1997	g
	Tajikistan	99.7	1997	m
	Turkmenistan	60.9	1997	m
	Ukraine	139.5	1997	g
	Uzbekistan	72.2	2003	m
Oceania	Australia	336.1	2005	w
	Fiji	28.6	1987	b
	New Zealand	397.0	1995	I
	Papua New Guinea	801.0	1987	b
	Solomon Islands	44.7	1987	b

(*Source*: Gleick *et al.*, 2011)
Compiled by Gleick, P.H., Cooley, H and L. Allen, Pacific Institute

The Sources for the data given in the abvoe table is as follows

a. Total natural renewable surface and groundwater. Typically include flows from other countries (FAO: "Natural total renewable water resources.").

b. Estimates from Belyaev, Institute of Geography, USSR (1987).

c. Frenken, K., ed. Estimates from FAO (2005), Irrigation in African in Figures- AQUASTAT Survey 2005, Rome, Italy: Food and Agriculture Organization.

d. Estimates from WRI (1994). See this source for original data source.

e. Estimates from Goscomstat, USSR (1989). as cited in Gleick. 1993. Table A16.

f. Estimates from FAO (1997), Water Resources of the Near East Region: A Review, Rome, Italy: Food and Agriculoutre Orgfanisation, United Nations.

g. Estimates from FAO (1997), Irrigation in the Countries of the Former Soviet Union in Figures .Rome, Italy: Food and Agriculture Organisation. United Nations.

h. UNFAO. 1999. Irrigation in Asia in Figures. Rome, Italy: Food and Agriculture Organisation, United Nations.

i. Nix, H.1995. Water/Land/Life: The Eternal Trinagle . Canberra, Australia: Water Research Foundation of Australia.

j. UNFAO.2000. Irrigation in Latin America and the Caribbean. Rome, Italy: Food and Agriculutre Organisation , United Nations.

k. AQUASTAT website as of January 2011.

l. Margat, J./OSS.2001. Les resources en eau des pays de l'OSS. Evaluation, Utilisation et ges-tion. UNESCO/ Observatoire du Sahara et du Sahel (updating of 1995).

m. Estimates from FAO (2003), Review of World water Resources by Country, Rome, Italy: Food and Agriculture Organisation , United Nations (see specific references in this document for more information).

n. United States Geological Survey revised: conterminous U.S (2071); Alaska (980); Hawaii (18).

o. EUROSTAT,U.Wieland.2003. Water resources in the EU and in the candidate countries. Statistics in Focus, Environment and Energy, European Communities.

p. Margat, J, and Vallee, D.2000. Blue Plan- Mediterranean Vision on Water, Population and the Environment for the 21st Century. Sophia Anti polis, France.

q. Geres, D. 1998. Water Resources in Croatia. International Symposium on Water Management and Hydraulic Engineering . Dubrovnic, Croatia, September 14-19, 1998.

r. AQUASTAT website as of November 2005.

s. EUROSTAT.2005.

t. Pearse, P.H., Bertrand, F., and MacLaren, J.W,1985. Currents of Change. Final Report of Inquiry on Federal Water Policy . Ottawa, Canada: Environment Canada.

u. China Ministry of Water. Annual Report 2007-2008. Ministry of Water Resources People's Republic of China, 2007-2008 Annual Report. www.mwr.gov.cn/english/2007-2008 .doc

v. EUROSTAT, accessed 2010.

w. Australian Government National Water Commission .2005. Australian Water Resources 2005.

x. Frenken, K., ed. Estimates from FAO (2005). Irrigation in the Middle East region in Figures. AQUASTAT Survey-2008.

A quick glance at the table shows that highest annual renewable water resources (km^3/year) are recorded by Brazil (8233) followed by Russia (4498), Canada (3300) ,USA (3069), Indonesia (2838) and China (2838). The data from FAO'S AQUASTAT was updated to include the most recent data in the data base.

Data in the above table comprises both renewable surface–water and ground water supplies, including surface inflows from the neighbouring, countries. In the data, flows to other countries were not subtracted from these numbers. Since the data was collected from different sources and estimated over different periods, the data must be viewed and used with caution. Many countries do not directly measure or report internal water resources; some of the data were produced using indirect methods (Gteick et al. 2011).

6.3 Surface Water over India

India is bestowed with rich surface water resources. All the water systems in northern part of the country are perennial which are fed from both monsoon rainfall and melting snow from Great Himalayas. The rest of the river systems in the country depend on monsoon rainfall and the water availability varies with the quantum of monsoon rainfall. A glance on the land and water resources in the country provided by National Institute of Hydrology, Roorkee is as follows (Table 6.3).

Table. 6.3: Land and water resources of India

Particulars	Quantity
Geographical Area	329 million ha.
Flood prone Area	40 million ha
Ultimate irrigation Potential	140 million ha.
Total Cultivable Land Area	184 million ha.
Net Irrigated Area	50 million ha.
Natural Runoff (Surface Water and Ground Water)	1869 Cubic km.
Estimated Utilisable Surface Water Potential	690 Cubic km.
Groundwater Resource	432 Cubic km.
Available Groundwater resource for Irrigation	361 Cubic km.
Net Utilisable Groundwater resource for irrigation	325 Cubic km.

(*Source*: www.nih.ernet.in)

The above table shows that the total water availability due to run off is estimated as 1869 Km^3 out of which utilizable surface water potential is only 37 percent of (690 Km^3). This was attributed to the fact that 1) over 90 percent of the annual flow of the Himalayan rivers occur over a four month period and 2) potential to capture such resources is not possible due to limited suitable storage sites in the river course.

The balance of the water resources over the country as a whole is given below (Table 6.4).

Table 6.4. Annual water balance for India

S. No.	Parameters	Volume in BCM	Precipitation (%)
1.	Precipitation	3838-4000	100
2.	Stream flow	1953	50.1 - 48.8
3.	Ground water	433	11.3-10.8
4.	Evapotranspiration	3838-(1953+433) = 1452	37.8
		4000-(1953+433) =1614	40.3

(*Source*: S.K. Jain, 2012)

Two values of volume of precipitation in the above Table. 6.4 were obtained 1) by multiplying the geographical area of the country 3.28 x 10^6 sq.km with average annual precipitation of 1170 mm (3838 BCM) and the value of 4000 BCM given by Planning Commission of India. A review of India's water balance and evapotranspiration by Jain,S.K (2012) critically examined the procedures to compute evapotranspiration and found that they are much lower than the what may be expected given under India's climate and land use. It was reported that discrepancy in water balance of India (Table 6.4) is because of the major rivers like, Indus, Ganga, Brahmaputra and their tributaries bring in large flow to the Indian portion from across the border and mean annual flow values for these basins include this flow. But precipitation, ground water and ET for the areas are not included in the water balance computations. This results in low estimates of ET for these basins and hence the overall estimate of ET for India is low. Basin wise water availability of the major river systems is given in the following Table 6.5.

Table 6.5: Basin-wise water availability (Km^3/year) of the Major River systems in India

Sl No.	Name of the River Basin	Average Annual Surface Water Potential	Estimated Utilisable Surface Water	Estimated Replenishable Ground Water Resources	Total Utilisable Water
1.	Indus (up to Border)	73.31	46.00	26.49	72.49
2.	a) Ganga	525.02	250.00	170.99	420.99
	b) Brahmaputra, Barak Others	585.60	24.00	35.07	59.07
3.	Godavari	110.54	76.30	40.65	116.95
4.	Krishna	78.12	58.00	26.41	84.41
5.	Cauvery	21.36	19.00	12.30	31.30
6.	Subernarekha	12.37	6.81	1.82	8.63
7.	Brahamani & Baitarni	28.48	18.30	4.05	22.35
8.	Mahanadi	66.88	49.99	16.46	66.45
9.	Pennar	6.32	6.86	4.93	11.79
10.	Mahi	11.02	3.10	4.20	7.30

(*Contd.*)

Sl No.	Name of the River Basin	Average Annual Surface Water Potential	Estimated Utilisable Surface Water	Estimated Replenishable Ground Water Resources	Total Utilisable Water
11	Sabarmati	3.81	1.93	3.00	4.93
12	Narmada	45.64	34.50	10.83	45.33
13	Tapi	14.88	14.50	8.27	22.77
14	West Flowing Rivers from Tapi to Tadri	87.41	11.94	8.70	20.64
15	West Flowing Rivers From Tadri to Kanyakumari	113.53	24.27	9.00	33.27
16	East Flowing Rivers Between Mahanadi & Pennar	22.52	13.11	9.00	22.11
17	East Flowing Rivers Between Pennar and Kanyakumari	16.46	16.73	9.20	25.93
18	West Flowing RiversOf Kutch, Saurashtra including Luni	15.10	14.98	11.23	26.21
19	Area of Inland drainage in Rajasthan desert	N.A Neg.	N.A	N.A	
20	Minor River Basins Draining into Bangladesh & Myanmar	31.00	N.A	18.80	18.80
	Total	1,869.35	690.31	431.42	1,121.73

Source: Central Water Commission for Col. 3 & 4 and Central Ground Water Board/report of the standing sub-committee for assessment of availability and requirement of water, Aug - 2000 for Col. 5.

The table also gives information on the estimated utilizable surface water along with estimated replinshable ground water resources. It can be observed that the surface water potentials of Brahmaputra and the River Ganga far exceed the other river basins. However its utilizable surface water against the potential are low compared with other rivers *viz.,* Godavari, Krishna, Cauvery, Mahanadi, Pennar etc. Though the rivers Brahmaputra, Baraka and others has highest surface water potential, the replenishable ground water resources are not highest. The replinshable ground water recorded in Ganga basin followed by Godavari, Krishna and Indus up to the border is the maximum over the other river basins in the country.

6.4 Ground Water Potentials in India

Groundwater is the most important water resource which was instrumental in achieving self sufficiency in food production by increasing the minor irrigation command area across the country. The reliability of ground water resources for irrigation is very high as surface waters cannot be taken to all areas and fields, which involves huge expenditure. Also the ground water quality is excellent in most parts of the country and provides a relatively safe source of drinking water for rural and urban populations. The availability of ground water varies greatly with changes in topography, sub surface geology and the climate. In some areas of the country the ground water is available in deep aquifer while in some parts it is stored close to the surface. The location of aquifers influences the recharge rate.

The recent estimates of the total annual replenishable ground water resources assessed by National Institute of Hydrology (NIH) as 433 Billion Cubic Meter (BMC) and the net annual ground water availability is estimated as 399 BMC/ year (**www.nih.ernet.in**) and the existing ground water draft in recent years for all the use is about 231 BCM/year. The spatial distribution of zones for exploitation ground water resources are given in the following Figure 6.4. It is observed that in general, ground water availability is mostly confined to valleys and other low lying areas in the mountainous and hilly regions in the North which don't allow adequate infiltration. The underlying geology in the peninsular part limits the formations of large continuous aquifers. Therefore it is scattered where fissures permit adequate storage. This has resulted in low yields in this region. However, in some areas medium to high yield potentials were noticed which depend upon local hydrogeology.

Coastal region are generally rich in ground water due to alluvial terrain but the aquifer can easily be contaminated with sea water incursion due to over pumping. The alluvial tract of Gangetic plains across central and North India has the best potential for ground water extractions as large areas in their region has many favourable characteristics for ground water storage and recharge. Basin wise assessment of total replenishable ground water resources is as follows (Table 6.6).

The total ground water potential is estimated at 431.4 km^3/year. The data showed that the Ganga basin has the maximum ground water potential and accounts for nearly 40 per cent of the total ground water potential in the country. Godavari basin is next with a record of 9.4 per cent. After subtracting the provision for the domestic and industrial purpose, the potential available for the irrigation is 360.35 km^3. However, there is still a scope for exploitation of 245.13 k^3 of water. Perhaps proper ground water resources development and utilization will effectively mitigate the challenges posed by drought conditions.

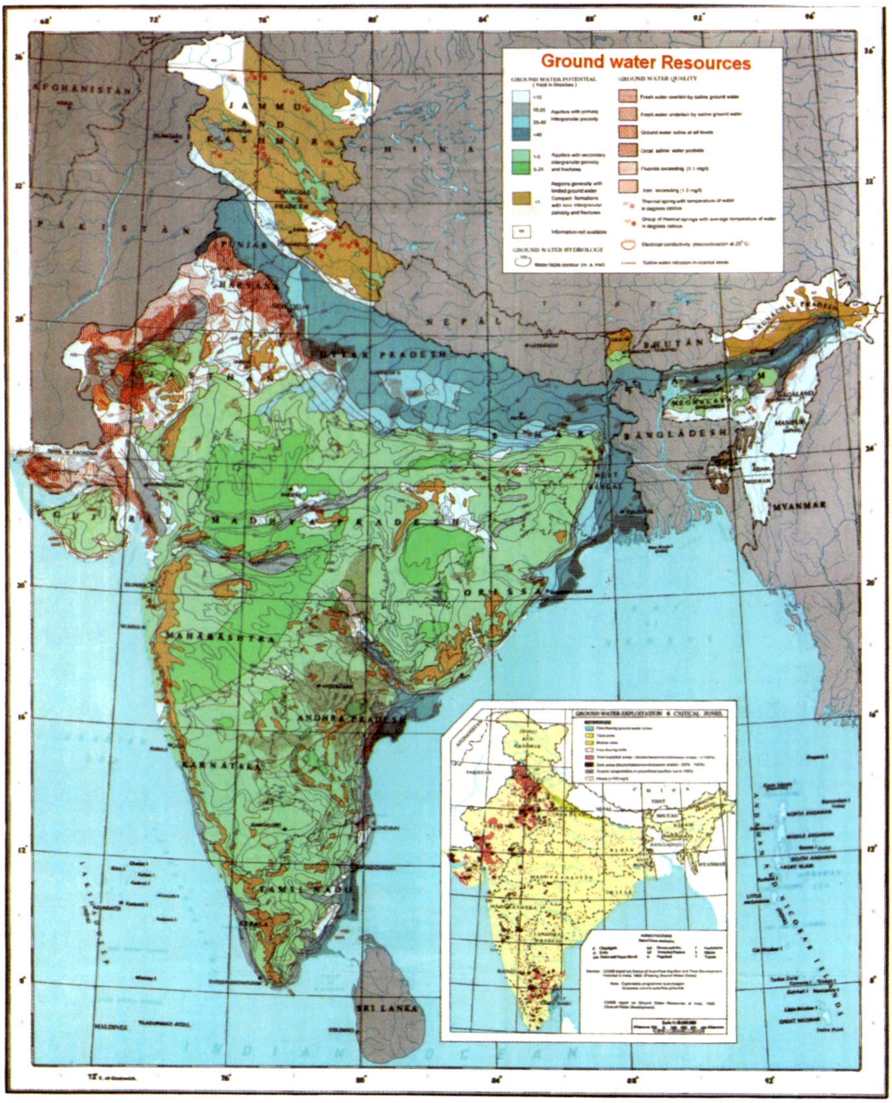

(*Source*: www.nih.ernet.in)

Fig. 6.4: Ground water resources of India

Table 6.6: Basin Wise groundwater potential, available for current irrigation, net draft and percentage development (Km^3/year). (*Source*: S K Gupta *et. al.*, 2000)

Basin	Total replenishable ground water resources	Provision for domestic, industrial and other uses	Available ground water resources for irrigation	Net draft	Balance ground water potential available for exploitation	Level of ground water development (%)
Brahmani with baitarni	4.05	0.61	3.45	0.29	3.15	8.45
Brahmaputra	26.55	3.98	2.56	0.76	1.80	3.37
Cambay composite	7.19	1.08	6.11	2.45	3.66	40.09
Cauveri	12.30	1.84	10.45	5.80	4.67	5.33
Ganga	170.99	26.03	144.96	48.59	96.37	33.52
Godavari	40.65	9.66	30.99	6.05	24.94	19.53
Indus	26.49	3.05	23.43	18.21	5.22	77.71
Krishna	26.41	5.58	20.83	6.33	14.50	30.39
Kutch & Saurashtra Composite	11.23	1.74	9.49	4.85	4.64	1.14
Madras & south Tamil Nadu	18.22	2.73	15.49	8.93	6.55	57.68
Mahanadi	16.46	2.47	13.99	0.97	13.02	6.95
Meghna	8.52	1.28	7.24	0.29	6.95	3.94
Narmada	10.83	1.65	9.1	1.99	7.18	1.74
N-E composite	18.84	2.83	16.02	2.75	13.26	17.20
Pennar	4.93	0.74	4.19	1.53	2.66	36.60
Subernarekha	1.82	0.27	1.55	0.15	1.40	9.57
Tapi	8.27	2.34	5.93	1.96	3.97	33.05
Western Ghats	17.69	3.19	14.50	13.32	11.18	22.88
Total	431.43	71.08	360.35	115.22	245.13	31.97

Similarly, the state wise ground water resources and irrigation potential of India is given in table 6.7. The total ground water potential estimation is 43.2 M. ha. and among the states, Uttar Pradesh which is located in the Ganga basin has the highest potential followed by Madhya Pradesh, Maharashtra, Andhra Pradesh and Bihar. Himachal Pradesh has the least potential amongst the states. In terms of area, the irrigation potential of the ground water has been estimated as 64.05 M. ha.

Table 6.7: State wise Ground water resources (Km³/years) and irrigation potential in India

S. No	States/Union Territories	Total Replenshable Ground Water Resource	Provision for Domestic, Industrial & Other Uses	Available Ground water Resource for Irrigation in net terms	Net Draft	Balance Groundwater Resource for Future use in net terms	Utilizable rrigation Potential for Development
1.	Andhra Pradesh	3. 52916	0.52938	2.09978	0.70922	2.29056	3.96000
2.	Arunachal Pradesh	0.14385	0.02158	0.12227	-	0.12227	0.01800
3.	Assam	2.47192	0.37079	2.10113	0.09418	2.00695	0.90000
4.	Bihar	3.35213	0.50282	2.84931	0.54676	2.30255	4.94763
5.	Goa	0.02182	0.00327	0.01865	0.00164	0.01701	0.02020
6.	Gujarat	2.03767	0.30565	1.73202	0.71702	1.01500	2.75590
7.	Haryana	0.85276	0.12792	0.72484	0.60798	0.11686	1.46170
8.	Himachal Pradesh	0.03660	0.00731	0.02929	0.00530	0.02300	0.00850
9.	Jammu & Kashmir	0.044257	0.06639	0.37618	0.00500	0.37118	0.70795
10.	Karnataka	1.611857	0.24279	1.37578	0.43010	0.94568	2.57281
11.	Kerala	0.79003	0.13135	0.65868	0.10062	0.55806	0.87925
12.	Madhya Pradesh	5.08892	0.76332	4.32560	0.71312	3.61249	9.73249
13	Maharashtra	3.78673	1.23972	2.54701	0.77403	1.77298	3.65197
14.	Manipur	0.31540	0.04730	0.26810	Neg	0.26810	0.36900
15.	Meghalaya	0.05397	0.00810	0.04587	0.00182	0.04405	0.06351
16.	Mizoram	Not Assessed					
17.	Nagaland	0:07240	0.01090	0.06150		0.06150	-
18.	Orissa	2.00014	0.30002	1.70012	Neg	1.55699	4.20258
19.	Punjab	1.86550	0.18652	1.67898	0.14313	0.10322	2.91715
20.	Rajasthan	1.27076	0.19945	1.07131	1.57576	0.52893	1.77783
21.	Sikkim	Not Assessed			0.54238		
22.	Tamil Nadu	2.63912	0.39586	2.24326	1.35578	0.88748	2.83205

(Contd.)

S. No	States/Union Territories	Total Replenshable Ground Water Resource	Provision for Domestic, Industrial & Other Uses	Available Ground water Resource for Irrigation in net terms	Net Draft	Balance Groundwater Resource for Future use in net terms	Utilizable rrigation Potential for Development
23.	Tripura	0.06634	0.00995	0.05639	0.01885	0.3754	0.08056
24.	Uttar Pradesh	8.38210	1.25743	7.12467	2.68354	4.44113	16.79896
25.	West Bengal	2.30923	0.34642	1.96281	0.47452	1.48829	3.31794
	Total for States	**43.14769**	**7.007414**	**36.07355**	**11.50055**	**24.57300**	**64.04513**
1.	Andaman & Nicobar Islands	Not Assessed					
2.	Chandigarh	0.002968			0.00245	0.000512	
3.	Dadra and N. Haveli	0.004220	0.000633	0.003587	0.00046	0.003130	0.00504
4.	Daman & Diu	0.001300	0.000200	0.001100	0.00090	0.000200	
5.	NCT Delhi	0.029154	0.017832		0.01180		
6.	Lakshadweep	0.000243			0.00016	0.000088	
7.	Pondicherry	0.002877	0.000432	0.002445	0.00059	0.001850	
	Total for UTs	**0.040760**	**0.019197**	**0.007132**	**0.016362**	**0.005780**	**0.00504**
	Grand Total	**43.18850**	**7.093337**	**36.08062**	**11.516912**	**24.57878**	**64.05017**

(*Source* : S K Gupta *et. al.*, 2000)

6.5. Influence of Drought on Water Resources, its Fauna and Flora

Drought has significant impacts on various resources both in developed and the developing countries. It is more felt in the latter regions due to higher population density, which had increased exploitation of water resources is of great concern. The impacts usually lead to economic losses and social problems. In some cases, the impacts may reach areas outside the region experiencing drought. For example the limited urban water supply may be the result of the receipt of inadequate rainfall in the reservoir area and not in the area where city is located. Drought severely affects the water resources and maintaining the supply of drinking water assumes top priority at the management level. Drought conditions may lead to adoption of costly emergency measures like transportation of water for drinking and fodder to animals. Under extreme conditions, it becomes essential to restrict the supplies and also the supply time. This may affect the recreation and tourism industries where rivers are mostly used for transport purposes. Low level waters impair the navigability of rivers and increase the transportation cost as the essential commodities must be transported by road only.

6.5.1 Effects on Fauna and Flora

Droughts on both standing and flowing water systems can have major ecological effects. Stress and depletion of both fauna and flora were studied by various scientists and the results are as follows. Fall in surface level in water bodies occurs due to decline in surface runoff and stream inputs due to drought. Usually the air temperature and evaporation increase sharply during this period. As drought progresses, water level recedes from normally highly production littoral zone and stranding of some of the fauna (eg. Mussels, snails) and flora, such as rooted aquatic macrophytes are observed (Furey, *et al.,* 2006).High water temperature may be accompanied by stratification and increasing conductivity (salinity) along with decreasing levels of oxygen, severely stress the aquatic fauna. The combination of high temperature and low oxygen levels may eliminate some fish species. With stratification and stagnant conditions, nutrients may build up resulting in increasing the risk of algal blooms during drought, in particular blooms of toxic blue green and golden algae (Ha *et.al* 1999, Colley , 2004), which can cause extensive damage to fish and may be harmful to live stock and domestic animals (Davis, 1978, Briand *et.al* 2003) . The initial flows breaking the drought period have been observed to spread out breaks of some forms of toxic microalgae down the stream (Brazos River Authority, 2002).

In large flood plain rivers, drought prevents seasonal inundation and extends the duration of dry spells during which flood plains are not flooded. The reduction in flooding for longer durations. caused either by droughts or river regulation or by both may greatly deplete the invertebrate egg bank of flood plain, so that when flood plain inundates again, the expected micro-invertebrate "boom" gets greatly diminished (Jenkins & Bowlton, 2007). As the drought continues, fish and invertebrates may move into pools and some of which may live as refugee, while drying up in other pools may kill all their inhabitants. Very noticeable with flow cessation is the decline in fauna, strongly dependent on flowing water such as riffle dwelling insects and other invertebrates such as mussets (Golladay *et.al* 2004). A noticeable change that can occur in isolated riverine pools is the switch from a "flowing water invertebrate fauna to one of typical standing water bodies" largely composed of air breathing predators *viz.,* hermipteroms (bugs) and coleopterans (beetles) which are well adapted to severe conditions (Lake 2003 and Boulton & Lake 2008) In pools with high densities of trapped fauna, the levels of parasitism and disease notably of fish can rise (Medeiros & Maltchik, 1999).

6.6 Influence of Changing Climate on Droughts and Water Resources

6.6.1 Global warming and droughts at global scale

Global warming processes have notable impact on droughts, by increasing the severity of the incidents as a result of more water loss from the soil and the plants due to high vapour pressure deficits. Arnel (2004) summarized the current understanding of the potential implications of the climate change on droughts and mentioned that the fourth Assessment Report of IPCC (2007) have concluded that altered frequencies and intensities of extreme weather events are expected to have mostly adverse effects on natural and human systems and more specifically that the area affected by droughts is likely to increase. Early Part of 21 century already witnessed significant drought conditions in many parts of world including Southern Europe, SE parts of USA, Australia, East Africa and large parts of Asia including India and China. Though low precipitation is the main cause for droughts, the importance of temperature was noticed in recent studies on trends in water resources (Nicholls 2004, Cai & Cowan 2008, Lorenzo – Lacruz *et al.,* 2010). Ciais *et al.* (2005) has shown how warming processes in 2003 summer in Europe increased the severity of droughts. In USA, Walter *et al.* (2004) quantified the warming effect by means of catchment balances in areas characterized by low anthropogenic fluctuations and reported an increase in water losses by evapotranspiration from 1950 onwards. To fit into the latest trend in drought studies, a new term *viz.*, "global warming- type drought" was coined by Adam's *et al.* (2009). A study on the drought indicators to assess whether the temperature increase during

1930-2006 in NW / Iberian peninsula in Spain has any impact on characteristics of drought experiencing in their region and on the water resources was conducted by Vicente - Serrano *et al.* (2011) and observed that the warming process did not have a noticeable influence on the stream flow variability in the study period.

Assessment of trends in drought occurrence over globe by employing Palmer Drought Severity Index (PDSI) was reported by (Dai *et al.,* 2004: Trenberth *et al.* 2007). The index over a time period is basically a function of the difference between AET and the amount of precipitation required to maintain normal water balance level, where the amount required is dependent on evaporation, soil water recharge and runoff as computed using a simple hydrological model. The Fig 6.5 shows the spatial pattern of PDSI over the period 1900-2002 years computed from observed precipitation and temperature data.

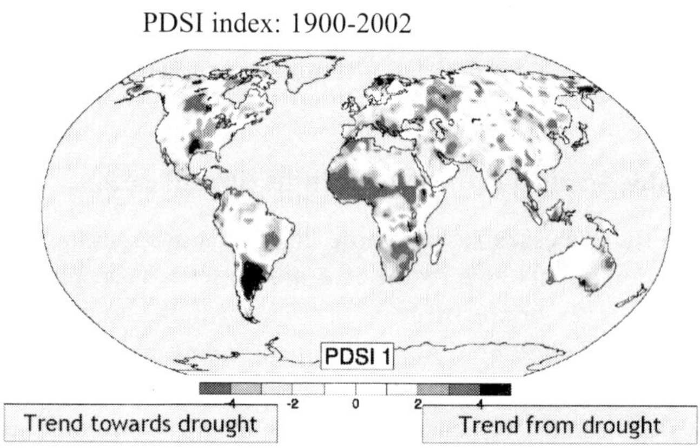

Fig. 6.5: Spatial pattern in the monthly PDSI over the period 1900–2002
(*Source*:Trenberth *et al.*, 2007)

The analysis has brought out that very dry areas (with PDSI less than-3) increased from 12 percent of the global land surface to 30 percent since the year 1970. Drought occurrence increased particularly in western and southern Africa, Southern Europe and Canada, Brazil, South Asia and Eastern Australia.

Burke *et al.* (2006) simulated changes in the PDSI over 21 century using Had CM_3Climate model drivers by A2 Scenario. The results projected are a overall drying trend throughout 21 century with more intensity over Amazonia, the US, North Africa, Southern Europe and Eastern Asia. The results further reveal that percentage land under severe drought increases from around 10 percent at the beginning of 21 century to around 40 percent at the end. The frequencies of severe drought events doubles by the end of century and their mean duration increases by a factor of around five (Burke *et al* 2006). However, there are considerable uncertainties in these projections like how the climate changes in the future and the indicators used to characterize drought.

6.6.2 Global Warning on Water Resources at global scale

The availability of fresh water in future to the humans is likely to shrink due to increase in population from 6.5 billion at present to 8.9 billion by 2050 and water use has been growing at more than twice the rate of population increase in the last century (World climate change-2008 **www.climate.org**). Increased urbanization demand more water among the thickly populated region around the world, and the consumption pattern of domestic water use is likely to increase significantly. All the changes that likely to occur with respect to rainfall pattern due to global warming would affect the fresh water resources especially in the water scarcity zones. The world wise water requirement pattern for domestic, industrial, agricultural and the total with respect to population for the years 2000, 2025, 2055 and 2075 is given in table 6.8.

Table 6.8: Global Water requirement (km^3/ year) for various purposes under A1b scenario

Year	Domestic	Industries	Agriculture	Total	Population (million)
2000	389.9	776.6	2657.8	3824.3	6132.0
2025	651.4	1243.2	3388.0	5282.5	7938.0
2055	903.1	2047.2	3683.8	6634.2	8722.7
2075	912.3	2867.1	3538.2	7317.6	8310.6

(*Source*: http: hydro.iis.u-tokyo.ac.JP/GW)

With the increase in population by approximately 2.6 billion from the year 2000 to 2055, the water demand as per A1b scenario increased by 232, 264 and 38 percent in respect of domestic, industrial and agricultural purposes respectively.

The Walker Institute for Climate System Research, University of Reading, U.K studied the impact of rise of global temperature by 4°C on the river flows and the availability of water resources in different countries over globe (Fig. 6.6). It observed a decrease of average run off by 70 percent over the current level at the Mediterranean, South Africa, Central Asia and large areas of North and South America. In some parts of the world such as high latitudes and wet tropics, run off would likely increase which might lead to increased flood risks in these parts.

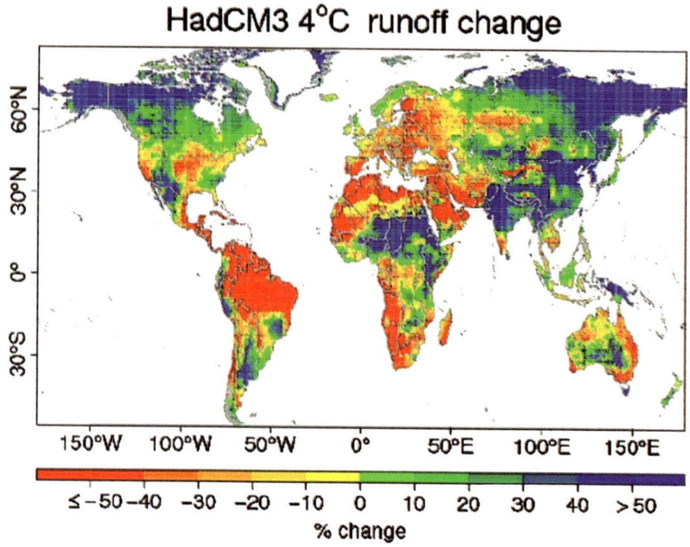

(*Source*: Walker Institute for Climate System Research)

Fig. 6.6: Change in river flow for an increase in global average temperature of 4°C

Several indicators have been developed over the years to identify water stress over the region. The indicators include the amount of water availability per person and the ratio of volume of water withdrawn to the volume of water potentially available. When withdrawls are more than 20 percent of the total renewable resources, water stress will be a limiting factor for development (Falkenmank and Lindh, 1976). If withdrawls are more than 40 percent or more, it may lead to high water stress. Also it was further reported by the authors that if a country has less than 1700m^3/ year of fresh water per person, water stress shall be major issue (Falkenmark and Lindh, 1976), which may hinder the overall development of the region. Simple numerical indices may provide only partial indications of water stress experienced by various regions because, the consequences of water stress depend on how efficiently the water is managed. At the global scale, usually assessment of water stress is made at country level because that is the unit at which water use data are available. During 1990, it was estimated in a WMO report of 1997 that approximately one-third of worlds human population lived in countries using more than 20 percent of their water resources and by 2025 it is estimated about 60 percent of population would be living in such stressed countries in the absence of climatic change (WMO, 1997). Based on the water withdrawal as percentage of total available, fresh water stress during 1995 and the projection for 2025 for different countries as reported by UNEP (1999) is given in the following Figure 6.7.

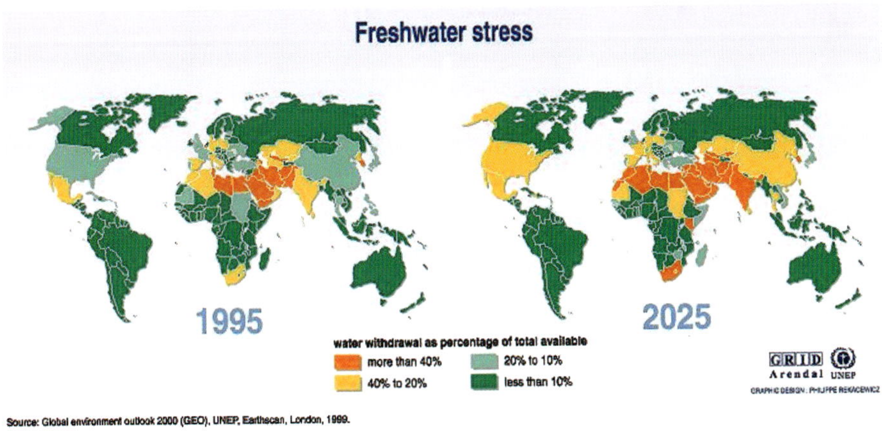

Fig. 6.7: Global fresh water stress during 1995 and 2025

It is seen that water stress by 2025 increases in almost all the countries except in Middle East and North African countries which are identified chronically water stressed regions. Arnell (1996, 2000) estimated the effect of different climate change scenario on National water resources availability and compared this with estimated future demand of water. The results of the study reported significant geographical variations. For all the business as usual Scenarios considered viz. HadC$_{M2}$, HadC$_{M3}$, ECHA$_{M4}$, CGCMI, CSIRO, CCSR, GFDL, NCAR, Climate change increases water stress in many countries, in southern and western Africa and the Middle east and in parts of Asia. Summary report on the impact assessment of challenges of water scarcity and droughts in the European union presented to Commission of European Communities during 2007 revealed that 33 river basins affected by water scarcity have been identified which represents an area of 460,000 Km^2 (10% of total EU area) that has a population of 82 million (16.5% of total EU population).

6.6.3 Global Warming and Droughts at National Scale in India

The Ministry of Environment and Forests of Govt. of India has launched Indian Network of Climate Change Assessment (INCCA) in 2009 to enhance the knowledge on impacts of climate change at the National and at Regional level. As a part of the climate change program under India's Initial National Communication to UNFCCC, impacts of climate change on water resources, agriculture, natural ecosystem, forestry, human health, infrastructure and energy were assessed for the years 2050 and 2080 and reported to IPCC. The climate change scenarios presented in the IPCC 2007 document were derived from a

Regional Climate Change Model PRECIS (a down scaled global climate modal version of Had RM3) developed by Hadley Center, UK with a resolution of 50 x 50 km and forced by a Green House Gas (GHG) emission scenario from A1B. All the projected changes are with respect to the average for the period 1961-90 (base line). The projected changes in summer monsoon rainfall in 2030 with respect to 1970 are given in the fig 6.8 (INCCA, 2010).

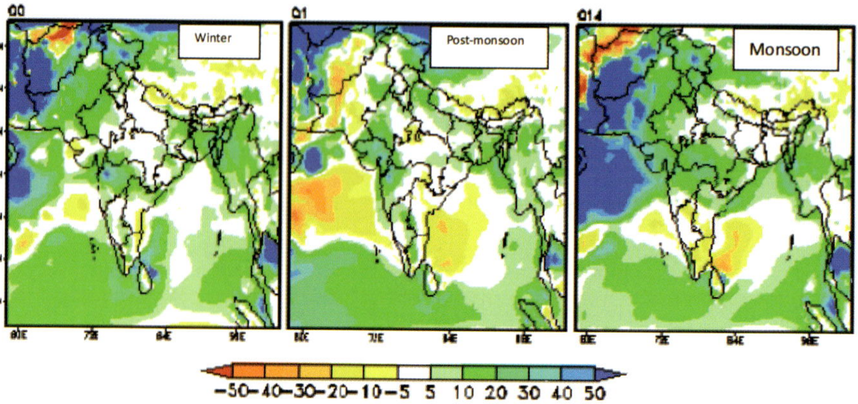

(*Source*: INCCA, 2010)

Fig. 6.8: Projected changes in winter, pre-monsoon and summer monsoon rainfall in 2030s with respect to 1970s

The projection indicates 3 to 7 percent overall increases in summer monsoon rainfall in 2030 with respect to 1970. However on a seasonal scale, except for Himalayan region all other regions *viz.,* coastal regions of east and west, Northeast region are likely to have lower rainfall in the winter period as well as pre-monsoon period. The spatial pattern of monsoon rainfall indicates a significant decrease in some parts of the southern peninsula.

The impact of global warming on the projected spatial distribution of percent changes in drought weeks during monsoon period June-Sept across 12 major river basin in India for the period Mid Century (2021-2050) and End Century (2071-2098) were studied by Gosain *et al.* (2011). The details of methodology and the highlights of the work are as follows. In this study, the Soil Moisture Deficit Index (SMDI) developed by Narasimhan and Srinivas (2005) was used to monitor the agricultural drought severity on weekly basis using SWAT output to incorporate the spatial variability across various river basins. Two scales *viz,* 1 and 2 representing the severity of drought wherein scale 1 (SMDI value 0 to -1) represents the drought development stage and scale 2 (SMDI value between -1 to-4) represents mild to moderate and extreme drought conditions. Weekly SMDI values were calculated using simulated soil moisture data of 30 years each under Baseline (BL), Mid Century (MC) and End Century (EC) climate change scenario. The spatial distribution of percentage change in drought weeks for both the scales from BL to MC and EC respectively was worked out using the SWAT output for drainage basins (Fig 6.9).

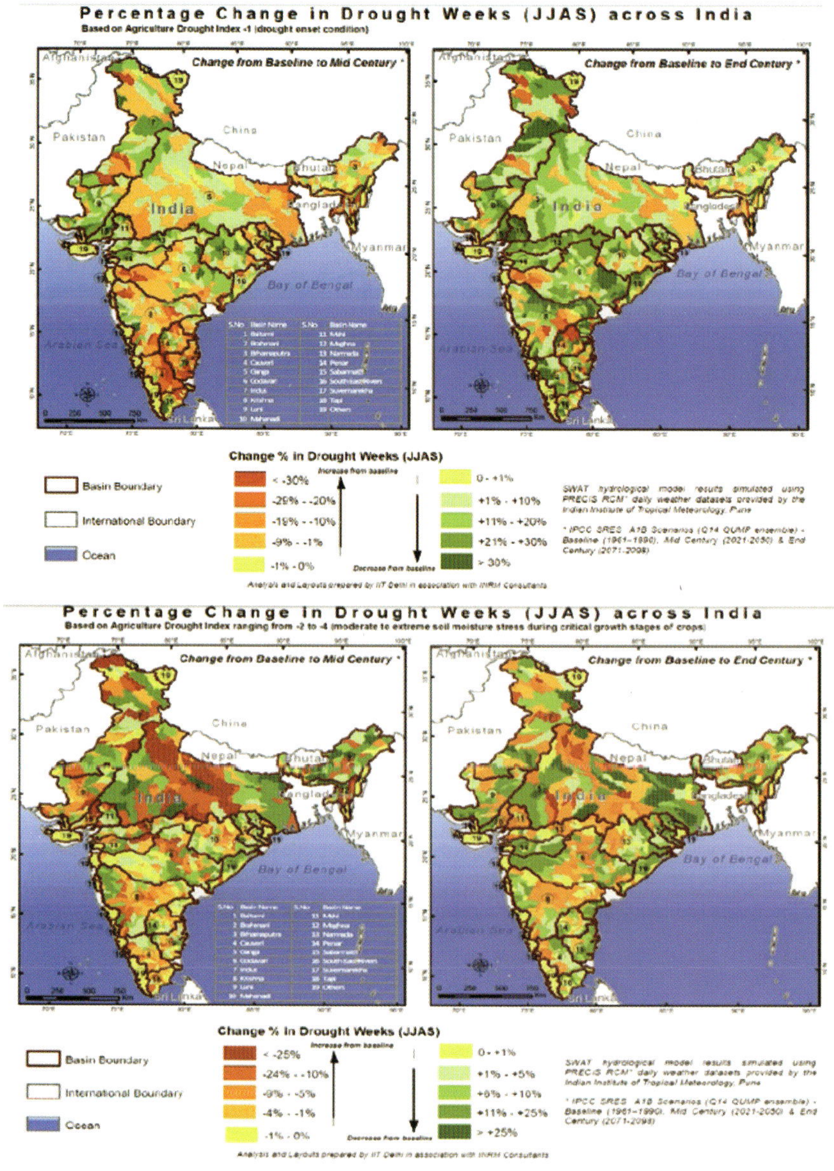

(*Source:* Gosain *et al.,* 2011)

Fig. 6.9: Spatial distribution of percentage change in drought weeks (JJAS) based on Agricultural Drought Index-1 & -2 to -4 for baselines to Mid & End Century across India

It is seen that an increase in moderate drought development (scale-1) for Krishna, Narmada, Pennar, Cauvery and Brahmani basins which have either predicted decrease in precipitation or have increased rates of ET for MC scenario. For scale- 2, moderate to extreme drought have been projected for Baitarani, Sabarmati, Mahi and Ganga river systems. The expected increase in drought is in the order of 5 to 20 percent for many areas in spite of increase in precipitation.

Regional level impacts of climate change scenarios of 2021-2050 and 2071-2098 on drought probabilities for the state of Andhra Pradesh were studied in which 8 of 23 total districts were identified as chronically drought prone (Rao AVMS *et al* 2010). This study was taken up under Network Program on Climate Change (NPCC) at CRIDA Hyderabad an ICAR sponsored project during X and XI Plan periods. The daily rainfall data of baseline and projected scenario for 120 gird points that are covered by the state was obtained frcm NATCOM database and meteorological drought analysis was carried out as per the criteria followed by India Metrological Department and the probabilities of moderate and severe droughts at 120 gird points for the baseline and for 2021-2050 and 2071-2098 periods and presented in the following Fig 6.10.

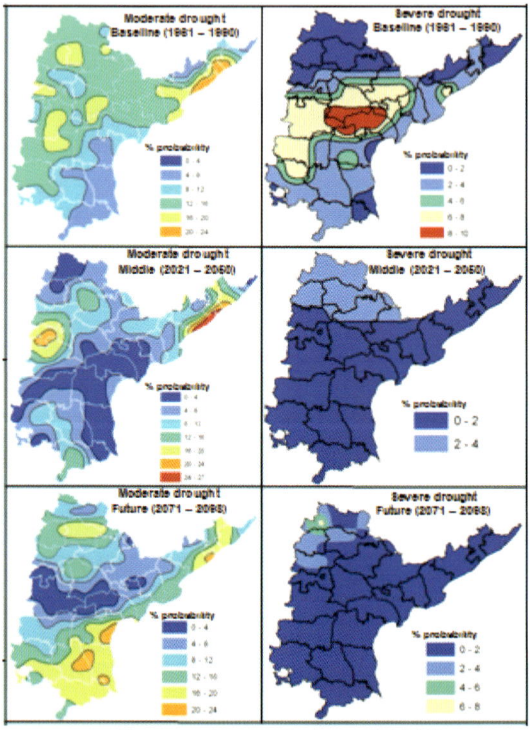

(*Source*:. Rao, AVMS. *et.al.* 2010)

Fig. 6.10: Probabilities for moderate and severe drought for the periods of Base line, 2050 and 2080 in Andhra Pradesh

The study found an increase in moderate drought probability in Ranga Reddy and Medak districts from 12-16 percent to 12-24 percent. However, in the north coastal districts, the probability increased from 16-24 percent to 20-27 percent during the period 2021-2050. Severe droughts have shown a decline in trend in all the districts from the baseline except in north Telangana districts for the period 2021-2050 and 2071-2098.

6.6.4 Global Warming Impacts on Water Resources at National Scale

The total water resources of India account only for 4 percent of global fresh water resources though it shares 16 percent of total global population. (INCCA-2010). The, available water resources and demand from various sectors *viz.*, domestic, irrigation, industry energy and other are shown is Table 6.9.

Table 6.9: Water Resources of India

Water Resources	Volume (Km^3)
Estimated annual precipitation (including snowfall)	4000
Run-off received from upper riparian countries (say)	500
Average annual natural flow in rivers and aquifers.	1869
Estimated utilizable water	1123
(i) Surface	690
(ii) Ground	433
Water demand _utilization (for year 2000)	634
(i) Domestic	42
(ii) Irrigation	541
(iii) Industry, energy & others	51

Source: National Water Mission under National Action Plan on Climate Change, MoWR,GOI,2010

It is observed that irrigation sector uses roughly 83 percent of water in India followed by industry and domestic sector. Due to increased population from 361 million in 1951 to 1.13 billion in 2007, the per capita availability of water decreased from 5177 m^3/y to 1654 m^3/y in 2007 (INCCA, 2010). Gupta and Deshpande (2004) estimated that that the gross per capita water availability in India will decline from about 1820m^3/y in 2001 to as low as about 1140$m^3/$year by 2050. According to India Initial National Communication (NATCOM) to United Nation Frame Work convention on climate change, is likely to adversely affect the water balance in different parts of India due to changes in rainfall and sea level rise.

The studies on impacts of climate change on river flows in the major river basin in India (Gosain *et al.*, 2004, Gosain *et al.*, 2006 and Gosain *et al.*, 2011) have quantified the effect on Indian water resources The results showed that the amount of surface run off due to climate change vary across river basins as well as in sub- basins by the year 2050 (INCCA-2010). This study uses the PRECIS daily weather data to determine the spatio-temporal water availability in the river system across the country. The hydrological model *viz.*, SWAT was used to simulate major water balance components. The model generated detailed out put at daily interval at the sub base level *viz.*, out flow, actual ET and soil moisture status etc. Surface and sub-surface run-off and recharge to ground water were also available as sub components. The effect of climate change on the water balance components analyzed by Gosain et al 2011 for each basin and the long term variations in percentage in these basic water balance elements for various river basins is shown in the following Fig 6.11.

The figure describes the percentage change in major components of water balance *viz.*, precipitation, water yield and ET from Baseline (BL) to Mid century (MC) and End century (EC). Positive change indicates decrease from BL and negative values shows increase from BL.

It can be noticed from the figure that majority of river systems under both the scenarios showed increase in precipitation except in Brahmaputra, Cauvery and Pennar basins. Accordingly, the water yield changes were noticed. Under MC scenario, the water yield levels for all the river basin were less when compared with EC Scenario. The situation under EC improves wherein almost all the river systems exhibit increase in precipitation and associated increase in water yield for the respective river systems. The Change in ET under MC scenario exhibits increase up to 10% for the Brahmaputra, Indus and Luni River basins. The rest show marginal increase/ decrease. In the Sabarmati and Luni river basins, the west flowing rivers showed a drastic reduction in precipitation. Consequently decrease in total runoff to an extent of two thirds of prevailing runoff may lead to severe drought conditions in future. Further, climate change may increase the severity of drought and intensity of floods in various parts of the country (Gosain *et al.*, 2011).

Thus these studies established a comprehensive understanding of the possible impact of climate change on water resources at global and national level for preparation of long term plans for managing the water resource efficiently.

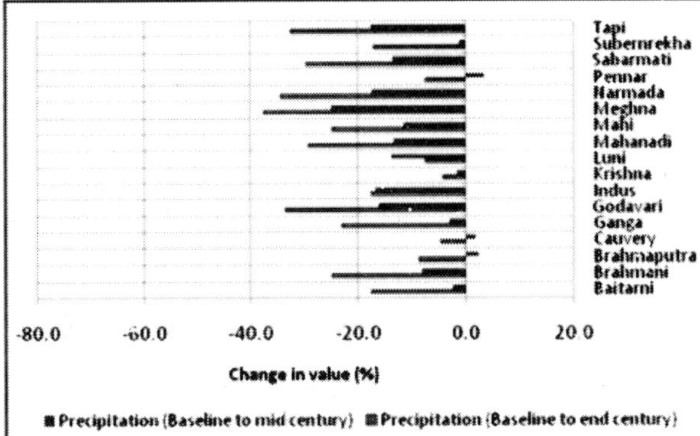

Change in precipitation

Basin	MC	EC
Baitarni	-2.5	-9.9
Brahmani	-8.1	-8.9
Brahmaputra	2.3	-20.7
Cauvery	1.7	-10.2
Ganga	-2.5	-23.0
Godavari	-16.1	-8.0
Indus	-16.6	-9.3
Krishna	-1.5	-10.9
Luni	-13.8	-2.5
Mahanadi	-13.3	-9.0
Mahi	-11.5	-3.9
Meghna	-25.0	-11.6
Narmada	-17.4	-6.8
Pennar	3.5	-6.2
Sabarmati	-13.7	-2.7
Subernrekha	-1.1	-8.3
Tapi	-17.5	-5.7

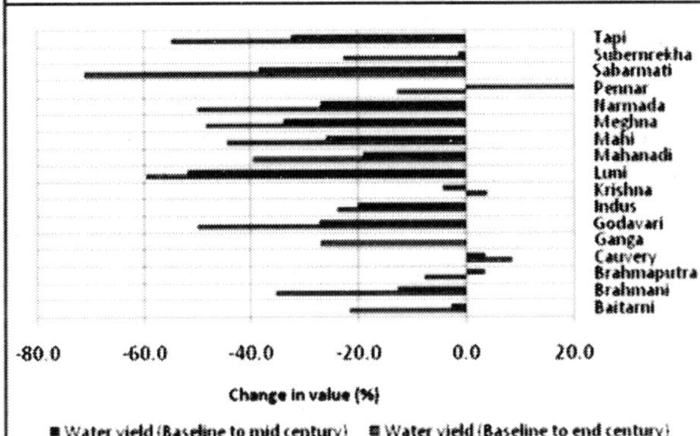

Change in water yield

Basin	MC	EC
Baitarni	-2.8	-17.6
Brahmani	-12.6	-24.9
Brahmaputra	3.5	-8.7
Cauvery	3.4	-4.7
Ganga	0.5	-27.0
Godavari	-27.2	-33.6
Indus	-20.0	-17.5
Krishna	-4.3	-4.4
Luni	-51.9	-7.5
Mahanadi	-19.0	-29.3
Mahi	-26.0	-25.0
Meghna	-33.8	-37.8
Narmada	-27.0	-34.5
Pennar	30.9	-7.4
Sabarmati	-38.6	-29.8
Subernrekha	-1.4	-17.1
Tapi	-32.5	-32.6

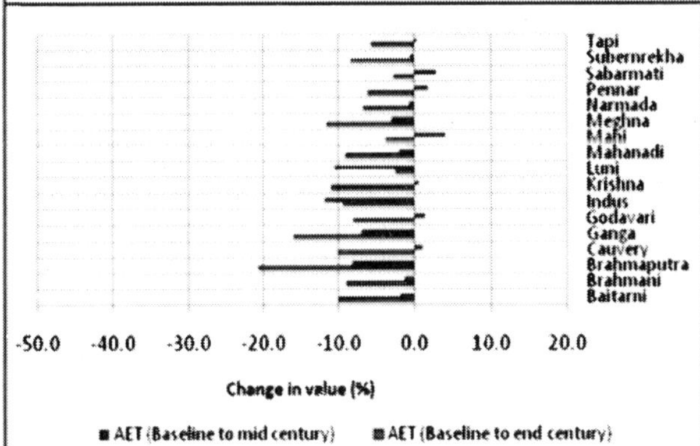

Change in actual ET

Basin	MC	EC
Baitarni	-1.9	-21.6
Brahmani	-1.2	-35.3
Brahmaputra	-8.1	-7.7
Cauvery	1.1	8.5
Ganga	-7.0	-16.0
Godavari	1.4	-50.0
Indus	-11.9	-23.9
Krishna	0.7	3.8
Luni	-10.5	-59.7
Mahanadi	-2.0	-39.6
Mahi	4.1	-44.3
Meghna	-3.0	-48.4
Narmada	-0.8	-50.1
Pennar	1.9	-12.8
Sabarmati	2.8	-71.1
Subernrekha	-0.6	22.6
Tapi	0.3	-54.9

Fig. 6.11: Change in water balance components towards 2030s (MC) and 2080s (EC) with respect to 1970s (BL) (*Source*: Gosain *et al.*, 2011)

References

Adams, H.D., Guardiola-Claramonte, M., Barron-Gafford, G.A., Villegas, J.C. et al., 2009. Temperature sensitivity of drought-induced tree mortality portends increased regional die-off under global-change-type drought. Proc Natl Acad Sci USA 106: 7063"7066.

Arnei, N.W. 2004. Climate change and global water resources: SRES emissions and socio-economic scenarios. Global Environmental Change 14 : 31–52.

Arnell, N.W. and N.S. Reynard. 1996. The effects of climate change due to global warming on river flows in Great Britain. Journal of Hydrology 183: 397–424.

Arnell, N.W. and N.S. Reynard. 2000. Climate change and UK hydrology. In: The Hydrology of the UK: A Study of Change [Acreman, M.C. (ed.)].Routledge, London, United Kingdom, 3–29.

Boulton, A. J. and P. S. Lake. 2008. Effects of drought on stream insects and its ecological consequences. In Lancaster, J. & R. A. Briers (eds), Aquatic Insects: Challenges to Populations. CAB International, Wallingford, UK.

Brazos River Authority, 2002. Brazos River Authority's Annual Water Quality Report. http://www.brazos.org/crpPDF/FY02BasinHighlightsExecutive%20Summary.pdf. Accessed 28 October 2007.

Briand, J., S. Jacquet, C. Bernard & J. Humbert, 2003. Health hazards for terrestrial vertebrates from toxic cyanobacteria in surface water ecosystems. Journal of Veterinary Research 34: 361–377.

Burke, E. J., Brown, S.J. and Christidis, N. 2006.Modelling the recent evolution of global drought and projections for the twenty –first century with the Hadley – Centre climate model. Journal of Hydrometeorology 7 (5) : 1113 - 1125.

Cai, W. and Cowan, T. 2008. Evidence of impacts from rising temperature on inflows to the Murray-Darling basin. Geophys Res Lett 35: L07701.

Ciais, P., Reichstein, M., Viovy, N., Granier, A. et al. 2005. Europe-wide reduction in primary productivity caused by the heat and drought in 2003. Nature 437: 529"533.

Colley, J. 2004. Drought Preparedness Council Biennium Report 2003–2004. Governor's Division of Emergency Management, Austin, Texas.

Critchfield, H.J.1983. General climatology. Fourth edition, Prentice Hall, New York. 453p.

Dai, A., Trenberth, K. E. and Qian,T. 2004. A global data set of Palmer Drought Severity Index for 1870-2002: Relationship with soil moisture and effects of surface warming. J. Hydrometeorology 5: 1117-1130.

Davies, A. W. 1978. Pollution problems arising from the 1975–76 drought. Proceedings of the Royal Society of London. Series A, Mathematical and Physical Sciences 363: 97–107.

Falkenmark, M. and G. Lindh, 1976: Water for a Starving World. Westview Press, Boulder, CO, USA.

Furey, P. C., R. N. Nordin & A. Mazumder, 2006. Littoral benthic macroinvertebrates under contrasting drawdown in a reservoir and a natural lake. Journal of the North American Benthological Society 25: 19–31.

Golladay, S. W., P. Gagnon, M. Kearns, J. M. Battle & D. W. Hicks, 2004. Response of freshwater mussel assemblages (Bivalvia: Unionidae) to a record drought in the Gulf Coastal Plain of southwestern Georgia. Journal of the North American Benthological Society 23: 494–506.

Gosain, A. K. and, Sandhya Rao. 2004. Impacts of climate change on water sector. Climate change and India: Vulnerability assessment and adaptation [Eds. Shukla et al], University Press, New Delhi, 159-191.

Gosain, A.K., Sandhya Rao and Basuray Debajit. 2006. Climate change impact assessment on hydrology of Indian river basins, Current Science 90 (3): 346-353.

Gosain, A.K., Sandya Rao and Anamika Arora. 2011. Climate change impact assessment of water resources of India. Current Science. 101 (3) : 356 - 371.

Gupta, S.K. and Deshpande, R.D. 2004. Water for India in 2050: first order assessment of available options. Current Science 86: 1216-1224.

Gupta,S.K., Minhas, P.S., Sondhi, S.K., Tyagi, N.K. and Yadav, J.S.P. 2000. Water resource management. In : Natural resource management or agricultural production in India (Eds. J S P Yadav and G B Singh). Indian Society of Soil Science, New Delhi , 137- 244.

Ha, K., E.-A. Cho, H.-W. Kim & G.-J. Joo, 1999. Microcystis bloom formation in the lower Nakdong River, South Korea: Importance of hydrodynamics and nutrient loading. Marine and Freshwater Research 50: 89–94.

INCCA. 2010. Climate change and India: A 4X4 assessment : A sectoral and regional analysis for 2030s. Ministry of Environment & Forests, New Delhi. 160p.

IPCC. 2007. Contribution of Working Group I to the Fourth Assessment Report of the Intergovernmental Panel on Climate Change (IPCC). Summary for Policymakers. Geneva, Switzerland.

Jain, S.K. 2012. India's water balance and evapo- transpiration. Current Science. 102(7): 964-967.

Jenkins, K. M. & A. J. Boulton, 2007. Detecting impacts and setting restoration targets in arid-zone rivers: Aquatic micro-invertebrate responses to reduced floodplain inundation. Journal of Applied Ecology 44: 823–832.

Lake, P. S. 2003. Ecological effects of perturbation by drought in flowing waters. Freshwater Biology 48: 1161–1172.

Lorenzo-Lacruz J, Vicente-Serrano SM, López-Moreno JI, Beguería S, García-Ruiz JM, Cuadrat JM. 2010. The impact of droughts and water management on various hydrological systems in the headwaters of the Tagus River (central Spain). J Hydrol (Amst) 386: 13"26.

Medeiros, E. S. F. & L. Maltchik, 1999. The effects of hydrological disturbance on the intensity of infestation of Lernaea cyprinacea in an intermittent stream fish community. Journal of Arid Environments 43: 351–356.

Narasimhan, B. and Srinivasan, R. 2005. Development and evaluation of soil moisture deficit index (SMDI) and evapotranspiration deficit index (ETDI) for agricultural drought monitoring. Agric. Forest. Meteorol., 133 : 69–88.

Nicholls N (2004) The changing nature of Australian droughts. Climate Change 63: 323-326.

Peter Gleick, with coauthors Lucy Allen, Juliet Christian-Smith, Michael Cohen, Heather Cooley, Matthew Heberger, Jason Morrison, Meena Palaniappan, and Peter Schulte. 2011. The World's Water Volume 7: The Biennial Report on Freshwater Resources. Island Press, Washington. 440p.

Rao, A.V.M.S., V.U.M. Rao, Satyanarayana, T., Madhu Malathi, E. and Manikandan, N. 2010. Drought situation under projected climate change scenarios in Andhra Pradesh. Proceedings of the International Conference on Climate Change Perspectives and Projections – A Systems Approach . December 9-11, 2010. Osmania University, Hyderabad, 106p.

Trenberth K.E., Jones P.D., Ambenje P., Bojariu R., Easterling D., Klein Tank A., Parker D., Rahimzadeh F., Renwick J.A., Rusticucci M., et al. 2007. Observations: surface andatmospheric climate change. In : Climate Change 2007. The Physical Science Basis, Contribution of WG 1 to the Fourth Assessment Report of the Intergovernmental Panel on Climate Change, Solomon S., Qin D., Manning M., Chen Z., Marquis M.C., Averyt K.B., Tignor M. and H.L., Miller (Eds.), Cambridge University Press, Cambridge, New York, 235–336.

UNEP. 1999. Global environment outlook 2000 (GEO). UNEP, Earthscan, London.

Vicente-Serrano, S.M., López-Moreno, J.I., Anita Drumond., Luis Gimeno., Raquel Nieto., Enrique Morán-Tejeda., Jorge Lorenzo-Lacruz., Santiago Beguería and Javier Zabalza. 2011. Effects of warming processes on droughts and water resources in the NW Iberian Peninsula (1930"2006). Climate Research. 48 : 203-212.

Walter, M.T., Wilks, D.S., Parlange, J.Y., Schneider, R.L. 2004. Increasing evapotranspiration from the conterminous United States. J Hydrometeorol 5: 405-408.

WMO. 1997: Comprehensive Assessment of the Freshwater Resources of the World. World Meteorological Organization, Geneva, Switzerland, 34p.

CHAPTER 7

Drought Management Strategies

The capacity of droughts in ruining the nation's economy irrespective of its size and status is well understood in recent times. The impact is more felt by the nations that are located in the tropical regions which are prone to various intensities of droughts during the rainy season. With the advancement in atmospheric sciences, better understanding of weather could be achieved in recent period by developing models and advanced computational procedures. Also greater cooperation in exchange of information among the neighboring countries and data sharing by the international organizations like WMO, FAO, UNDP etc, made it possible to develop suitable plans before and during the season for managing the climate risks associated with droughts. Over centuries, human settlements on this earth have evolved techniques for managing water deficits and still many of them are relevant even today. Initially, the early civilizations efficiently setteled only along the river banks where assured water supply round the year was available. Similarly construction of tanks, spill over or earthen dams, flood water diversion canals, digging of wells etc., are some of the measures taken by the early settlers to manage droughts. In recent years, based on long term meteorological data analysis many countries have evolved drought management plans and drought proofing technologies that are tested at several locations. The respective governments have responded to droughts by providing emergency, short and long term assistance to the affected regions. Some of the palaces, storage water reservoirs and roads were constructed by the earlier rulers in India as drought relief measures.

Good Practices for Drought Management

Based on the experiences gained from the countries in Africa and Asia, the following issues viz. Good practice for Drought Risk Management were suggested under UNDP program by Paul Venton (2012).

- There is a need for an integrated development approach centered around sustainable land /water resource management practices.
- The importance of indigenous knowledge (not only as a specific area of importance, but as a thread running throughout all community-based considerations) in drought management strategies should be given priority.
- The importance of numerous methods of raising local community awareness about drought-related issues should be given wider publicity
- The importance of a community-led participatory approach and use of community organization, (e.g., farmer's groups and water user association), especially in relation to sustainable local natural resource management may be encouraged.
- A need is felt for a stronger focus on diversification of livelihoods, including crop and livestock varieties and other income-generating activities.

This chapter briefly deals with various methods/technologies adopted in drought management studies by different countries including India.

7.1 Drought Planning and Preparedness

To face any eventuality arising due to climate hazards like drought, a sound planning based on available resources is essential. Similarly, people in the drought prone regions must be alerted before the occurrence of drought and actions to be initiated should also be finalized. Drought planning can be defined as actions taken by Government, industry, individual citizens and others in advance of drought for the purpose of mitigating some of the impacts associated with its occurrence (Wilhite – 1987). He has indicated the following points for planning for drought.

1. Development of early warning system for droughts consisting of its onset, continuation, termination, severity and spatial extent for the decision makers at all levels to take appropriate steps.
2. Assessment of the reliability of likely impacts of drought on various sectors.
3. Establishment of an authority for coordinating the government actions and to liaise within and between various organizations including drought declaration and revocation criteria.
4. Development of appropriate drought assistance program, determine eligibility criteria and implementation procedures.

5. Identification of financial resources to maintain operational programs and to strengthen the existing Research activities for accurate assessments and development of appropriate technologies.
6. Conducting drought awareness campaign for the farmers to promote drought mitigation strategies.
7. Reassessment of drought prone regions in view of changes observed through extensive use of Remote Sensing Applications and GIS.
8. Involvement of Stake holders at different levels in preparation of drought management plans that are viable and economically beneficial and last for longer periods.

Drought management is often viewed by many nations as crisis management rather preparing for long term permanent plans. Many countries in the world concentrated mostly on developing short term strategies only. Except very few, long term crop plans pertaining to weather and soil have not yet been initiated. Crisis management often results in implementation of a hastily prepared procedures which may lead to ineffective, poorly coordinated and untimely response, (Wilhite, 1987). If planning were initiated between the periods of droughts, the problems related to droughts can effectively be addressed with limited financial resources.

The drought assistance measures that have been evolved over many years and majority of the decisions are politically motivated rather than of economic reality. Such measures taken by the countries are found to be ineffective. Hence, for the purpose of drought planning, the objectives must be stated explicitly. Drought preparedness policies were initiated by Australia and South Africa and they have formulated plans and programs to initiate the work. Wilhite (1987) proposed three objectives for preparation of national drought policy.

i) Assistance measures should not discourage agriculture producers and other groups from the adaptation of appropriate management practices that help to alleviate the effects of droughts.

ii) Assistance should be provided in an equivatable, consistent, and in predictable manner to all without regard to economic and geographic region.

iii) Importance of protecting the natural and agricultural resource base must be recognized.

To improve drought assessment and response capability, the experiences of Australia, South Africa, Canada and United States have been reviewed and the following recommendations were made (Wilhite, 1987).

1. Reliable and timely information products (advisories, reports, forecasts) and information about dissemination plans must be developed.

2. Impact assessment techniques must be improved and new tools be developed to provide decision makers in government and business with the types of information necessary to identify the onset and termination of drought and to understand the severity and to its likely impacts. These tools shall help to identify periods of abnormal risks to initiate the assistance measures.

3. Criteria must be determined in advance of drought, well publicized when drought occurs and applied consistently to all affected localities. Revocation procedure must be similarly defined.

4. Assistance measures must be developed in advance of drought that is, a proactive approach should be taken to avoid the delays in program formulations and approval.

5. The degree of fiscal and administrative responsibility that the states should have in support of assistance measures may follow the Australian approach of cost sharing. Such approaches would allow the states to have greater control over assistance measures.

6. More attention should be directed to the development of assistance measures that encourage producers to incorporate appropriate levels of risk management in individual farm plans. The recipients would benefit from the advanced information on drought aid that they are likely to receive for planning appropriate actions.

Wilhite *et al.* (1984) recommended a greater involvement of State Governments in Drought planning to complement and facilitate the government efforts. In Canada the organizational structure set up to deal with emergencies such as droughts was ignored during 1976-77 (Liverman, 1980) because of different jurisdictional units administered by various government agencies that had different boundaries. Wilhite (1987) observed that the Australian constitution delegated specific powers to federal government and took more active role in preparing appropriate action to plan for drought, while in US, the powers are with the State governments. The number of states with drought plans in USA had increased from 3 in 1982 to 28 in 1996 (Wilhite-1997). In addition to the 28 states that already have plans, five more states viz., Alabama, Oklahoma, Louisiana, Texas and New Mexico have prepared plans totaling to 33 states. The pattern of drought planning is so complex that it cannot be explained adequately on the basis of drought climatology alone. Therefore, drought plan should be prepared based on climatic, political, economic, environmental and demographic factors. The situation has not improved over the globe even by the year 2012. As per UN weather agency, Sivakumar, Director, WMO observation, only Australia has a national policy towards drought in the world (**world.time.com**, 23 August 2012). Therefore there is a need for all the countries in the world to gear up action plans to manage climate risk efficiently. Climate change is projected to increase the frequency, intensity and duration of droughts, which have impacts on many sectors in particular, food, water, health and energy – W.M.O. Secretary–General, Michel Jarrund observed that "we need to move away from piece meal crises driven approach and develop integrated risk-based nation drought policies" (world.time.com 23/8/2012).

7.1.1 Drought Planning in India

In India, 21 famines occurred between1770-1880 and a Commission was set up by the British Government under the Chairmanship of Colonel Baird Smith to look into the problems of 1860-61 famine of Orissa and Bengal and recommended development of irrigation systems, improvement in communication and remission of land tax in 1860-61. His observations on famines are relevant even today, "Indian famines are famines of work rather than of food, when work can be had and paid for, food is always forth coming". Thus the genesis of famines in India in the past did not lie in food shortage but in the shortage of purchasing power (Sinha *et.al,* 1987). On the basis of the Famine enquiry commission under General R. Stratchey in 1878, observed that "The paramount duty of the state was to offer protection to the people of India from the effects of uncertainty of the seasons". Though the commission recommended measures to improve production and assistance, the Government did not set up food supply measures and regulations of trade and prices during famines.

A famine code was promulgated during 1883 and was adopted by several provincial governments. The main points of the code were:

1. Opening of relief works in famine area
2. Prescribing the Scales and Systems of wages for labor employed on relief works.
3. Providing for the treatment of destitute women, children, ill and elderly victims.
4. Grant of "taccavi" loans to the farmers for resuming agricultural activities at the end of famine.
5. Remission and suspension of payment of land revenue demand.
6. Supply of grain and fodder the cattle.

The famine relief measures were to start only after the crop damage, since the drought declaration by the Government followed Annawari system which is expressed as the loss in crop production. Famine codes were used in post-independent India also during 1965-1966 drought periods.

In India, drought management responsibility lies with the Central Government for providing the necessary financial aid to the State Governments that were affected by drought. Though the State Governments declare the drought affected regions based on the criterion of distribution of rainfall, area sown and the water levels in the reservoirs, the Central Government's sanction of financial aid depends on the reports submitted by the expert team's visit to affected regions. The State Agricultural Department in collaboration with SAU and ICAR institutions suggest the measures likely to be taken immediately based on the contingency crop plans prepared by the respective Agricultural Universities of the state.

National level drought management policies with respect to agricultural crop planning, alternate sources of livelihoods, water management issues, infrastructure development, literature awareness, relief to farmers against the impacts of droughts have yet to be formulated by the GOI. The recommendations of workshop on setting up a "Mechanism for Continuous and Integrated Management of Droughts "by the Ministry of Agriculture (MOA) of GOI during 21st September 2007 has highlighted the need for preparation of a National policy on drought for its better management. The recommendations of the workshop with reference to policy and themes are as follows.

Policy Imperatives / Themes of GOI

1. Department of Agriculture & Cooperation, being the nodal Department on drought, as recognized by the National Executive Committee of Ministry of Home Affairs, under the Disaster Management Act 2005, should evolve a common policy to dovetail short-term relief measures into long-term interventions being handled in different Departments / organizations for comprehensive and all inclusive drought management.
2. Policy on Drought Management should be simple, dynamic and actionable in planning right interventions at different stages of drought evolution to help mitigate the impact of drought on the affected population. The policy should use science and technology inputs, and community based practices to promote better coping strategies.
3. There is a need to create trigger points and an appropriate response matrix for implementing various short-term and long-term interventions. Objectivity should be part of District and State drought plans that would

be drawn up on the model guidelines prepared by the IDMC (India Drought Management Centre).

The Drought Management Division (DMD) of MOA of GOI has brought out a crisis management plan 2011 to deal with the drought situation by the Central and State Governments to minimize its impacts. The early warming indicators of drought for *kharif* and *rabi* season as identified by DMD are as follows:

For *Kharif* season (June to August)

i. Delay in onset of southwest monsoon
ii. Break in monsoon
iii. Insufficient rains during July
iv. Rise in fodder price
v. No rise in reservoir levels
vi. Drying of tanks used for rural drinking water supply
vii. Fall in the progress of sowing over successive weeks compared to figures of normal years.

For *Rabi* Season (November to January)

i. Deficiency in Southwest monsoon rainfall values
ii. Depletion of ground water levels compared to normal years
iii. Fall in reservoir level compared with normal years
iv. Indication of marked soil moisture stress
v. Price rise of fodder
vi. Increased deployment of water through tankers.

An expert committee on "Drought Prone Areas Program (DPAP) and Desert Development Program (DDP) headed by Hanumantha Rao (1994) delineated desert and drought prone districts from the arid, semi-arid and dry sub humid regions based on percent area under irrigation. In order to reduce the biotic pressure, the committee felt the need to improve the soil moisture conditions by developing watersheds at appropriate places to minimize the impacts of drought within a 10 year period as a policy to be taken up by GOI. The committee suggested the funding patterns for a 500 ha watershed area and the ratio of sharing of expenditure between Central and State governments to encourage the development of assured moisture supply in different agro climates as follows:

For DPAP the funding parttern is 50% by central and 50% by state governments. Incase of DDP it vary 75 to 100% funding by Central Government

Table 7.1: Suggested funding patterns for watershed programs

Ecosystem	Major problem	Funding pattern		
		Per hectare (average) (Rs)	Micro watershed of 500 ha (Rs. in lakhs)	GOI : SHARE
Hot sandy arid (DDP)	i. Sandy movement ii. Degraded pasture iii. Receding ground-water iv. Extreme moisture stress	5000	25.0	100
Hot arid (DDP)	i. Receding ground –water ii. Degraded pastures iii. Extreme moisture stress iv. Soil erosion	4500	22.5	75:25
Cold arid (DDP)	i. Deforestation ii. Poor water management iii. Lack of efforts on vegetable and fruit production	5000	25.0	100
Semi-arid (DPAP)	i. Receding ground-water ii. Increased moisture stress iii. Soil erosion iv. Lack of vegetal cover	4000	20.0	50:50
Dry sub-humid (DPAP)	i. Moisture stress ii. Soil erosion iii. Lack of vegetal cover	3000	15.0	50:50
Dry sub-humid (DPAP) (Hill region)	i. Moisture stress ii. Soil erosion iii. Lack of vegetal cover	4000	20.0	50:50

Source: Hanumantha Rao (1994)

7.2 Global Experiences in Drought Planning

Increased occurrence of droughts in recent years in Africa, Asia, Australia, Brazil, China, India and SE Asian countries demanded comprehensive plans at national level for managing droughts efficiently. Drought plans prepared by different nations are briefly discussed here.

7.2.1 USA

In the US, the states are clearly the policy innovators for drought management (Wilhite, 1991 a). The basic goal in State Drought plans is to improve the effectiveness of the state response by increasing the monitoring status of drought and developing drought preparedness plans, recovery and mitigation program at micro regional level. These plans improve the coordination within the agencies of State Government and between State and Federal Government (Wilhite, 1997). Municipal and Agriculture sector are primarily the principal sectors under State level drought plans while others are viz., migration of both human and livestock population, development of water resources and the necessary infrastructure.

As an example, Colorado State level drought plan was developed under the Authority of the Colorado Natural Hazards mitigation council (Truby & Bonlas, 1994). The State initiated development of comprehensive plans in 1981 and the plan was later updated during 1986 and 1990. The plan is Successfull as it incorporates three primary components viz., 1) Monitoring System 2) Impact assessment system and 3) Response system. The responsibility for monitoring the availability of water resources is with the Water Availability Task Force. They assess the monthly projection of snow packs, soil moisture levels, reservoir and ground water levels, precipitation, temperatures and stream flows collected by different agencies. This information can provide early warning signals to help the state to gear up for drought situation. Activation of drought plan is triggered by the assigned values of three indices viz. modified PDSI, water supply index and the SPI. The assessment system of drought plan in the above has eight different task forces covering the water related areas viz., 1) Municipal water, 2) Wild fire protection, 3) agricultural industry 4) tourism, 5) wild life, 6) economic, 7) energy loss and 8) health. The goal of each task force is to identify the existing and potential drought related problems and assess possible impacts on society. A Final Task Force shall review all assessments of impacts of different task forces and report information to policy makers, media and others. The Response System will deal with the unfinished needs identified by the Special Task Force. The agency is most closely associated with appropriate response and is assigned the responsibility to take action and enlist the cooperation of other agencies as necessary. These lead agencies are identified and for complex emergency responses, an inter-agency coordinating group is established. As drought begins as per the criteria, the state can immediately start mitigation measures to cope up with drought impacts.

7.2.2 South Africa

In South Africa, the responsibility of management of Agricultural drought lies with the National Department of Agriculture (NDA) and its regional branch at provincial and district level. Before the year 1980, Agricultural Drought in South Africa was managed as an abnormal disaster event that required emergency government assistance. The crisis management focused mainly on providing relief to livestock sector in areas declared as Disaster Drought (DD) areas with little provision extended to assist crop farmers. (Rashid Hassan, 2010). It was further reported by him that major changes in government policies since 1994 have emphasized a shift towards a more proactive approach that adopts risk reduction rather than the reaction to drought relief strategy. The country developed comprehensive inter-basin water infra-structure to move water from surplus to deficit regions. The Department of Water Affairs (DWA) is responsible for managing Hydrological drought. Guidelines for managing bulk water supply system were as per the guidelines stipulated by National Water Policy.

The Water Policy places higher priority to meet the basic requirements for human needs and ecological demand, where lowest curtailments are imposed. The National Water Act of 1998 gives priority in water allocation to strategic users such as power generation and key industries. Irrigation suffers greatest reduction in water supply during drought period. One of the agricultural drought management policies is to provide subsidy to encourage stock reductions among live stock producers. Plans for access of assured water supplies to the rural poor and their live stock is the major mandate of the government polices.

7.2.3 African Nations

A UNDP/UNSO project was initiated in 1997 to assess the status of drought preparedness and mitigation activities, constraints for developing drought plans, primary drought policies and planning needs in selected Sub–Saharan African nations viz., Angola, Botswana, Lestho, Malawi, Mauritius, Mozambique, Namibia, South Africa, Swaziland, Zambia and Zimbabwe. Botswana and South Africa clearly stand apart from the other countries in this assessment. Common themes on current status of drought preparedness in the above countries as quoted by Wilhite (2003) are as follows:

- No permanent Government body to deal with drought issues.
- Drought response is often coordinated through natural disaster authorities.
- Formal drought plans are rare and mainly directed as response actions.
- Drought policies are usually lacking.

The status of some of the constraints in preparing drought policy and plan are as follows:

- Lack of institutional capacity.
- Poor quality of meteorological network.
- Limited coordination between governmental agencies.
- Low level involvement of NGOs in drought management.

7.2.4 Namibia

The Government of Namibia developed a series of drought policy recommendations based on the 10 step drought planning process developed by Wilhite and Svoboda (2000). Namibia has large number of livestock, therefore a three-pronged survival strategy consisting of anticipatory measures, drought crisis management and drought after-care for livestock management, was proposed by Axel Rothauge (2001). As a drought preparedness measure the farmers were advised to follow forecasts issued by meteorological services, accumulate fodder and establish pastures of indigenous grasses and drought resistant fodder plants. Drought crisis management consists of progressive destocking in accordance with the fodder availability and its reserves. Drought after- care consists of no grazing until it starts raining and restocking progressively in accordance with improving carrying capacity

7.2.5 Ethiopia

The Ethiopian Government has taken important steps for better management of droughts, with the establishment of productivity safety net program (PSNP) in 2005 by the donors. The Government has initiated program beyond the "Food first" culture to ensure higher levels of livelihood support (Humanitarian Exchange Magazine, March 2012). A Disaster Risk Management Technical Working Group (DRMTWG) was established to coordinate Sectorial Task Forces including Agricultural Task Force (ATF) which covers hazard management for drought, flood, livestock disease, crop pests etc. The government also developed a draft Disaster Risk Management policy and Strategic Policy Investment Framework (SPIF) which were meant to manage droughts better than even before.

7.2.6 Botswana

Drought Relief Planning in Botswana was evolved as a result of critical evaluation of 1979-80 drought relief program as reported by Tswelopele & Moremi (1987) has the following components.

a) Human relief

It has four elements viz.,

1) Supplementary food distributions for vulnerable groups on a monthly basis
2) Primary school feeding program
3) Direct feeding for malnourished persons
4) Logistic support to transport food supplies to remote areas

b) Labour based relief

Temporary employment of rural people, selected by their village development committees for construction of structures viz., (dams, airstrips, buildings, roads, storage structures)

c) Water relief

Allocation of special funds for repairing of existing water systems, transport water to the settlements, to livestock and to create small scale irrigation facilities.

d) Agricultural relief

Free distribution of seeds to farmers and subsidy on stock food, cash loans to farmers and heavily subsidized draft power for households.

Drought management plans in Europe

Drought management plan of a few countries in Europe as given in the EU report (2008) are briefly given in the following pages.

7.2.7 Spain

The earlier Water Act (1985) developed by the Government of Spain gave certain responsibilities to Reservoir Committees of River Basin Authorities to manage water shortages in agreement with the rights. Modifications were brought in the original Water Act as it was found inadequate to take drought risk management measures. It was recommended that the River Basin Authority may set up Water Interchange Centre to enable the user rights to be waived by voluntary agreement. The measures to be initiated after the declaration of drought depend upon its Severity. The measures are grouped into structural measures (New pumping wells, new pipes and use of desalination plants) and the non-structural measures (Water saving by applying restrictions to the users, increase the use of ground water). It was further mentioned that the Directorate

General for water coordinates and joins with River Basin Authorities and approves the drought plans. Based on the Hydrological Indicator System (HIS) threshold values, monthly maps of drought situation shall be developed and drought plans activate immediately and serve as a reference framework for specific urban supply plans. The urban water supply authorities have to prepare drought emergency plans and implement the same after drought warning is declared by River Basin Authority.

7.2.8 England and Wales

The Department of Environment, Food and Rural Affairs in England and the Welsh Assembly Government in Wales have policy responsibility for the legislation that controls water resources, which includes the law relating to hose pipe ban, drought permits and drought related work orders. It works closely during drought with the Environmental agencies and the water companies to ensure proper water supply, without any compromise on an environmental safety.

The environmental agency has to prepare drought plans based on the following points

- Role of environmental agency in managing water resources during drought period.
- Requirement of additional environmental monitoring system.
- Development of frame work for liaison with water companies, awareness campaigns and determination of drought permits.
- Advise the Senior Government officials on the prospects and possible actions.

The water companies in England and Wales have produced drought plans since 1999. With the approval of Water Act-2003, it is a statutory requirement for water companies to prepare, maintain and publish the drought plans. Depending upon severity of drought, campaigns to reduce consumption by the public and others shall be conducted. With further increase in severity, the water companies issue drought orders for change in discharge rates that could restrict the non-essential uses of the public supply. The trigger curves for drought management in England are shown in the following Figure 7.1.

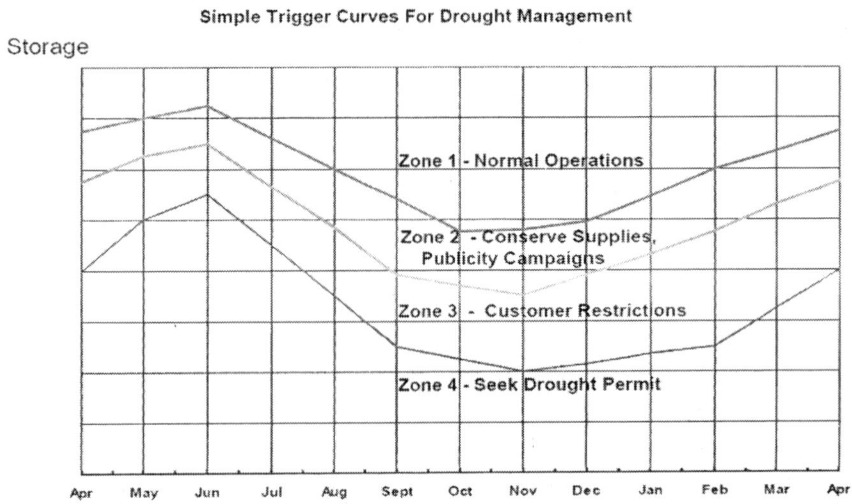

(*Source*: EU report-2008)

Fig. 7.1: Simple trigger curves for drought management in England

7.2.9 Portugal

The Water Institute in Portugal is the National Body that manages water resources in which drought is included. The Drought Commission prepares drought plans for the political and strategic issues and the Technical Secretariat for the technical and operational issues. The main activities of Drought Commission are:

- Development of communication system
- Identification of agencies responsible for implementing measures for drought management based on the recommendations by National Program for efficient water use
- Preparation of budget requirements and submit for approval by the legislative body
- Identification of measures that prevent forest fires
- Activation of emergency management group
- Elaboration of the Drought Balance report after the Drought for the activities of Technical Secretariat
- Fortnightly monitoring of drought situation
- Asses the technical and financial requirement of agencies associated with the program.

7.2.10 Australia

The earlier drought policies developed in Australia were based on providing direct subsidies and other forms of support. Such methods were phased out in 1992 and a new National Drought policy was developed by Australia's Common Wealth and State Governments. The core principle of managing droughts in Australia moved the drought policy away from a subsidy-based to reactionary or crisis driver approach. In this process drought is considered as a normal part of the Australian environment and Government developed a policy of Drought Exceptional Circumstances (DEC) allowing welfare support for affected producers which includes the welfare of animals and protection to environment. The plan was formulated on the following three principles (Greg Laughlin and Anthony clark-2000).

1. Encourage primary producers and other sections of rural Australia to adopt to self-reliant approach.
2. Maintain and protect Australia's agricultural and environmental resource base during periods of extreme climate stress and.
3. Ensure early recovery of agricultural and rural industries, consistent with long term sustainable level.

7.2.11 China

Primary responsibility for drought management and flood control lies with Flood Control and Drought Relief Headquarters (FCDRH) established in 1955 under the Ministry of Water Resources. It also implements China's drought relief policy which deals with drought prevention, relief and mitigation (ABARES-2012). Drought Relief Planning takes into account the needs of economic and social development, comprehensive development of water resources and the water needs of urban and rural population, agricultural and industrial production. The government at the district level is responsible for strengthening of water conservation works, rural drinking water projects.

7.2.12 Palestine

Droughts continue to threat the livelihood of Palestinian both urban and rural population. Strategies for drought management and prevention includes development of sustainable agricultural practices, soil conservation, crop diversification and integrated watershed management. Qasen Abdon (2007) proposed the following policies to improve the capabilities of the Palestinians to respond to different intensities drought.

- Preparation of risk and vulnerability profile maps
- Development of mitigation and management plans to combat droughts
- Strengthening and developing technical and institutional capabilities to reduce the impacts of droughts.

7.2.13 Global Initiatives

The UN Convention to Combat Desertification (UNCCD) emphasized the importance of improving the early warning systems and seasonal climate forecasts and developing drought preparedness plans. In this endeavor, the National Drought Mitigation Centre (NDMC) at the university of Nebraska has been working in partnership with the United Nations Secretariat of the International Strategy for Natural Disaster Reduction and other organizations to develop Regional Networks on drought preparedness and then to link these networks into Global Drought preparedness network (GDPN) Wilhite (2003). The GDPN provided the more opportunities for nations and regions to share experiences and lessons learned on drought policies, emergency response measures, methodologies, stakeholder involvement and development of early warning system, impact assessment methods and procedures for addressing environment conflicts through virtual networks.

7.3 Drought Preparedness Action Plans in India

Drought preparedness mechanism consists of 1) Drought monitoring, 2) Assessment of vulnerability of regions to various intensities of drought and their frequencies, 3) Agencies responsible for coordinating and implementation of the drought activities at regional and national level, 4) Availability of Regional level plans and relief measures, 5) Development of climatic risk management strategies to minimize crop loss, 6) Crop contingency plans for different weather conditions, 7) Development of Decision Support Systems (DSS) for management of drought, 8) Awareness of droughts and the local climate knowledge among the farming community *etc.* Since many of these topics have already been covered in earlier sections of this book, the preparedness plans followed by various agencies in India are briefly described. Drought preparedness plans and their management in India is addressed by the following mechanisms and sectors (Singh, 2001).

Governmental level

- Policy issues, national, regional, and district level
- Rural development infrastructure

- Input supply, marketing and farm advisory services.

Non-Governmental level

- Implementation of drought management strategies in the farmer's fields by various organizations such as rural institutions, local self-government, Private sector, Philanthropic organizations etc.

Research and Development Institutions

- Best practices for rainwater and soil management through linking on-station and on-farm research
- Weather forecasts
- Contingency crop planning / mid-season corrections.

Drought declaration being the state subject in India, the State Government usually gears up the district administration to procure adequate quantities of seeds, fertilizers and diesel for pumpsets to derive benefits from the early onset of monsoon rains. De-silting of tanks, ponds, repairing, and renovation by utilizing the opportunities under National Rural Employment Guarantee Act (NREGA) and other schemes are taken up before the commencement of rainy season. The district administration would be also advised to keep seeds of short duration varieties to cope up with delayed sowing or re-sowing due to poor germination or failure of crop due to early season droughts.

The Drought prepared mechanism and various actions to be taken by Relief Commissioner and District Collectors that are chronologically reported in the Manual for Drought Management (2009) is given below:

7.3.1 Drought Preparedness Checklist

Drought preparedness and response checklist for Relief Commissioner and District Collector is as follows

To activate State/district disaster management authorities*:* Following action points need to be checked upon the occurrence of a drought situation:

- Meetings of State / district disaster management authorities have been convened.
- Information has been obtained on all the key indicators of drought.
- Proposal has been prepared for the declaration of drought.

- Relief Commissioner/District Collector has developed a strategy for drought management in active consultation with all the participating heads of departments.
- A contingency plan for drought management has been developed.
- A control room for drought management has been set up.
- All the departments / agencies have prepared their own contingency plans.
- Memorandum has been prepared and submitted to the Government of India for assistance from the NCCF.
- Necessary funds/ authorization for drought relief measures have been obtained.

The relief measures to be taken up by monitoring the drought indicators by different agencies as given in Mannual for Drought Management (2009) are summarized in the Table 7.2.

As a strategy to improve production under all-weather situations, Agricultural Universities located in each state generally prepare a crop calendar for each year which contains the cropping strategies and the recommend new varieties suitable for different regions in the state. The publication is mostly released before the onset of the season. The GOI decided to prepare District level contingency agricultural plans to gear up District administration to meet any climate situation *Viz*: droughts, floods, heat, cold waves and hail etc. CRIDA on behalf of ICAR was given the task of preparing District level Contingency plans in consultation with the respective SAU. The program was initiated during 2009 and initially the plans for 209 districts were ready by June 2011 and are made available to the district authorities, which had helped them to take appropriate actions to meet the drought conditions occurred in various parts of the country during 2012 season.

Dr. Swaminathan in his drought management and rural livelihood security program plans developed in August 2009 identified the following actions to be taken to manage phenomenal drought occurred during 2009. He has classified the drought affected areas into two categories *viz:* Most Seriously Affected (MAS) areas and Most Favorable Areas (MFA). Apart from relief farm operations, the urgent needs of MSA are 1) Saving farm animals from distress sale by arranging animal camps near water sources, where animals can be fed with agricultural residues. 2) Growing short duration crops like Saathi maize (60 day maize), sweet potato, pulses, oilseeds and other less water requiring crops. 3) A Pond in Every Farm using National Rural Employment Guarantee Act (NREGA). Atleast five cents in every acre should be reserved for

Table 7.2: Drought relief measures to be implemented by different agencies

Drought indicators	Agencies responsible for monitoring	Indices to be monitored	Drought relief measures to be organized
Monitoring Water Resources:	Departments of Revenue, Irrigation, water supply and water conservation.	Daily, weekly and monthly rainfall, ground water sources, water storage levels in reservoirs/ ponds/lakes	• Reservation orders have been issued for drinking water • Water is being equitably distributed for multiple purposes: drinking, commercial, industrial and agricultural. • Temporary water supply schemes have been sanctioned. • Supply of drinking water is being organized through tankers/bullock carts/ trains. • A ban on digging deep bore-wells has been enforced.
Crop Management	State departments of agriculture and revenue, state agricultural universities	Soil moisture, area under sowing and type of crop, status of growth, crop yield, alternative cropping possibilities. •	• Seeds for second sowing have been secured and supplied. • A crop contingency plan (less- water consuming, drought-resistant crops) has been developed. • Crop insurance premium has been paid by the Government. • Micro-irrigation equipments (sprinkler and drip irrigation systems) are available to the farmers for using water optimally.
Fodder management	State departments of animal husbandry, agriculture and revenue	Availability of fodder, fodder prices, fodder cultivation, availability of water for fodder cultivation.	• Supply of fodder increased through getting fodder from surplus States /districts.

Drought indicators	Agencies responsible for monitoring	Indices to be monitored	Drought relief measures to be organized
			• Coordination mechanism set up with the forest department and Agriculture University. • Farmers to get surplus fodder. • Fodder cultivation encouraged and incentives provided through Government schemes. • Ban imposed on taking fodder from the State. • Fodder depots set up and the prices fixed at levels, which farmers can buy. • Cattle camps set up through the Government, NGOs and cooperative societies. • Water supply arranged for cattle camps. • Vaccination and other health measures organized for cattle.
Relief Employment	Department/ agencies implementing the National Rural Employment Guarantee Schemes (NREGS) and other schemes such as Pradhan Mantri Gram SadakYojana (PMGSY) and Swaranjayanti Gram Swarozgar Yojana (SGSY)	Demand for relief employment, number of relief works on shelf, number of works in progress, attendance of workers, wages distributed.	• Demand for relief employment assessed. • Job cards available with the people. • Relief employment plan prepared in consultation with the agencies. • Number of sanctioned works available on shelf. • People informed of their rights and entitlements under the NREGS.

(*Contd.*)

Drought indicators	Agencies responsible for monitoring	Indices to be monitored	Drought relief measures to be organized
			• Relief works started in response to the people's demands. • All the amenities organized on the sites of relief works. • Distribution of wages supervised to ensure it is timely and fair. • Attendance on relief works reported on a daily/weekly/monthly basis.
Food Security:	Food and Civil Supplies Corporation/Department, Food Corporation of India (FCI)	Availability of food grains, food grains prices in open market, free movement of food grains, practice of hoarding food grains, stock of food grains in FCI and Government warehouses, functioning of Public Distribution System (PDS) shops, and PDS price of kerosene oil.	• Enough food grains available in the open market. • Drive against hoarding food grains initiated. • Stock of food grains available in FCI and Government warehouses adequate. • PDS shops lifting food grains from availability of kerosene oil, open warehouses as per the number of ration card holders. • PDS shops selling food grains at the rates fixed by the Government. • Access to food grains for all ration card holders. • Food grains available for distribution as wages on relief employment works.

(Contd.)

Drought indicators	Agencies responsible for monitoring	Indices to be monitored	Drought relief measures to be organized
			• Kerosene oil available through the PDS system. • Food supplies augmented through monitoring of other social security schemes: Antyodaya Anna Yojana, National Old Age Program, Annapurna Scheme, Integrated Child Development Services, Mid-day Meals for school children.
Distribution of Relief Assistance	State revenue and agriculture departments	Funds available through CRF and NCCF, funds allocated for distribution of inputs subsidy, distribution of input subsidy and gratuitous relief.	• Seeds stock available for distribution among the farmers. • Enough seeds available in the open market for agricultural operations

(*Source*: Mannual for Drought Management, 2009)

construction of pond. 4) Wherever ground water resources are adequate in MSA area, subsidized power and diesel should be made available on priority basis. In the MFAs, where there is enough moisture for good crop, a compensatory production program can be launched by taking appropriate steps to increase the productivity of already sown crops. The following are some of the actions suggested by him.

- Drought, flood and good weather code should be prepared for every agro climatic region.
- Identification and training of two members from each Panchayat as climatic Risk Mangers.

Midterm actions

- Build a national grid of ultra-modern grain storage structures all over the country
- Establishment of village level seed, grain and water banks
- Enlarge food security basket by including wide range of millets and grains like Ragi (finger millets) in Public Distribution System (PDS).

The India Meteorological Department (IMD) gives its first forecast about the monsoon rainfall in the last week of April and revises it in 3rd week of June after evaluation of the weather conditions that existed after first forecast. These forecasts are helpful for the senior Government official to take up immediate actions in regions that are likely to receive deficit rains and prepare crop plans accordingly. It is the constant endeavor of Government of India to develop crops that are tolerant to droughts and to higher temperatures. Under their National Program on drought management, the Crop based Research Institutes of ICAR have a mandate to develop drought resistant and thermal tolerant varieties to cope up with the seasonal and intra seasonal weather aberrations. New technologies Viz. advanced breeding techniques are being employed to develop crops that are genetically modified to withstand droughts.

Other agencies *viz.*, Central Water Commission (CWC) that monitors the water levels in all the reservoirs and regulates water supplies at their respective irrigation commands in the country watches closely the availability of water for raising crops before and during the season. Farmer's Awareness program on crop management practices under the Climate Change initiatives taken by GOI; one day training camps for farmers have been organized regularly by ICAR and SAU through the Farmers Day celebrations at the respective university and at Krishi Vigyan Kendras (KVK). In addition to the above, about 25-50 Research Centers of All India Coordinated Research Projects on

Agrometeorology and of Dryland Agriculture of ICAR, combined together have been conducting climate awareness and technology demonstration programs of about 6 to 7 per year at each centre to combat weather extremes since 2010 under ICAR sponsored Research program on National Initiative on Climate Change Resilient Agriculture (NICRA).

Above all, the Integrated Agromet Advisory Services(IAAS) in India, a joint collaborative program of IMD, ICAR and SAU, weather based Agro Advisories at weekly interval are issued to the farmers from 130 Agro Met Field Units(AMF) fairly covering all the agroecological regions in the country. These advisories are useful in improving yields through appropriate agronomic, water management and pest/disease interventions. The information thus generated is disseminated through mass media communication systems like Newspapers, Television, internet and mobile services. Each SAU has its own website and it is mandatory to exclusively devote a page for weather forecast and agro advisories. With the exponential growth of cell phone technology even in remote rural areas, the weather forecast and related advisories are reaching the farmers at a greater speed to manage the climate aberrations effectively.

Few Governmental agencies *viz*., National Rainfed Area Authority (NRAA), National Institute of Drought Management (NIDM) and Drought Prone areas Program (DPAP) and Desert Development Program (DDP) are responsible for preparing drought preparedness program viz. to prepare perspective plans, outlining the national strategy and a road map for holistic and sustainable development of drought prone regions and to analyze the impact assessment of drought on agricultural crops and water resources.

The recommendations of the policy paper on Drought preparedness and management by Ministry of Agriculture (2011)GOI based on one day Brainstorming session at NIDM suggested the following.

- Interventions in emergencies in terms of sequence of events could be distinctly identified for planning and executing different actions. The eight types of actions or phases of emergency sequence are 1) prevention, 2) preparedness, 3) early warning, 4) impact, 5) relief, 6) rehabilitation, 7) reconstruction and 8) sustainable recovery.
- The existing response programs for drought prone area tend to be limited in scope and depth. They focus on immediate needs such as drinking water, food crops and livestock loss prevention and short term employment. New approaches, strategies and methods to be included are farming systems, resources and improved livelihood in drought prone region program.

- Integrated water shed management coupled with enterprise diversification based on carrying capacity and product value addition should be the major thrust that would ensure optimum resource use, sustainability and income maximization.
- Establishment of water (for human and livestock) seed, fodder and feed bank assumes higher priority in drought prone areas.
- The harvested rain water and augmented ground water resources should be used only through efficient irrigation systems (Drip/sprinkler)
- Drought management plans for different locations should be developed by analyzing the resources.

7.4 Drought Preparedness Action Plans in Different Countries

7.4.1 Arizona State in USA

The Tucson city in the state of Arizona located in South-western United States is known for its arid type climate and drought occurrence is common and water availability is limited. Due to continuous drought conditions, the State of Arizona passed a law to prepare drought preparedness and response plan by 2007. Accordingly Arizona Department of Water Resources (ADWR), developed plans that include specific drought indicators and actions that the customers can take based upon the level of response needed. As an example, Tucson water plans are as follows. **(http://cms3.tucsonaz.gov/water/drougth-sum)**. The plan has four stages. The Stage 1 is declared for Tucson water's service area and the stages 2,3 and 4 will be declared based on the threats to Colorado River supplies to the Tucson city and fall in ground water supplies. The drought preparedness plan has the following recommendations.

Stage 1:

- Public education on drought issues
- Modify water system operations
- Identify water saving and water efficiency measures for all city facilities
- Voluntary self-audit programs for commercial, multifamily and industrial users.

Stage 2

- Additional public education and additional systems or operational actions
- Prohibit Public fountains and non-essential uses

- Implement water saving and efficiencies identified in stage-1
- All potable water users requested to make additional voluntary reductions.

Stage 3

- Potable water reduction by all water users
- Plumbing retrofit on release for residential, commercial, multifamily and Industrial users.

Stage 4:

- Implement city's emergency water conservation ordinance
- Non essential outdoor water use restricted / prohibited
- Restaurants to provide water on request only
- No filling of swimming pools, fountains and car wash only with recycled water.
- Additional response action to be developed if warranted.

7.4.2 Texas State in USA

To coordinate the preparedness and response to droughts in Texas State, the state legislature created Texas Drought Preparedness Council in 2006, that has four committees and a special task force in which the state drought response agencies like Health, Water, Agriculture, Forestry, Wild life etc. are the members.

The four committees are 1) Drought planning and coordination committee, 2) Drought monitoring and water supply committee, 3) Drought Technical Assistance and Technology Committee and 4) Drought Impact Assessment Committee and a Drinking Water Tank Committee.

The responsibilities of each committee are as follows:

1. The drought preparedness councils works together to find better methods of identifying and monitoring drought situation and advise the public with suitable methods to cope up with droughts.
2. Specific actions of the drought preparedness council are related to the energy management cycle functions such as mitigation, preparedness, response and recovery.
3. If drought situations degrades further, the council can immediately seek the support of state emergency management council for food and water supplies.

4. If the drinking water situation is likely to be effected. Emergency Drinking Water Contingency plan can be implemented.

7.4.3 Italy

Rossi, *et.al* (2010) identified the responsible authorities for preparation of plans for 1) Strategic water shortage preparedness plan, 2) Water supply system management plan and 3) Drought contingency plan as follows:

Table 7.3: Responsible authorities and main contents of drought management plans in Italy

Plan	Responsible Authority	Contents
Strategic Water shortage Preparedness Plan (SWSPP) to be annexed to the River Basin Management Plan	Basin Hydrographic District Authority	• Identification of drought vulnerable areas, though the analysis of past drought events and their impacts • Estimation of water shortage risk in water supply systems • Definition of appropriate long term actions to reduce water shortage risk; • Comparison and selection of alternative drought mitigation measures; • Procedures to ensure the transparency of information on drought situation and planning process and to allow public participation to the decisions • General indications for the drafting of Water Supply System Management Plan (WSSMP) • General indications for the drafting of Drought Contingency Plan(DCP)
Water Supply System Management Plan	Agencies for Water Supply Management	• Definition of indicators and their values to define normal, Alert and Alarm conditions with respect to drought for a given water supply system • Definition of the best mix of long term and short term measures to avoid emergency • Estimation of costs and financing sources for the chosen mitigations measures • Tools for fostering the stakeholder participation and exchanges

Plan	Responsible Authority	Contents
Drought Contingency Plan	Regions or Basin Authorities	• Definition of indicators and their values for declaration of a drought as natural calamity • Tools for an institutional participation through a devoted task force on drought • Definition of short term mitigations measures and their costs • Indications for the coordination of state, regional and local interventions during emergency • Tools to ensure the transparency of information on drought situation • Possible recovery from drought damages

Source: Rossi *et al*., 2010)

7.4.4 Mediterranean Region

Droughts are not uncommon in the Mediterranean countries and are a recurrent event in Middle East region where Jordan is a typical example. The droughts in Jordan, Syria, and Palestine are the worst and it was observed that they have been observed more frequently in recent times. In Mediterranean European countries like Cyprus, Greece, Spain and South France dry condition were observed after 1970. Traditionally all the countries in the region react to drought (Crisis management approaches) by providing relief or emergency assistance to the affected regions and the countries have no drought plans that can mitigate the drought conditions and elevate the economic status of rural poor community. Atef Handy (2004) suggested ten steps for planning and emphasized the need for a National Drought Management policy for the above countries. The steps are:

1. Appoint a Drought Task Force
2. State the purpose and objectives of drought plans
3. Seek Stake holder participation and resolve conflict
4. Take inventory of resources and identify groups at Risk
5. Develop organizational structure and prepare drought plan
6. Integrate Science and policy, and close institutional gaps
7. Publicize the proposed plan, solicit reaction
8. Implement the plan

9. Develop educational program
10. Post-drought evaluation.

Some of the short comings found out in the drought preparedness plans as reported by Altaf Handy (2004) in this region are

1. Inadequate data net works
2. Limited data sharing among different agencies
3. Early warning system products are not user friendly
4. Drought monitoring tools – indices are inadequate
5. Lack of impact assessment methodology and
6. Slow delivery of information about drought.

7.4.5 Pakistan

Monitoring of droughts in Pakistan are carried out by Pakistan Meteorological Department, Water and Power Development Authority, Provincial Irrigation and Drainage Authorities and District Governments. The major limitation in monitoring drought is that no single institution in responsible for monitoring of drought and their impacts on various systems in the country (IWMI-Working paper 1985, 2004). It is further reported in the above publication that through the institutional involvement for drought relief activities are well defined but the drought preparedness and mitigation mechanism are absent. Only ad-hoc measures were adopted between the years 1998-2004. The linkages between public sector institutions with non-governmental organizations, private sector and civil society are very limited. The commitment at the political level for the formation and to enforce the policy on drought need to be strengthened. The following points have been suggested by the author to prepare a drought policy at Regional / National level.

1. Review of the existing policies related to drought and water management.
2. Identify potential collaborators for initiating the policy research viz., re-assessment of resources, basin wise management plans, drought mitigation and water management interventions.
3. Formulate policies that link development of resources within the frame work of Integrated Water Resources Management (IWRM).
4. Subsidy on electricity and less water Tax on surface waters resulted in over exploitation of water resources. Hence there is a need to formulate policies for judicious management of resources.

5. Community participation in implementing the drought mitigation plans may be given priority.

7.4.6 Darfur in Sudan

The UNEP mission (2008), which reviewed Water Resources management at Internally Displaced Persons (IDP) camps and host communities in Darfur in has found no plans to manage drought in Sudan and suggested a four stage process for development of drought preparedness as follows:

1. Continue to improve the quality and coverage of hydrological data collection
2. Design and implementation of drought mitigation work such as construction of sand dams and check dams
3. Preparation of interim drought preparedness plans
4. Drought preparedness plans should address residual risks, which remains high in some cases.

7.4.7 Bangladesh

The most important disaster in Bangladesh is the tropical cyclones. However droughts do occur at some interval which are mostly confined to North-West part of the country. The Government of Bangladesh constituted various Disaster Management Committees at National and at filed level. Though there are seven National Committees headed by various ministers for managing disasters, there is no special committee exclusively to care for Droughts. Perhaps the National Disaster Management Committee headed by the Prime Minister of Bangladesh may be responsible to prepare policies and issues to all the Disasters and Drought being the one among them.

7.5 Indigenous Knowledge for Drought Management

Curiousity in watching the position and the movements of celestial bodies, lead to predict the occurrence of seasons. The prediction of weather in the coming season was reflected in the ancient Indian writings of the 3000 B.C *viz.*, Upanishads. Perhaps they were the first to use mathematical calculations to follow the planetary movement and there by relating the weather phenomena. Varahamihira's classical work "Brihatsamhita" written around 500 AD provide evidence of knowledge on Atmospheric process. Similarly Kautilya's Arthasastra mentions about measurement of rainfall and its application to revenue and relief work (Varshneya-2010). In addition to the Astronomical knowledge, the farmers experiences over the centuries has made them to develop empirical art to predict weather by closely observing the nature for example, the wind direction, animal, bird and insect behaviour, flowering of some wild plants, cloud type, hallow ring around sun and moon etc. Human settlements all over the globe have developed this folk-art in forewarning of weather mostly for initiation of agricultural operations. As an example, a local folklore in Marwar area of Rajasthan describes "A century is made up of seven years of famine , 27 years of plenty, sixty four years of droughts and two years of extreme drought" (Aparna, 2001). However there are some limitations in following such predictions as they may not able to withstand scientific logic and evaluation. Farmers' wisdom continues to play a decisive role in managing climate risks. Perhaps blending of traditional knowledge with modern science may bring necessary synergy in managing droughts efficiently. In the following section, traditional knowledge in predicting weather and their application to manage droughts at different location spread across the globe are briefly given.

7.5.1 Astrometeorological Methods

Varahamihira postulated different principles for prediction of rainfall (Tamboli and Nene, 2002). According to the ruling planet of the year, the overall rainfall of that particular year is anticipated as follows:

1. Sun - Moderate
2. Moon - Very heavy
3. Mars` - Scanty
4. Mercury - Good
5. Jupiter - Very Good
6. Venus - Good
7. Saturn - Very low (Stormy wind)

As per the astrological principles, sun is the main planet that influence weather and climate on the earth. Mars and Pluto are responsible for increase in temperature. Different aspects between Moon, Venus and Neptune are responsible for rains. Saturn and Uranus are responsible for decrease in temperature. Mercury acts on wind and is responsible for movement of clouds (Varshneya, 2010).

Ancient methods of weather forecasting may be divided in two categories (Mishra *et.al.* 2002) *viz.*, 1) Theoretical and 2) Observational methods. The Theoretical methods are based on computation of planetary positions, conjunctions of planets and stars. Observational methods include observations on atmospheric changes, bio indicators, chemical changes, physical changes, cloud forms and other sky features. Some of the biological indicators for prediction of monsoon rains are well documented and reviewed (Kanani *et.al.* 1995). Pisharoty (1993) observed that abundances of bunches of golden yellow flowers of Golden shower tree (Cassia Fistula) occurs just 45 days before the onset of monsoon. Bhadali, an eminent astronomer of 12 Century described 10 variables viz., wind, needle or sword shaped clouds (high altitude clouds), lightning, colors of the sky, rumbling thunder, dew, snow, rainbow and occurrence of halo around sun or moon for development of rain embryo. If any of these things appear in the sky, then rain is assured 192/193 days after. Thus observations taken for the period approximately six months between i.e., Ashwin Krishnapaksha (Dark half of Ashwin – Oct-Nov) of previous year to Vaisakha Pournima (April-May month) of the current year can be used to predict monsoon season rainfall of current year. Such study was conducted for Barshi region in Sholapur district of Maharashtra for the year 2004-2007 (Varshneya *et.al.* 2010). The results showed that on an average, the symptom of rainy clouds has the highest frequency, while the symptom of squalls has the lowest frequency. The Anand Agricultural University at Anand in Gujarat was the first to combine the Astrometeorology with Agrometeorology by bringing a "Monsoon Research Almanac 2006" which predicted daily rainfall of eight stations representing "eight agro climatic zones of Gujarat. The Calendar also provided information on mean monthly maximum and minimum temperature and crops to be grown on different climatic conditions, package of practices and farm operation to be carried out. Similar type of Calendar for the year 2007 was also prepared in local language and distributed among the farmers.

A one day workshop-cum-Group meeting on the "Relevance of Ancient wisdom on weather forecasting for improving Agro Advisories was held on 29th May 2006 at CRIDA, Hyderabad. Dr. Varshneya presented A "Nakshtra" based rainfall forecast calendar that provides the probability of occurrence of rainfall in different agro climate zones of Gujarat. The accuracy of rainfall forecast varies between 40 to 70 per cent across the state. According to Gayathri Vasudev an astrometeorologist from Bangalore, when sun ingresses in to Scorpio and Saturn in seventh position, below normal rainfall can be predicted. According to V.V. Ravi, an astrologer from Hyderabad, as the planet Mars is a head of sun in 2006, below normal rainfall was expected during 2006 which came true. Dr. Sudhakr Joshi advocated two methods viz., "Anthariksha Paddhati" and "Bhowma Paddhati" which can predict rainfall 5 to 10 days in advance. Dr. Dhansukh Shah from Pune demonstrated a model that predicts day to day rainfall one month ahead of monsoon of Saurashtra region of Gujarat.

7.5.2 Traditional Knowledge in Weather Prediction

The traditional knowledge available with rural population that was inherited from their ancestors in prediction of weather events at different locations in India are as follows:

North-West Himalayas

Based on the data collected from six districts of Himachal Pradesh, Mohan Singh Jangra *et.al.* (2010) compiled information from astronomical phenomena, bird activities animal activities, social beliefs and folklore. They are briefly presented below.

a) Astronomical phenomena

i) Corona (Halo ring) around sun with grater diameter is an indication of rain where as smaller corona around moon indicates dry periods.

ii) During rainy season more red color intensity in rainbow indicates fair weather whereas more green colour intensity predicts rainfall in the coming days.

iii) Red colour of sky at sunset predicts rain for next day.

iv) Bright sunshine at the evening after rains predicts more rains.

b) Bird activities

i) Peacock dancing in groups predicts good rainfall.

ii) Bathing of "Maina" in water indicates early rainfall.

iii) Frolicking of a group of sparrows in sand predicts rainfall, but playing in water predicts dry weather on the next day and rain after 2 or 3 days.

c) Animal activities

i) Jumping of Bull and cow with joy during grazing predicts rain and good weather.

ii) Seeing of snakes quite often during summer predicts rainfall occurrence.

iii) If sheep giving milk to the young at high hill ranges, early rains are predicted.

d) Social beliefs

i) If 15 days in the month of "Jeth" are very hot then good rainfall is predicted.

ii) If the "Khejari" tree bears good fruits, it predicts good rainfall and vice-versa

iii) When "tobacco bag" shows more moisture indicates rains in near future.

iv) People in Nauni area predict rainfall when grey colored clouds descent below the hill of "Doko" fort.

Rain forecasting in Andhra Pradesh

Identification of Traditional knowledge for prediction of rainfall and other weather parameters, proverbs and Folklores were collected and compiled by Ravi Sankar *et.al.* (2008). They have identified a total of 66 indicators out of which 24 were Bio-indicators, 40 Non bio-indicator and one Astronomical indicators. A few of the indicators along with explanations are listed here.

S.No	Indicator	Category	Explanation
1.	Movement of dragon flies	Bio-indicator	When humidity reaches saturation, a couple of hours before, dragon flies move in swarms indicating rain
2.	Jumping cattle	Bio-indicator	Cattle anticipate rain few hours early by sensing cool breezes developed before rain for a short period. Therefore, they jump joyfully.
3.	Flocking of Sheep and goats	Bio-indicator	When low clouds are formed, enough energy is released from water vapor resulting in formation of excess heat, which cannot be tolerated, by sheep and goats. Hence they form flocks.
4.	Chirping of Rishi pitta (bird)	Bio-indicator	When these birds move at heights above 0.5 kms, they sense the upper air winds and chirp. Usually rain bearing clouds occur at heights between 0.5 and 2.5 km. Because of the sensible response of these birds to the winds, farmers predict rain. It is also believed that if these birds fly in rows after the occurrence of rain, the possibility of further increase in rain is very high.
5.	Movement of Jangamaiyyalu (Red hairy caterpillar)	Bio-indicator	The adults of red hairy caterpillar become restless as the humidity increases beyond 90%. Their quick movement is considered as an indicator of rain.

(Contd.)

S.No	Indicator	Category	Explanation
6.	Squeaking of owls	Bio-indicator	Owls are blind to light but sensitive to humid conditions. The fear of increasing humidity and heat released by clouds instigates restlessness among owls. The peculiar squeaking sound of owls has been attributed to be an indicator of rain.
7.	Pink colored sky in the evening	Non-Bio-indicator	Mie scattering along with high relative humidity (more than 90) and low temperatures in the evening results in short duration high intensity rains.
8.	Rainbow in the West during SW monsoon	Non-Bio-indicator	Because of more depth and heavy coverage of sky, rainbow appears in west. This is an indication of onset of S-W monsoon.
9.	Dampness of jaggery, tamarind and salt	Non-Bio-indicator	Increasing humidity results in increase in water vapor content in the atmosphere. On saturation, water vapor gives rise to clouds bearing water, thereby rains. Since relative humidity and absolute humidity are considered as factors that determine the response of these materials, an indication about the possible occurrence of rain.
10.	Smell of dried fish	Non-Bio-indicator	When dried fish absorb moisture, some smell emanates. More than 80% of relative humidity has capability to give such smell that indicates nearby rain. In tropical conditions, 80% relative humidity coupled with air currents and gentle winds gives enough rain

Some of the beliefs in different districts in Andhra Pradesh based on Almanac Calendar is as follows

a) Visakhapatnam

Nela Kommu (Curve in a month)

Geographically the appearance of moon after every *amavasya* (Full moon day) looks like an arc. It looks very thin but angular(drooping) to one side. During rainy season if the moon appears evenly balanced on the third day after *amavasya* between 6-7 PM then, it is believed as auspicious and farmers expect good rains.

b) Ranga Reddy

Stiff jowar silage

When jowar silage is plucked in bundles from the large heaps located in the fields and is broken in the process, it is considered as still dry weather. If the bundles of dried leaves of jowar are intact along with the small bundle when plucked, it is considered as arrival of "*mrigasira*" nakshatra alongwith rains.

c) Anantapur

Conveyance mode of the Lord

It is believed that Lord Shiva arrives on earth by different vahanas (vehicles). Future rain depends on the vahana in which he arrives on January 15 (Sankranti festival).

Proverbs

Some of the proverbs mentioned by the farmers related to onset of rains were as follows:

1. If "*Hastha*" constellation (27September – 9 October) does not interfere, then it will rain copiously in "*Chitra*" (10-23 October).
2. Thunder in "*Makha*" (1-15September), rain in "*Uttara*"(16-30 September) is certain.
3. Rainbow in "*Arudra*" (22 June –5 July) constellation means heavy rain.
4. Wait and watch for half an hour in "*Hastha*" to experience heavy rain.
5. If when eyes are red or sky is red, be ready for down pour.

7.5.3 Traditional wisdom in Drought Risk Region (DRR) Rajasthan and Bundelkhand region in Uttar Pradesh

Jaisalmer in Rajasthan that receives rainfall of less than 200mm /year could able to meet its water requirements for centuries using traditional water conservation practices. This clearly indicates that if a society has good relationship with its water resources and knows how to store and does not allow to pollute it, there will not any shortage of water even in low rainfall region (Gupta, A.K and Anjali Singh, 2011). Some of the traditional wisdom and indigenous knowledge for controlling erosion, Water Conservation and Housing as reported by them are as follows:

1. Erection of obstacles at an angle to the wind that forces it to change the direction, this leads to deviation of the sand, which therefore ceases to accumulate.
2. In gullies, fixing in place of flat dry stones together with logs and large bunches; help to reduce the water velocity.
3. Collection of roof water and store it in tanks for drinking purposes.
4. Development of water harvesting structures (khadings) for recharging ground water provides drinking water to humans and livestock.
5. Construction of "qanats" underground dykes and tunnels for the transfer of sub surface water to the surface by gravity.
6. Application of dilute cow urine to plants and seeds as a pesticide to control pests/diseases.

7.5.4 Indigenous Technologies for Moisture Conservation in Drylands of Semi-arid Regions of India

Moisture conservation begins right from seed bed preparation to entire crop growth period in the semi-arid regions of India, which are characterized by erratic rainfall and frequent droughts that limit the agricultural productivity. Farmers over the centuries have evolved techniques to manage droughts by developing simple moisture conservation techniques. Mishra,(2002) compiled Indigenous Technical Knowledge (ITK) on Soil and Water Conservation in Semi-arid India under the National Agricultural Technology Project (NATP). The following are the few ITK collected by him

i) **Conservation furrow**: It is an age old practice for *insitu* moisture conservation in which furrow store more rain water. Water stress for crops is reduced during rainless period.

ii) **Inter Cropping**: This is an age old practice descended from the wisdom of ancestors. A method where two crops of different durations viz., short and long duration are sown. During a drought year, if one crop fails the other crop supplements the income.

iii) **Cultivation and sowing across the slope**: Cultivation and sowing of groundnut in Anantapur regions is done across the slope to reduce the runoff losses and increase soil moisture. It would reduce the top soil loss due to erosion.

iv) **Application of tank silt**: To increase the fertility and moisture holding capacity of soil and the crop yield, tank silt is applied over the field and mixed thoroughly.

v) **Rain water measurement by Indigenous raingauge:** A hole of approximately 20cm depth with 18 cm diameter made on a granite stone of 1x0.5m size is used to collect rainfall. When the hole is filled with rain water then the farmer goes for ploughing. To measure subsequent rainfall events, the water need to be removed.

vi) **Rain water harvesting from roof top and road surfaces:** To recharge wells and to store water in Kund / Tanks for drinking and establishment of trees in very low rainfall regions.

7.6 Traditional Wisdom for Drought Management in a few Selected Countries

Qashqai nomadic communities in Iran

Pastoral communities comprise about 2 per cent (1.3 millions)of total population of Iran and meet about 25 per cent of livestock needs of the country. The grazing lands are affected and the production falls due to frequent droughts. The largest group of migratory nomadic pastoralist in Iran is "Qashqai Confederation" livestock based livelihood system which depends on migration to make the best use of rainfall patterns. They usually migrate with their herds from summer to winter pastures twice a year. The "Koohi" subtribe, one of the dozens of subtribe of the confederation travel about 860 km in a year in search of best grazing areas. The traditional knowledge for drought management is a part of the knowledge related to Natural Resource Management as reported in CENESTA (2004) is briefly described below.

- Drought is predicted based on weather patterns usually around the fist of Esfand (21 Feb). Prediction is made about the year like "a good year is heralded by a good spring".
- The tribes would send scouts to the ranges 20 days ahead of the planned migration for assessing the conditions of the grazing land thereby to adjust the date and duration to avoid harshest impact of droughts. The same practice of assessing the ranges is done in normal rainfall year also.
- Another sub tribe of the Qashqai, the "Hoolegooh" were known "Monnajem" (astronomers). They specialize in predicting the chances and duration of drought. They use vegetation, wind and cloud patterns to predict rainfall and drought. Based on wind blowing pattern they predict rainfall 5 months in advance.
- Another sub-tribe "Ghareghani" classifies rainfall according to style, density, seasons and associated events.

- The tribal groups developed systems for rehabilitating the range lands and improving vegetation as a drought preparedness measure. They developed a system of planting seeds of wild plants along the migration route by tying small bags filled with seeds having small holes around the necks of sheep and goat. These seeds will spillout from these bags as the herds migrate across the mountains.

Indigenous knowledge systems in Swaziland

Swaziland in African continent is prone to disasters includes, droughts, occasional floods and devastating lightening during rainy season. Manyatsi (2011)collected information on indigenous methods at Ntndozi area of Swaziland to predict weather based on environmental cues and animal behaviors are as follows:

Table 7.4: Indigenous methods used to predict weather patterns in Swaziland

Predicator/ signal	Description	Implications
Nesting position of *Ploceus spp* birds	When floods are likely to happen, the *Ploceus spp* birds nest very high on the trees next to rivers between November and February When floods are not likely to occur, they nest low on the trees.	Planting of crop along flood plains is avoided.
Cry of *Cuculus solitaries* bird	The cry of *Cuculus solitaries* bird between August and November is a sign of the wet season.	Farmers prepare farming inputs upon the cry of the bird.
Chirping of *Centropus burc hellii* bird	The chirping *Centropus burchellii* bird during October to April is a sign that a thunderstorm is approaching.	Herd boys will bring livestock home for the safety of the livestock
The visit of *Scopus umbrella* bird in a homestead	The visit of *Scopus umbrella* bird in a homestead is associated with imminent strike by lightning in that homestead.	Some means of protecting the homesteads against the lightning is done.
Abundance of some wild fruits	The abundance of some wild fruits that include *Vanguer ia infausta, Englerophytum natalense* and *Sclerocarya birrea* during the period of December to February is taken as a sign of imminent famine.	Harvested products of previous season are preserved. Wild fruits and vegetable are also collected and preserved.
Cry of frogs	The cry of frogs during the summer season (September to March) is taken as a sign of approaching rainfall.	Farmers prepare their farming inputs in readiness to plough and plant.

(Contd.)

Predicator/ signal	Description	Implications
The position of the moon	When the moon is slightly tilted to the west and the crescent is facing down during the periods of August to December, it is taken as imminent sign of rainfall within a week	Farmers prepare their farming inputs in readiness to plough and plant.
Dark clouds on the west of the country	The appearance of towering scary darkened clouds on the west, of immediate fierce hailstorm accompanied by thunder and lightning.	Here boys bring livestock at home and every member of homestead leave fields and rush home for safety.
Abundance of certain insects	The abundance of butterflies, locusts and grasshoppers during the farming season is taken as a sign of imminent drought and famine	Harvest from previous season is preserved and preserved.

(*Source*: Manyasti, 2011)

Traditional community warning systems in Wajir and Turkana counties in Kenya

The Arid lands Resource Management Project (ALRMP), under the ministry of state for development of Northern Kenya' is the official government agency responsible for developing early warning signals at drought in the country. Most of the communities in Kenya have indigenous weather and climate prediction system that are based on indicators established over generations. The following are some few examples of indicators for Wajir and Turkana areas for predicting coming season was collected by Sammy Mutua (2011).

i. Early warning systems in Wajir communities

- Grunting of the camel bulls symbolizes a good rainy season
- When drought looms over, the cattle take excess water and reluctant to move away from watering point.
- If camels run away, browsing haphazardly and appear to have lost their hearing senses, it depicts bad season.
- When black ants make and follow only one route from their hole is an indicator of bad season. If they made more than four routes from their hole, it shows a good season ahead.
- "*Deyr bula kalaheys, rati iiriiulahaw*" – This somoli saying means, if the rains start early in the mornings, the seasonal rains are going to be good for browsers (camel & goats) and no grass for cattle.

ii. Early warning systems in Turkana community

- When livestock start feeding on dry bones and droppings of other animals, symbolizes a terrible upcoming drought that would kill the whole herd.
- When livestock become highly infested by animal pests especially licks and lice is a sign of drought on the way.
- When certain birds (Elelee) fly very high in the sky, is an indicator that the rains are far (drought) and opposite is true when they fly low.
- When the new moon rises, the shape of the moon and tilting positions, viz. symmetrical shape – symbolizes no rain during that month. When the crescent is directed to the north (left) is an indication of bad season/ sign of drought.

Indigenous knowledge for managing droughts in West Asia and North Africa (WANA) region

To cope with recurrent droughts, strategies was developed by pastoralists enabled them to survive and produce food was summarized by Hazel *et al.* (2001). These strategies were well adopted for both human and livestock population which were relatively low and the range lands are large in size and easily accessible.

i) Reciprocal grazing arrangements with distant communities for access to their resources in dry years.

ii) Adjust flock sizes and stacking rate to match available natural feed resources.

iii) Keep extra animals for easy liquidation during drought, either for food or for cash.

iv) Diversification of crops and storage of surplus grains, straw and forage during good rainfall years.

v) Diversification among animal species and breeds allow the vegetation recovery and survival of minimum number of livestock.

The Bethma practice during drought year in Sri Lanka

Bethma is a practice that temporarily redistributes plots of land among shareholders (paddy land owners) in part of the command area of tank (reservoir) during drought periods. It is an old indigenous method that is still practiced in some areas in Sri Lanka, which aims to provide opportunities for farmers to cultivate during drought where some water is available in the reservoir. It is practiced when there is not enough water available to cultivate the entire command area. It may be practiced in combination with field rotation and the farmers may cultivate either paddy or other field crops. The allocation of the plots is usually done by the farmers' organizations.

7.7 Role of Weather-based Agro-advisories

The importance of weather parameters on the crop growth, development and its production has been well recognized since the evolution of human history. Their wisdom and experiences gained over the period on farm management under all weather conditions are still relevant as they stood against rigorous evaluations over time. With the revolutionary developments and break through research in physical and atmospheric sciences in the last century, understanding of the weather and its variability over space and time has enormously improved. The modern science of weather played a significant role in achieving sustainable agricultural production by developing climate based agricultural planning and providing timely advisories relating to crop management practices viz., time of sowing, irrigation scheduling, fertilizer application, pesticides spraying etc., to the farmers. The India Meteorological Department (IMD) which was established in 1875 started issuing farmers weather forecasts way back in 1940. However, these forecasts were of climate oriented in nature and were not specific to any crop or to any particular area. A separate Division of Agrometeorology was established by IMD in 1930 to study the weather-plant interactions and to prepare a climate based agricultural planning. With the advent of modern computing facilities, regional weather forecast for next 3 to 4 days period is being issued by National Centre for Medium Range Weather Forecast (NCMRWF), from 1990 onwards. Using this forecast, Agro Advisory bulletins are issued from 130 Agromet field units (AMFU) covering all the 127 NARP (National Agricultural Research Project) zones of the country (Figure 7.2).

AMFU stations are mostly located either in the State Agricultural or Veterinary University Headquarters or at its regional stations, IITs, ICAR Research Institutes and in KVKs. These Units are responsible for preparation and dissemination of weather based Agro advisories twice a week for the districts falling under the agroclimatic zone representing each centre. Improvements in weather forecasts by using latest forecasting models like Meso- Scale Model - 5 (MM-5) and Weather Research Forecast (WRF) model enabled to provide forecasts for smaller grid regions *viz*., 50x50 km for MM-5 9x9 km for WRF respectively. These developments lead to provide Medium Range Weather Forecast at district level. Thus the district level Agromet advisories are being issued from 2008 onwards by the respective AMF units. These services are monitored by Agromet Division of IMD which is known as "Integrated Agromet Advisory Services"(IAAS) from 2008 onwards. The Agro Advisory Board

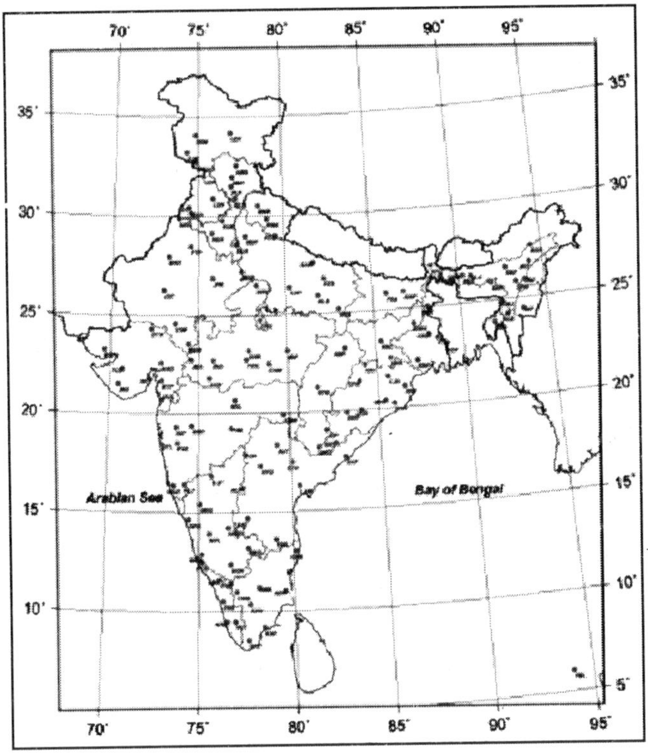

Fig. 7.2: Spatial distribution of AMFU stations in different states

(AAB) that prepares agro advisory bulletin consists of Scientists from different disciplines viz., Agronomists, Soil Science, Entomology, Pathology and Agrometeorology as it contains information on various aspects of crop management. Usually the AAB is headed by the senior most officer at AMF. The contents of Agro Advisory Bulletin in local language and the Dissemination System are as follows:

7.7.1 Strategies for dissemination of advisories

To ensure timely delivery of information to the farming community, a multi-mode dissemination system for agro-meteorological advisories is essential in which beside the conventional modes (radio, television & print media), the emerging modes of communication such as mobile phones and Internet have been deployed in India by IMD as well other organizations involved in issuing Agromet advisories. To improve the AAS further it requires back flow of information from the stakeholders on quality and relevance of information or demand for specific product.

The use of electronic media such as e-mail or the Internet for dissemination of AAS is picking up in India particularly the initiation taken by of Department of Information Technology. AAS is a scalable system which can be extended to cover all the farmers (crops) of India in a cost-effective manner. With the expansion of computers and Internet, emphasis is more towards electronic communication systems. However, TV and radio services are still the best among rural people. Under Integrated Agromet Advisory Service (IAAS) scheme efforts have been made by IMD to strengthen the Agromet Advisory as per the need of the farmers. In addition the old system of dissemination of AAS to the farmers of nearby villages through the messenger is still continuing at many of AMF Units.

7.7.2 Role of ICAR and AICRPAM on Agromet Services

The efforts of IMD in providing weather based agro advisories for sustainable agricultural production were supplemented by ICAR through development of teaching and research facilities on Agro-meteorology at State Agricultural Universities (SAU's) since 1970. Establishment of All India Coordinated Research project on Agrometeorology (AICRPAM) in 1983 at Central Research Institute for Dryland Agriculture (CRIDA) also strengthened AAS system. The research findings on crop-weather relationships and the crop-pest-weather relations obtained from experiments conducted by various Agricultural Research Institutes of ICAR and by the cooperating centres of AICRPAM at Agricultural Universities made it possible to support the efforts of IMD and NCMRWF to provide valuable weather-based advisories to the farmers. Number of training programs in techniques in agro climate analysis and other related topics were conducted for the agro meteorologists managing AMF units. Further, the AICPRAM project at CRIDA initiated a web site called **'www. cropweatheroutlook. in'** first of kind in the country to utilize the internet services for issuing weather based agro advisories since 2003 on weekly basis through its 25 centres across the country. It also prepares weekly contingency crop plans for the different meteorological sub- divisions in the country during kharif season. These contingency plans are also reflected in the ICAR and in AICRP on agro meteorology websites. Many of the SAU's and ICAR institutes have initiated issuing district level agro advisories in their states through websites of respective SAUs and the list of such websites (Table 7.5) are given below.

Table 7.5: Addresses of websites providing agro advisories in India

AMFU station	Website address
Basar, Arunachal Pradesh	https://sites.google.com/site/iaasbasar/
Rahuri, Maharashtra	http://mpkv.mah.nic.in
Ambikapur, Chhattisgarh	http://igau.edu.in
Namakkal, Tamil Nadu	www.tanuvas.ac.in
Jodhpur, Rajasthan	www.cazri.res.in
Delhi	www.iari.res.in
Solan, H.P.	www.yspuniversity.ac.in
Thrissur, Kerala	www.kau.edu.in; www.kauagmet.org
Jorhat, Assam	http://192.168.99.1/dee/aaservices.php http://www.aau.ac.in/dee/aaservices.php
Raipur, Chhattisgarh	http://igau.edu.in.
Junagarh, Gujarat	www.jau.in
Jabalpur, M.P.	www.jnkvv.nic.in.
Dantiwada, Gujarat	www.sdau.edu.in
Jorhat, Assam	http://192.168.99.1/dee/aaservices.php, www.aau.ac.in/dee/aaservices.php
Hyderabad, A.P.	www.agromet.ap.nic.in
Palampur, H.P.	www.hillagric.ernet.in/info/kisano_ke_liye_soochna/index.htm
Raichur, Karnataka	www.uasraichur.edu.in
Srinagar, J & K.	www.skuastkashmir.ac.in
Junagarh, Gujarat	www.jau.in
Hissar, Haryana	www.hau.ernet.in/coa/agromet.htm www.hau.ernet.in/coa/agrometpub.htm
Jagdalpur, Chhattisgarh	www.igau.edu.in
Kanniwadi, Tamil Nadu	www.oddanchatrammarket.com
Modipuram, U.P.	http://www.svbpmeerut.ac.in
Raichur, Karnataka	www.uasraichur.edu.in/info/infofile
Imphal, Manipur	www.mosdac.gov.in
www.icar.org.in	Barapani, Meghalaya www.indiaweatherwatch.org; www.meghalaya.nic.in www.icarneh.ernet.in; www.neigrihms.gov.in
Darisai, Jharkhand	www.baujharkhand.org; www.atmachaibasa.org; www.atmaseraikella.org; www.sameti.org
Chhindwara, M.P.	www.jnkvv.nic.in
Karimganj, Assam	www.aau.ac.in; www.karimganj.gov.in
CRIDA, ICAR	www.cropweatheroutlook.in (For 25 Units)
Pune, Maharashtra	www.mahaagri.gov.in

All these efforts have helped in improving the reliability and accuracy of agro advisory services in India to a great extent which resulted in improved production and rural economy. Its services have been extended from regional to district and now to Block level (Figure 7.3) with the availability of forecast to smaller region (9x9 km^2).

Under the National Initiative on Climate Resilient Agriculture (NICRA) project of ICAR a few AICRPAM centers have started issuing agro advisories for the blocks in few districts in Karnataka from 2011 rainy season onwards. It has now expanded further to all the AICRPAM centers located in different agro climate regions in the country by the year 2012.

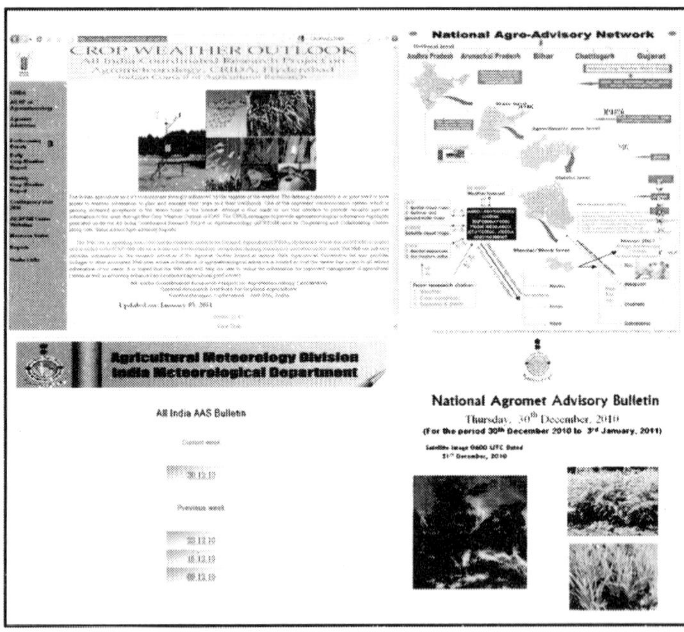

Fig. 7.3: Crop weather outlook website, proposed National Agro Advisory Network and IMD Agromet website

7.7.3 Agromet advisory services through mobile phone

In addition to different multi-channel dissemination system, Agromet advisories under the project Integrated Agromet Advisory Service (IAAS) are disseminated to the farming community in India through SMS and IVR(Interactive Voice Response) Technology. Under the SMS system an information platform has been created which allows the existing Agromet Field Units (AMFUs) located at State Agriculture Universities (SAUs), institutes of Indian Council of Agriculture Research (ICAR), Indian Institute of Technology (IITs) etc. to provide the information in a convenient and timely manner. Under IVR system the information from AMFUs for each state are collected and then stored, and converted into voice where the farmer would be calling and receiving the desired information.

A number of private firms namely Reuter Market Light, Handygo, Vritti Solution, IFFCO Kisan Sanchar Limited (IKSL) are disseminating the Agromet advisories generated under IAAS through SMS and Interactive Voice. At present 16 states namely Delhi, Uttar Pradesh, Punjab, Haryana, Rajasthan, Madhya Pradesh, Orissa, West Bengal, Gujarat, Karnataka, Kerala, Tamilnadu, Andhra Pradesh, Bihar, Maharashtra and Himachal Pradesh have been covered under this service. Advisories are also disseminated through SMS to the farming community of Maharashtra in collaboration with the State Department of Agriculture, Government of Maharashtra. Some of the agrometeorologists of AICRPAM centers, who happened to be the nodal officer under IAAS, are in constant touch with the farmers and advise them on likely weather in the next 4 days and the actions to be taken through their mobile phone.

International Horticultural Innovation and Training Centre (IHITC) in India has tied up with the Michigan State University (MSU) to explore and test the cost effectiveness of a cell-phone mediated personalized advisory service for growers cultivating horticultural crops in polyhouses, shade nets and open fields. The Tata Consultancy Services (TCS) provides the technical support for the period through Mobile-Agro-Advisory System (MAAS) which is called ***Mkrishi***. There are many other public private sector initiatives in India that are using mobile technology.

7.7.4 Status of Agromet Advisory Services in other Countries

USA

Specific farm advisories based on weather outlook are not issued to the users. However, weekly weather and crop bulletins containing weather and crop status are issued by US Department of Agriculture outlook Board. Under the weekly weather scenario it contains maps of different parameters for the current week viz., crop moisture index, palmer drought severity index (Fig. 7.4) and US drought monitor, US seasonal drought outlook, Extreme Maximum temperature, minimum temperature and temperature anomaly, average pan evaporation and total Growing Degree days.

It also provides the National Weather data collected at selected cities. Under the crop bulletin scenario, information on crop progress condition for selected states with respect to planting percentage, harvested percentage of various crops viz, soybean, wheat, corn, oats, sunflower and identifying the crop condition at different locations as very poor, poor, fair, good and excellent. The National Agricultural Summary and the State level agricultural summaries are incorporated in the bulletin. The state level summaries include crop and weather condition with limited advisories. As an example the state level crop bulletin contents of Hawai state is as follows:

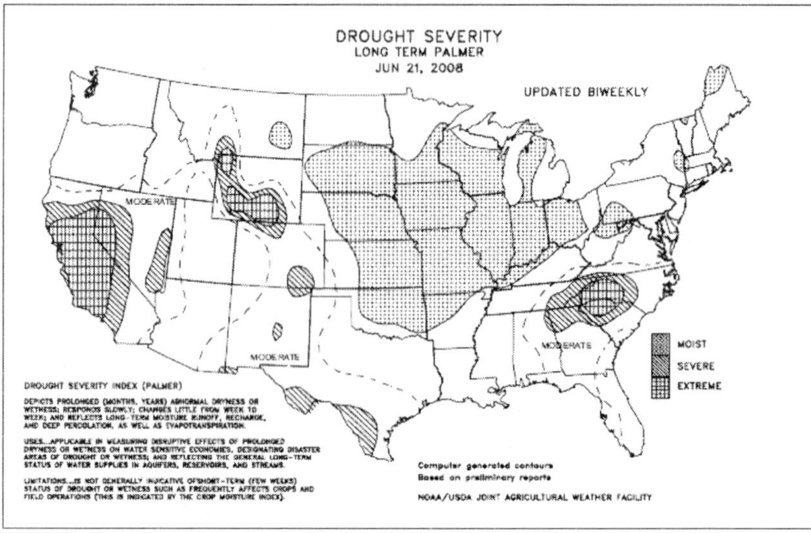

Fig. 7.4: Weekly Drought Severity Index (Palmer) over US

"Days suitable for fieldwork". Soil moisture was adequate with some windward areas receiving a boost from daily showers. Banana orchards were in fair to good condition. Soil moisture was raised in windward areas by light showers. Roguing of trees infected with the Banana Bunchy Top virus continued. Papaya trees were in fair to good condition. Light showers, warm temperatures, and long daylight hours were beneficial for the crop. Regular spraying was controlling insect infestations. Head cabbage plantings were in mostly fair to good condition. Increased insect pressure was noticed. Sweet corn fields were in good condition. However, water restrictions have affected crop development and planting schedules in the windward area of Oahu. Coffee fields benefited from light showers. Sugarcane fields on Kauai were in good condition. Building trade winds resulted in a mixture of sunshine and showers for the State. Shower activity increased over the weekend as moisture embedded in the trade wind flow brought daily showers to some windward areas".

International Weather and Crop Summary are also given in this Bulletin.

Bangladesh

Bangladesh Meteorological Department (BMD) through its network of weather observatories spread across the country issues weather forecasts regularly for the benefit of various user agencies. For agricultural crop planning, one month long-range forecast and for regular agricultural operations, a weekly agromet forecast is issued by BMD. The information provided under this weekly forecast includes the rainfall departure (%) maps of the previous week, cumulated rainfall forecast for the current week and the extended outlook for the next week. It also provides information on highlights of the weather during last week and advisory for the current week (Figure7.5). The advisory simply gives the information on the expected evaporation and sunshine in the current week. In addition, the information on expected variations in temperature is also available in the forecast. However, there is no specific information on the farm advisories (what the farmers should do) utilizing the forecast.

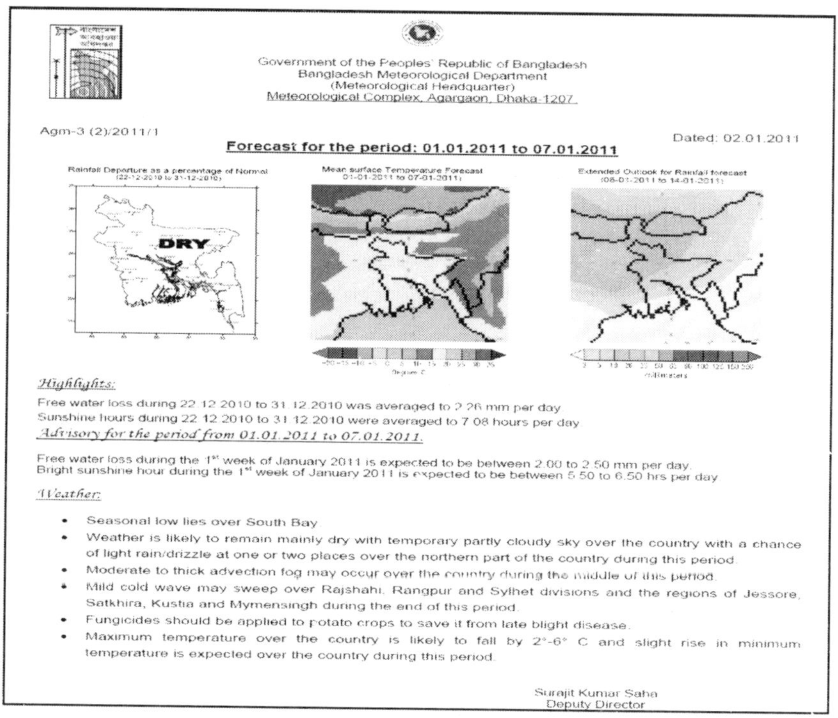

Government of the Peoples' Republic of Bangladesh
Bangladesh Meteorological Department
(Meteorological Headquarter)
Meteorological Complex, Agargaon, Dhaka-1207

Agm-3 (2)/2011/1

Dated: 02.01.2011

Forecast for the period: 01.01.2011 to 07.01.2011

Highlights:

Free water loss during 22.12.2010 to 31.12.2010 was averaged to 2.26 mm per day.
Sunshine hours during 22.12.2010 to 31.12.2010 were averaged to 7.08 hours per day.

Advisory for the period from 01.01.2011 to 07.01.2011.

Free water loss during the 1st week of January 2011 is expected to be between 2.00 to 2.50 mm per day.
Bright sunshine hour during the 1st week of January 2011 is expected to be between 5.50 to 6.50 hrs per day.

Weather:

- Seasonal low lies over South Bay.
- Weather is likely to remain mainly dry with temporary partly cloudy sky over the country with a chance of light rain/drizzle at one or two places over the northern part of the country during this period.
- Moderate to thick advection fog may occur over the country during the middle of this period.
- Mild cold wave may sweep over Rajshahi, Rangpur and Sylhet divisions and the regions of Jessore, Satkhira, Kustia and Mymensingh during the end of this period.
- Fungicides should be applied to potato crops to save it from late blight disease.
- Maximum temperature over the country is likely to fall by 2°-6° C and slight rise in minimum temperature is expected over the country during this period.

Surajit Kumar Saha
Deputy Director

(*Source*: Ramakrishna Y.S. 2010)

Fig. 7.5: Weekly Agromet Forecast in Bangaladesh

Pakistan

The Agromet service in Pakistan is in operation since 1988 with the establishment of National Agro Met Centre (NAMC) at Islamabad and five Regional Agromet Centres (RAMC) at Rawalpindi, Quetta, Faisalad, TanJodan and Usta Muhammad. The responsibilities of RAMC are (1) monitoring local agrometeorological conditions, (2) development of regional crop-weather studies, (3) liaison with Agricultural Institutions to determine the local requirements and research priorities, and (4) processing of Agromet information on regional scale for onward transmission to NAMC, Islamabad. Considerable progress has been made by PMD on the collection and archiving and analysis of weather and climatic data and their conversion into weather-based agro-advisories.

The webpage (Figure 7.6) of National Agro Met Centre (NAMC) provides information on current week's precipitation analysis *viz*., probability of occurrence of rainfall, percent departure. Also, the other information available are soil moisture analysis, region-wise Agromet advisories, seasonal precipitation and ETo climate outlook for the next 10 days with respect to temperature, rainfall and soil moisture, forecast to farmers, separate forecast to cotton crop, crop vegetation index maps, weekly forecast and crop report, crop calendars etc. are available. Based on the analysis with respect to rainfall and soil moisture, the advisories are issued and disseminated through official website of PMD, *viz*., **www.pakmet.com.pk** and other electronic and print media. Advisory bulletins at monthly, fortnightly and 10 days interval are issued each month.

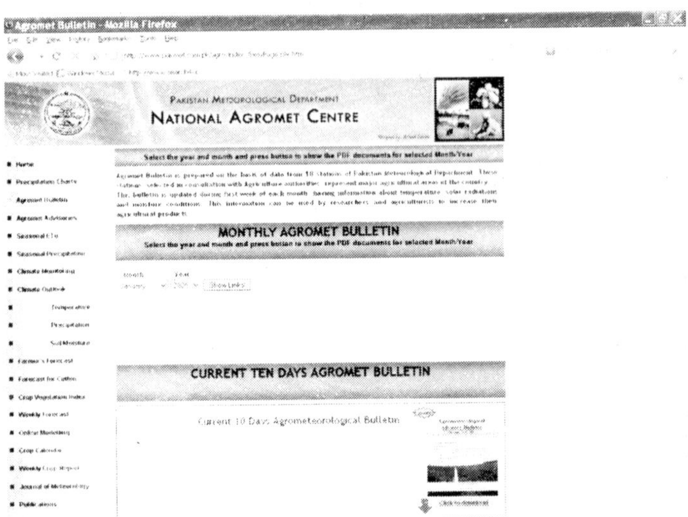

(*Source*: Ramakrishna Y.S. 2010)

Fig. 7.6: Webpage of Pakistan Meteorological Department

Sri Lanka

The National Meteorological Centre (NMC) is primarily responsible for collection and processing of weather data received from field units. Issue of regular weather warnings to the farmers, fishermen and to the general public is done by NMC. The Agromet Division collects Agromet data from 37 Agromet units and is subjected to quality checking and averages are worked out by them. NMC at present is not providing any specific weather based agro advisories to farming community except warning them on adverse weather conditions (Figure.7.7). Some private organizations and department of Agricultural extensions are issuing agro advisories on a limited scale to the farming community.

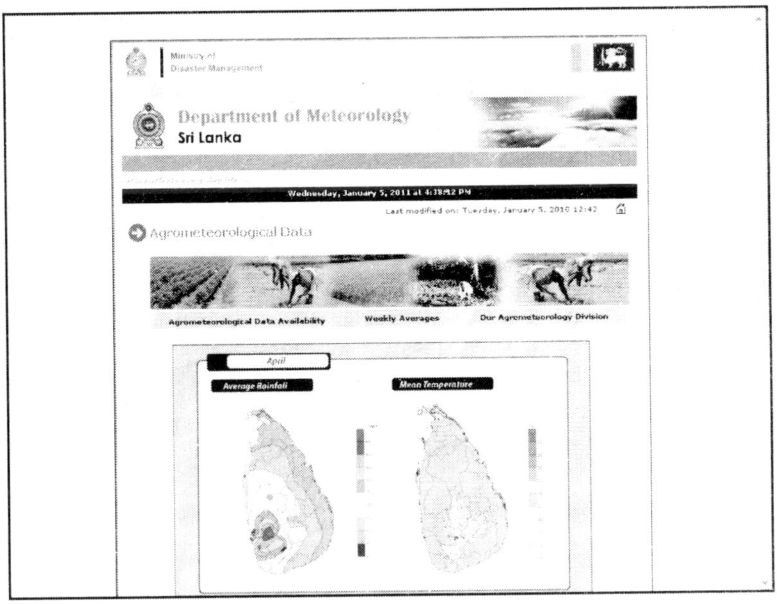

(*Source*: Ramakrishna Y.S. 2010)
Fig.7.7: Meteorological website of Sri Lanka

Nepal

The Department of Hydrology and Meteorology is responsible for the weather data collection and forecasting and there is no special division which deals with Agromet problems. No special bulletins or in the form of weather agro advisories are available for farming communities. However some organizations like Agro Enterprise Centre, in Khatmandu monitors all aspects of agricultural production right from production, market, financial and other relevant information to the farmers' community. A website **www.datameteo.com** has been providing 7 days forecast to parts of South Asia including Nepal, which generate agro-meteo-grams with respect to seven basic weather parameters which can be used to prepare weather based agro advisories by any local entrepreneurs. However, these agro-meteo-grams are priced and are available on demand (Figure.7.8). Another project which is operating under National

Information Technology issues agro advisory which has a tie up with GKP net projects. Under this project, Agro advisories are being currently issued through telephone or through internet in 21 villages located in Eastern Region of Nepal.

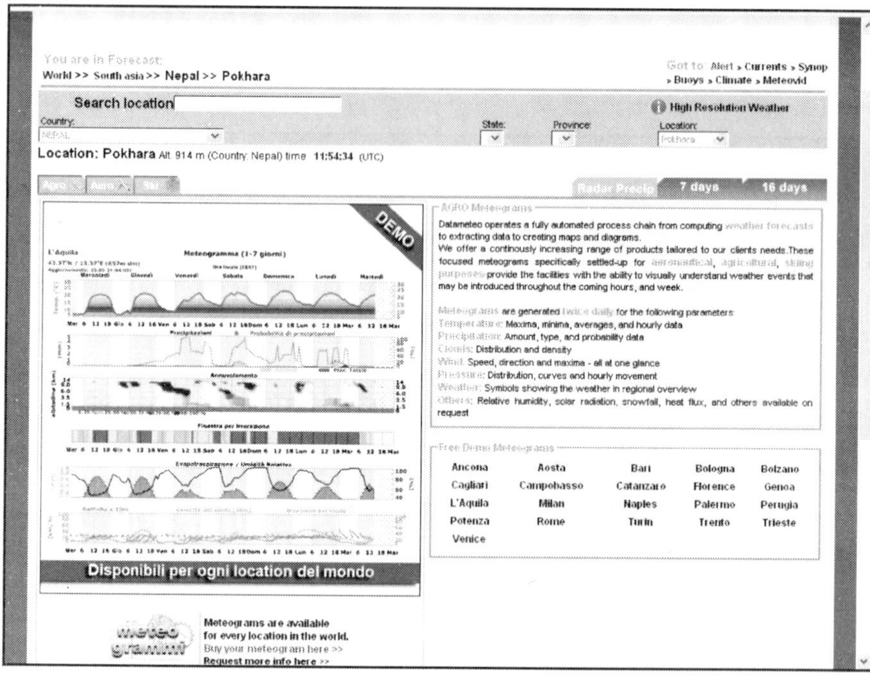

(*Source*: Ramakrishna Y.S. 2010)

Fig. 7.8: Agro-meteo-grams for Pokhara in Nepal (www.datameteo.com)

Multipurpose Community Telemeter

The Rural areas in Africa, Asia and Latin America have become a major hub for wide range of telecommunication services. They were designed to broad cast information regarding agriculture, health, education, environmental protection etc., to the rural population. Many countries disseminate the information on Agro Advisories using ICT that have changed the activities of farmers. Following are some of the websites in different countries which are related to Agrometeorological information (Table 7.6)

Table 7.6: Websites providing Agromet information in various countries

Country	Institution / University	Website
Austria	University of Agricultural Sciences	http://www.boku.ac.at/imp/agromet/agrarle.htm
Australia	Agriculture Western Australia Service Unit	http://www.agric.wa.gov.au/climate/
	Agricultural Production Systems Research Unit	http://www.apsru.gov.au
Belgium	UCL-FAOAGROMET Project (Spatial interpolation for Agrometeorological variables)	http://www.agro.ucl.ac.be/biom/recherche/projets/agromet/agromet.htm
Brazil	CEPAGRI/UNICAMP	http://orioncpa.unicamp.br/
Canada	Drought Watch Canadian Society of Agrometeorology.	http://www.agr.ca/pfra/drought.htm http://www.oac.uoguelph.cal/~csam
China	Agrometeorological Center of Jiangsu Province in P.R.China	http://www.angelfire.com/nj/jgWU/
Hungary	Meteorological services	http://www.met.hu/eao/eao-e.htm
Italy	Italian Agrometeorological Association	http://wwwfisbat.bo.cnr.it/AIAM/
Kenya	Agrometeorological Services	http://www.meteogo.ke/obsv/agro.html
Malaysia	Agrometeorological Services	http://www.kjc.gov.my/people/agromet/agromet.htm
Namibia	Meteorological services	http://weather.iafrica.com.na/

7.8 Value Addition to Agromet Information

The importance of climatic information for agricultural production and planning is well recognized and the weather forecasts valid for different durations play significant role in crop management measures such as scheduling irrigation, application of fertilizers and pesticides, weeding, inter-cultural operations etc. However, simple forecast for rainfall and other weather parameters for a region many not be much useful as many of the small and marginal farmers don't have access to forecasted information and may not be using information for agricultural operations. Therefore it would be desirable if the forecast is converted into agro-advisories using the relationship between crop and weather etc. Already a good progress has been made worldwide in preparing the agro advisories using the forecasted weather and the research results of Crop-weather and pest/diseases parameters. Before discussing further about the value addition to Agromet information for managing the climate risks, let us have a look at the Agrometeorological information requirement for the user community as reported by Ramakrishna *et.al.* (2006) is given below.

- Rainfall is the main source of water which causes instability in agricultural production. Therefore, information on the quantum of precipitation at various crop phenophases and at different probability levels enables the farmer to manage the risk involved in crop management. The information on the probability of occurrence of dry and wet spells during particular week is beneficial in preparation of the agricultural crop planning for a region.

- Monitoring of soil moisture through water balance computation enables the planners to monitor the drought situations over a region. Similarly, the same information can be used by the farmers for scheduling of the irrigation effectively.
- High and low temperatures due to heat and cold waves significantly affect the crop growth and yield by increasing the evapotranspiration, induction & sterility, destruction of cells and stunted growth. Therefore, the probability of occurrence of the above situations can help in preparation of crop calendars circumventing such events.
- Strong winds associated with weather disturbances can cause physical damages to the plants and also increase the water requirements considerably. The application of pesticides is also affected by the wind speed and direction. Therefore, information on the probability of occurrence of high windy days and the methods to reduce the impacts of strong winds on agricultural crops is very much essential.
- Crop pests and diseases are influenced by the prevailing weather conditions. Forewarning of onset and the spread of the pests and diseases is needed for timely management of these biotic stresses in a cost-effective manner and increase profitability. Hence, current efforts for development of forewarning models should be given priority.

To achieve the farmer's requirement, collection, scrutiny and statistical analysis of weather data from a large area is the first step that any genuine researcher attempts. Collection of long term weather data and the relevant crop and soils data over a region and its scrutiny for missing period and for typographical errors is essential before initiating the actual analysis. India Meteorological Department (IMD) is the repository of weather data for the country as a whole through its large network of surface observatories. In addition, to above the State owned Rain gauge stations, the State Agricultural University Research Farms, Zonal Research Centres, Krishi Vigyan Kendras (KVK), ICAR Research Institutes and their sub centres have been recording weather data for varied periods that is being widely used in agricultural planning programmes. ISRO and IMD have established a large number (above 1000) of AWS in the country. Some of the states like Andhra Pradesh, Karnataka, Tamil Nadu established a net work of AWS at their mandal, taluka level. In order to make available the weather data free of cost to the researchers, Agromet Databank facility has been created at CRIDA, Hyderabad during 1998 with financial support from DST. It is being operated through its coordinating unit of AICRPAM Agrometeorology project located at CRIDA, Hyderabad. The weather data collected at Research centres of AICRP on Agrometeorology and that of Dryland Agriculture and ICAR institutes since their inception has been collected and checked for its quality. Also the experimental data of 25 centres of AICRPAM with respect to date of sowing, duration of various pheno- phases, dry matter production, crop yields, varieties, irrigation etc have also been uploaded in the Databank. The data base management is being done using ORACLE database, where the climate derivates viz., weekly and monthly total, their probabilities, water balance parameters are generated and is supplied to the user for interpretation and its use in crop planning. Some of the screen shots of Agromet Databank are given below.

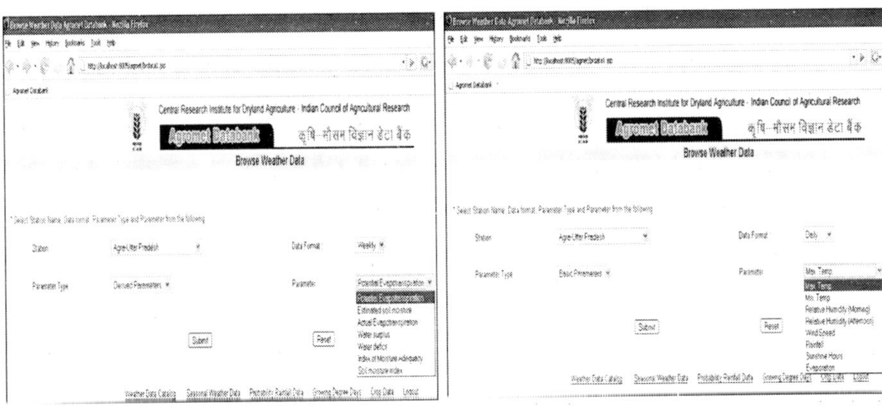

(Agromet Databank, AICRPAM, CRIDA Hyderabad)

Fig. 7.9: Screen shots of Agromet Databank at CRIDA, Hyderabad

7.8.1 AICRP on Agrometeorology

One of the mandates of AICRP on Agrometeorology project operating since 1983 at CRIDA, Hyderabad is the climate characterization of a region for agricultural planning. The project through its network of 25 centres has contributed significantly in collecting and analyzing the regional data to identify the climate risks during crop season and suggest measures to enhance crop production. Some of the significant achievements which are relevant for agricultural planning and climate risk management are as follows (Rao *et.al* 2010).

- Agroclimatic Atlas showing the spatial variation of weather parameter for Gujarat, Chhattisgarh, Haryana, Andhra Pradesh, parts of Karnataka,

Kerala were prepared. The spatial distribution of weather parameters and their derivatives in preparing thematic maps using advanced spatial interpolation techniques in Geographic information System (GIS) has been proved to be the most useful information for agricultural planning.

- Crop-weather calendars for crop planning purposes showing the optimum periods for agricultural operations for important crops grown in various agroclimatic zones were prepared.
- Weather-based relationships of pests/diseases and some thumb rules were developed and are being used in Agro Advisory Services at all the centres of AICRPAM.
- Micro level rainfall data analysis in Chhattisgarh identified regions suitable for direct seeding and for transplantation of rice crop under rainfed conditions. Vulnerability of upland paddy to droughts was assessed using water balance computations and management plans were developed.
- Studies on agro climatic constraints for rice productivity in rainfed rice growing regions of eastern India, comprising the states of Orissa, West Bengal, Chhattisgarh, Jharkhand, Bihar, East Uttar Pradesh and East Madhya Pradesh showed that the quantum of dependable rainfall (Rainfall 60% probability) expected during September month which is coinciding with reproductive stage is insufficient to meet the crop water requirements (Figure 7.10). In these regions, the probability of terminal droughts for rice crop is about 70 per cent, hence manipulation like advancement of sowing date, use of short duration varieties and fertilizers management has been suggested.

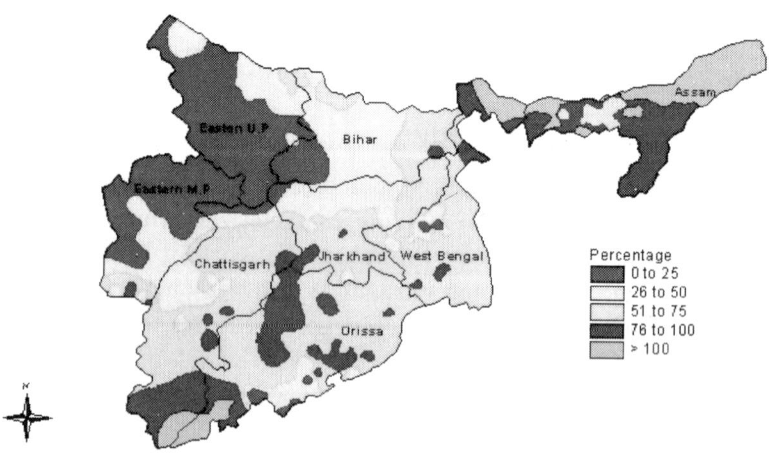

Fig. 7.10: Rainwater availability (as percentage of requirement) at 60 % probability during the maturity period for rainfed rice in Eastern India

- Crop contingency plans for various climate aberrations in different agro-climatic regions and Development of DSS for managing climate risk for major corps grown in different locations were developed and put into use by the respective centres in its Agro Advisory Services.
- The Water Balance parameters viz., Rainfall (P), Potential Evapotranspiration (PET), Actual Evapotranspiration (AET), Water Deficit (WD), Water Surplus (WS) and Soil Moisture (SM) computed using climate Water Balance procedures by Thornthwaite & Mather (1955) has been widely used in applied climatology studies like determining Length of Growing Period (LGP), irrigation scheduling, computations of droughts, climate shifts and climate classification etc.
- The distribution and duration of wet spells and dry spells for different weekly rainfall amounts during the crop growing period affect the crop growth and its production. Hence the probabilities of duration of wet and dry spell, and the conditional probabilities such as wet week followed by dry or wet period and dry week followed by dry or wet period provides the necessary input for preparation of suitable crops and the cultivars that can escape the unfavourable conditions at critical growth stages. As an example the following table provides the weekly probabilities of dry and wet spells.

Table 7.7: Weekly probabilities of dry and wet spells at Akola

	LIMIT=10 mm Consecutive Dry Probabilities			DATABASE: 1970-2008 Consecutive Wet Probabilities		
WeeK	P(2D)	P(3D)	P(4D)	P(2W)	P(3W)	P(4W)
1.	0.82	0.77	0.77	0.03	0.00	0.00
2.	0.85	0.83	0.78	0.00	0.00	0.00
3.	0.95	0.87	0.78	0.03	0.00	0.00
4.	0.90	0.80	0.80	0.00	0.00	0.00
5.	0.82	0.82	0.72	0.03	0.00	0.00
6.	0.87	0.76	0.74	0.00	0.00	0.00
7.	0.87	0.85	0.71	0.00	0.00	0.00
8.	0.85	0.71	0.69	0.00	0.00	0.00
9.	0.82	0.79	0.77	0.03	0.00	0.00
10.	0.79	0.77	0.71	0.00	0.00	0.00
11.	0.95	0.87	0.85	0.00	0.00	0.00
12.	0.90	0.87	0.87	0.00	0.00	0.00
13.	0.90	0.90	0.90	0.00	0.00	0.00
14.	0.97	0.97	0.97	0.00	0.00	0.00
15.	1.00	1.00	0.97	0.00	0.00	0.00
16.	1.00	0.97	0.95	0.00	0.00	0.00
17.	0.97	0.95	0.85	0.00	0.00	0.00
18.	0.95	0.85	0.78	0.00	0.00	0.00
19.	0.87	0.80	0.70	0.00	0.00	0.00
20.	0.82	0.72	0.43	0.05	0.03	0.03

- Rabi season crops are sensitive to temperature fluctuations due heat and cold waves. Identification of optimum temperatures for critical growth stages and the impact of departures of temperatures from its mean value and its duration are significant in estimating production. The analysis of temperature departures during rabi season for wheat crop grown in NW India (Ludhiana) have quantified that (Saikia et.al 2009) departure of average daily temperatures by 2°C for a period 5 days during reproduction stage will reduce the wheat yield by 10 per cent over its average. This type of quantification is useful in preparing weather insurance products.
- The Agromet Field Units (AMF)receives Medium Range Weather for the districts falling in its agro-climatic region. As the forecast accuracy especially during rainy season is around 60 to 70 per cent at many locations, the forecast need to be upgraded and modified with the climatic normals of the region, local forecasts. Satellite map, synoptic and upper air charts. Thus this type of the value addition to the forecast improves its quality and accuracy considerably and the weather based advisories becomes more relevant for the benefit of farming community.

7.8.2 Thumb Rules for Weather-pest/diseases

Next to weather, pest/diseases play a dominant role in agricultural crop production. On average about 10 per cent losses in crop production can be attributed to damages caused by pests and diseases; however on individual farm basis it may even go up to 30 per cent (MMP-17 project report). The possibility of damages from sucking pests like jassids, thrips, grasshopper, milibugs etc., to already stressed plants during drought condition is likely to be more, which results in greater fall in crop production during a drought year. Therefore it is necessary to develop relationships / thumb rules and forewarning models and Decision Support System for controlling of pests damages using weather variables. This information plays greater role in advising the farmers about the action to be taken through its weather based agro advisory services, operating through 130 AMF in the country. The India Council of Agricultural Research through its Research Institutes and coordinating centres of the project at SAUs developed relationships between weather-pest/diseases. Some of the thumb rules and relationships developed in respect of *Helicoverpa* in pulses, pink bollworm in cotton, aphids, leaf minor, rust and leaf spot in groundnut, aphids in mustard under National Agricultural Technology Project No. MMP-17 is as follows. (Fig.7.11a)

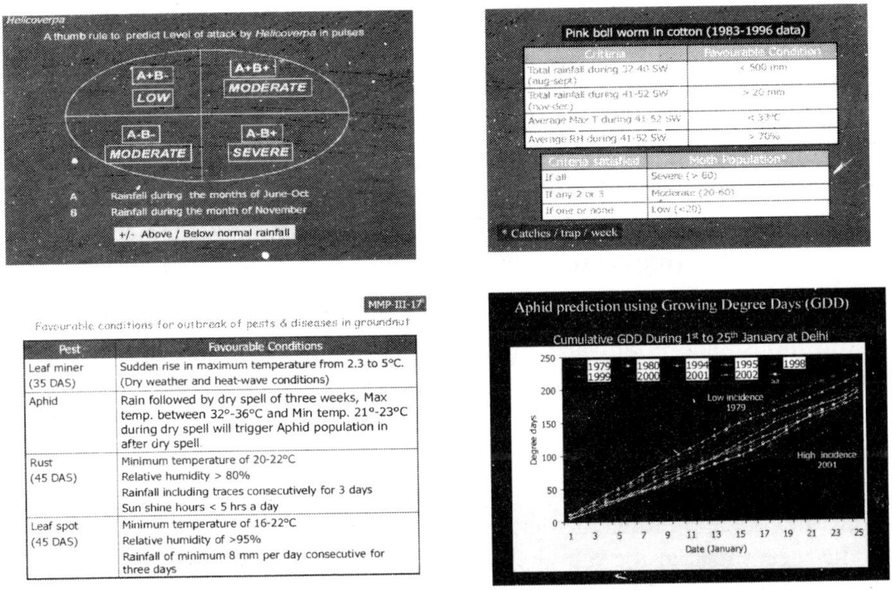

Fig. 7.11a: Thumb rules for important pests in pulses, cotton, groundnut and mustard.

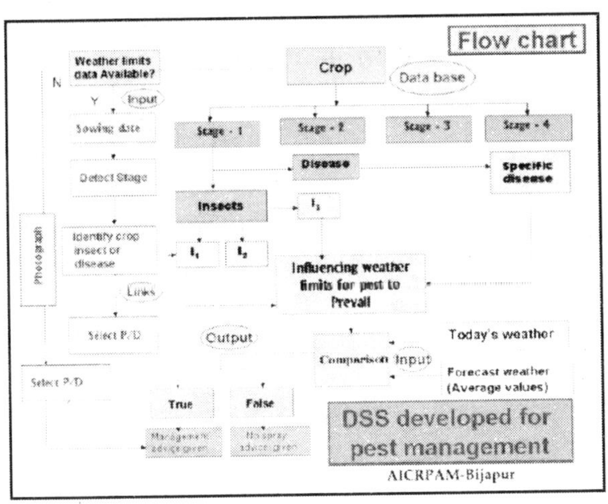

Fig. 7.11b. DSS Developed for pest management by AICRPAM-Bijapur.

A Simple Decision Support System for management of insect pest/disease was developed for NW Karnataka region by the AICRPAM centre at Bijapur and is being tested for its use in AAS (Figure 7.11b). The input to the system are crop details, stage of the crop, current days weather and forecasted weather. From the relationships between weather parameters and the population growth of pest and intensities of diseases and the current weather, the DSS gives an output whether management of pests/diseases to be activated or not be activated.

7.9 Contingency crop planning for minimizing risks

Crop production in the country is dependent on the timely onset of monsoon season, receipt of normal rainfall with good distribution during monsoon period. The impact of the temporal distribution of monsoon rainfall on food grain production from 1999 onwards reveal a close positive relation between these two parameters (Figure 7.12). The Kharif Season rainfall is characterized with breaks in rainfall (break monsoon) is common over greater part of the country.

Being located in the tropical region, the receipt of rainfall over the country during monsoon is erratic and unpredictable over 40 per cent of the time, which implies that in 4 out of 10 year period, there would be an adverse impact on crop production if appropriate strategies or crop plans are not available. (Venkateswarlu *et.al.* 2011). The production from rainfed regions plays an important role in achieving sustainable agricultural production in the country even after reaching fullest irrigation potential. Increasing variability of rainfall in the recent past at the regional levels has forced the authorities to improve their preparedness to face any water shortages during the crop growing period even in irrigated regions. In addition to the above, the frequencies of extreme weather events like heat waves, cold waves, untimely rainfall, hail storms and frost have increased over many locations.

It is customary for each State Agricultural University (SAU) to prepare crop calendars for every year(crop plans) indicating suitable crops and varieties for different agro-climatic zones falling under its jurisdiction. It also recommends the crops and varieties under delayed sowings due to delayed onset of rainy season and also suggests contingency measures to be taken up. As an example,

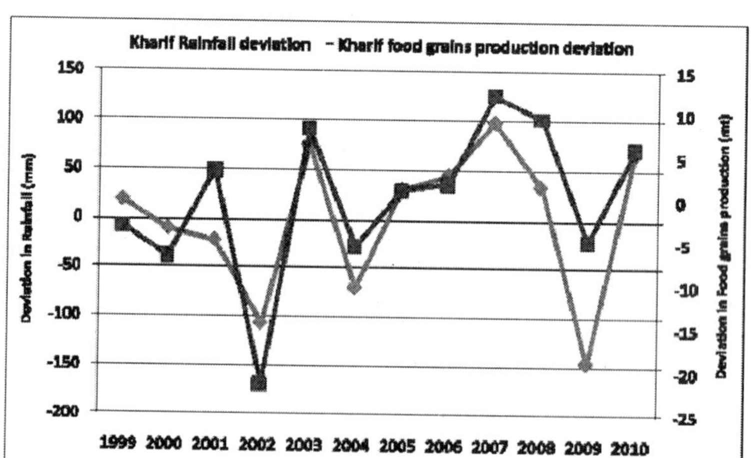

(*Source*: Venkateswarlu, *et al.* 2011)

Fig. 7.12: Rainfall vs Food grains production in India during Kharif season

the contingency plans for Rainfed chalka soils of southern Telangana region for delayed of monsoon and dry spell at different periods after sowing as suggested by Acharya NG Ranga Agricultural University (ANGRAU), Hyderabad for the year 2008-09 is given in Table 7.8

7.9.1 Contingency Crop Planning for Rainfed Regions of India

During the crop season, the important characteristics of the rainfall are 1) Early commencement of rains or considerable delay in onset of monsoon rains, 2) Intermittent breaks during crop growing period, 3) Variations in spatial and / or temporal rainfall, 4) Early cessation of rainy season or continued wet spells for extended period (Subba Reddy *et.al.* 2008). These situations call for the attention of scientists to prepare contingency measures to save the rainfed crops. Also there is a need to select crops and varieties matching with the Length of Growing season (LGP). In this endeavor, the All India Coordinated Research Project on Dryland Agriculture (AICRPDA), at CRIDA, Hyderabad which is operating at 25 cooperating centres and eight Operation Research Project (ORP) centres since 1972 have evolved plans over the years to manage droughts occurring at different stages of crop. A comprehensive plan for different centres falling under crop based production system viz., rice, maize, groundnut, soybean, cotton, rabi sorghum, pearl millet and finger milled were developed for normal as well delayed/failed rains during kharif season (June-September) and for Rabi season (November - March) (Subba Reddy et.al. 2008). Details of the production system and the locations are given in the Table 7.9.

Table 7.8: Contingency plans for Rainfed red chalka soils in Southern Telangana

	Monsoon rainfall condition			
Zone	Normal monsoon (2nd week of June)	Delayed in onset of monsoon by 2 to 3 weeks	Delayed in onset of monsoon by 4 to 5 weeks	Delay in onset of monsoon by 6 to 8 weeks
Southern Telangana	Sorghum,bajra, finger millet, fox tail millet, sorghum or bajra intercropping with redgram in 2:1 ratio, castor intercropping with Abhaya redgram in 1:1 ratio, castor intercropping with cowpea (1:1), groundnut, redgram, greengram	• Sorghum (short duration) bajra, finger millet, foxatail millet, castor intercropping with redgramin1:1ratio, castor intercropping with cowpea (1:1), groundnut, redgram, greengram • Plant protection against shoot-fly in sorghum is needed. Endosulfan spray @ 2ml/l of water two to three times at weekly interval.	• Avoid sorghum crop since shoot-fly damage will be severe· • Intercrop bajra, finger millet, foxtail millet, castor with Abhaya redgram in 1:1 ratio, • Castor intercropping with cowpea in 1:1 ratio • Avoid groundnut since the crop will be severly damaged by bud necrosis • Cultivate redgram and green gram	• Avoid sowing of long duration ragi varieties • Sow early varieties of castor with close spacing • Redgram and greengram to be avoided • Horesgram, foxtail millet, bajra can be taken up
	Dry spells	2 to 3 weeks after sowing	6 to 8 weeks after sowing	10 to 12 weeks after sowing
		• Due to sever moisture stress if crops have been damaged, re- sowing with short duration varieties/alternate crops can be taken up as suggested above	• Soil moisture conservation practices like hoeing, weeding and mulching to be taken up. • If dry spell continues, thinning of plants/rows to be taken up (1:4 of the population in jowar can be removed) • Spray 2% urea on crops like groundnut. "In case crops are damaged, alternate crops can be grown as suggested above.	• Harvest crops like jowar, bajra for fodder purposes and ratoon if sufficient rains are received later apply 40 kgN/ha for ratoon crops.
	Early with –drawl of monsoon	• Harvest the crops for fodder purpose		

Source: ANGRAV, 2008)

Table 7.9: Details of the rainfed production system and the locations

	Introduction
1. Rice based production system	
1.1	*Kharif – Rabi.* Sub-humid Deep Inceptisols in Eastern Plain Zone of Uttar Pradesh
1.2	*Kharif – Rabi.* Sub-humid Deep Inceptisols in Vindhya Zone of Uttar Pradesh
1.3	*kharif – Rabi.* Oxisols in Eastern Ghat Zone of Orissa
1.4	*Kharif.* Sub-Humid Alfisols in Centraland North Eastern Plateau Zone of Jharkahand
2. Maize based production system	
2.1	*Kharif.* Semi Arid Shallow to Deep Vertisols in Sub-Humid Southern Plain and Aravalli Hill Zone of Rajasthan
2.2	*Kharif – Rabi.* Semi Arid Inceptisols in Undulating Plain Zone of Punjab
2.3	*Kharif* and *Rabi.* Sub-Humid Deep Inceptisols in Low Altitude Sub-Tropical zone of Jammu and Kashmir
3. Oilseed based production system	
3.1	Groundnut based production system
3.2	*Kharif.* Arid Shallow Alfisols in Scarce Rainfall Zone of Rayalaseema of Andhra Pradesh
3.2.1	*Kharif.* Arid Deep Vertic/Vertisols in North Saurashtra Zone of Gujarat
3.3	Soybean based production system
3.3.1	*Kharif – Rabi.* Semi-Arid Medium to Deep Vertisols in Malwa Plateau Zone of Madhya Pradesh
3.3.2	*Kharif – Rabi.* Sub-Humid Medium Deep Vertisols in Keymore Plateau and Satpura
4. Cotton based production system	
4.1	*Kharif.* Semi-Arid Vertic Inceptisols/Vertisols in Central Maharashtra Plateau zone
of	Maharashtra
4.2	*Late Kharif-Rabi.* Semiarid Deep Vertisols in Southern Zone of Tamil Nadu
5. Nutritious Cereal based Production System	
5.1	*Rabi.* Sorghum based production system
5.1.1	*Rabi.* Semi Arid Vertisols in Northern Dry Zone of Karnataka
5.1.2	*Kharif – Rabi.* Semi-Arid Medium/Deep Vertisols in Northern Dry Zone of Karnataka
5.1.3	Semi-Arid Shallow to Deep Vertic Inceptisols/Vertisols in scarcity zone of Maharashtra
5.2	Pearlmillet based production system
5.2.1	*Kharif – Rabi.* Semi-Arid Inceptisols in South Western Semi-Arid Zone of Uttar Pradesh
5.2.2	*Khaiif.* Arid Deep Aridisols in North Zone of Gujarat
5.2.3	*Kharif.* Arid Deep Aridisols in Western Zone of Haryana
5.3	Fingermillet based production system
5.3.1	*Kharif.* Semi-Arid Deep Alfisols in Eastern Dry Zone of Karnataka
6.	**Basic Concepts and Practices for Crop Production in Rainfed Regions**
7.	**Livestock management In Rainfed Agriculture**

(*Source*: Subba Reddy *et al.*, 2008)

It would be difficult to provide the details of plans for all the agro-climatic zones in this book. Therefore, crop contingency plans for kharif season for a few important agro-climatic zones representing crop based regions are given as below.

1) Rice based Production System

Recommended kharif contingency plans for delayed monsoon in sub-humid Deep Inceptisols in Eastern Plain Zone of Uttar Pradesh are as follows

Month	Fortnight	Crop/Intercropp-ingsystems	Cultural operations including Mid season corrections	Remarks
June	I Fortnight	-	Summer ploughing	
	II Fortnight	Rice(NDR-97)	Dry sowing	
July	I Fortnight	Rice(NDR-97) Maize (Naveen) Pigeonpea (Bahar) Maize+Pigeonpea	Dry sowing	
	II Fortnight	Rice(NDR-97) Pigeonpea (Bahar)	Dry sowing	
August	I Fortnight	Pearlmillet (BJ-560) (BJ-104 Hybrid)	Sowing	If rains occurs
	II Fortnight	Pearlmillet (BJ-560) (BJ-104Hybrid) Blackgram (T-9) Greengram (Jyoti,Jagriti)	Sowing	
September	I Fortnight	Maize Sorghum Pearlmillet	Harvesting for fodder	
	II Fortnight	Maize Sorghum Pearlmillet	Harvesting for fodder	
October	I Fortnight	Rice Sorghum Maize Pearlmillet	Harvesting of various crops	
	II Fortnight	-	Land preparation for rabi crops	

2) Maize based Production Systems

Recommended kharif contingency plans for delayed monsoon and other aberration in Semi-arid inceptisols in undulating plains of Punjab are as follows.

Month	Fortnight	Crop/Inter cropping system (Varieties)	Cultural operations mid-season corrections	Remarks
June	I Fortnight	-	-	
	II Fortnight	-	-	
July	I Fortnight	-	-	
	II Fortnight	Maize Greengram, blackgram Sowing pearlmillet (f),clusterbean(f)	Re-sowing	Due to dry spell Late onset of monsoon
August	I Fortnight	Maize	Weed control, intercultural operations, application of 2nd dose of Nf bearthing-up	July II fortnight Sown maize
		Greengram, blackgram, pearlmillet (f), clusterbean (f)	Sowing	Late onset of monsoon
		-do-	Intercultural operations	July II fortnight Sown crops
	II Fortnight	Pearlmillet(f)	Harvesting	July II fortnight sown
		Pearlmillet(f)	Sowing	Late onset of monsoon
		Greengram, blackgram	Pest control	July II fortnight sown
September	I Fortnight	Greengram, blackgram, pearlmillet (F), clusterbean (F)	Intercultural operations	August I fortnight Sown crops
		Greengram ,blackgram	Pest control	August I fortnight sown
	II Fortnight	Pearlmillet(f)	Intercultural operations	August II fortnightSown crops
		Pearlmillet(f)	Harvesting	
October	I Fortnight	Maize, greengram, blackgram &clusterbean(f)	Harvesting & threshing	July II fortnight Sown crops
		Pearlmillet(f)	Harvesting	August II fortnight Sown crops
		-	Soil moisture conservation Practices for sowing of Rabi crops	

Month	Fortnight	Crop/Inter cropping system (Varieties)	Cultural operations mid-season corrections	Remarks
	II Fortnight	Greengram, blackgram & clusterbean(f)	Harvesting & threshing	August I fortnight Sown crops
		-	Soil moisture conservation practices for sowing of Rabi crops	

3) Groundnut based production systems

Recommended kharif contingency plans for delayed monsoon in the scarce rainfall zone of Rayala seema of Andhra Pradesh is as follows.

Month	Fortnight	Crop/Intercropping system (Varieties)	Cultural operations including mid-season corrections	Remarks
June	I Fortnight	Groundnut		—
	II Fortnight	Groundnut+pigeonpea/castor	Top dress of urea to castor	—
July	I Fortnight	Groundnut+pigeonpea/castor	,,	—
	II Fortnight	Groundnut+pigeonpea/castor	,,	—
August	I Fortnight	Groundnut+pigeonpea/castor	,,	—
	II Fortnight	Foddersorghum,horsegram, sunflower	Application of urea to fodder jowar Whenever rainfall is there	—
Sept.	I Fortnight	Fortnight	Horsegram	—
	II FortnightI	,,	—	—

4) Rabi Sorghum Based Production Systems

Recommended kharif contingency plans for delayed monsoon for the arid vertisols in the northern dry zone of Karnataka is as follows.

Normal rains with timely onset of monsoon in September

Sorghum (SPV86,CSH7R/CSH8R,5-4-1) sunflower and field beans (CO.7)	:	Complete sowing within one week after the receipt of I[st] soaking rains.
Safflower (A300)	:	Completesowingwith15-20 days after first soaking rains
Chickpea (A-1,N-52)	:	Complete sowing within a month after soaking rains

Delayed onset of monsoon in October		
Sorghum, field beans and sunflower	:	Sow SPV86 upto mid October. Beyond mid October, sorghum may be sown only for fodder. However sorghum M-35-1may be sown for grain upto first fortnight of November
Safflower and chickpea	:	Sow upto early November and complete sowing within a fortnight after the receipt of first soaking rains

5) Pearl millet based Production Systems

Recommended kharif contingency plans for delayed monsoon in deep aridsols in Western zone of Haryana is given in page 293.

These above crop contingency plans are for an agro-climatic region which consists of number of districts with varied soils types, social structure of farming community and the irrigation facilities etc. They may provide a broad outlook for delayed monsoon which may not be of much use for the district administrators to take up corrective measures based on the available resources. Therefore the Parliamentary Consultative Committee on Agriculture (PCCA), Food, Civil Supplies and Consumer Affairs, Government of India (GOI) suggested the preparation of district level contingency plans for the entire country for droughts, floods and other extreme weather events, covering crops, livestock and fishery sector based on the available research information. The Indian Council of Agricultural Research (ICAR) was entrusted with the responsibility to prepare the district wise contingency crop planning. This was carried out by Central Research Institute for Dryland Agriculture (CRIDA) during 2010 with the collaboration from all the State Agricultural Universities and other Research Institutes of ICAR. Initially, by the end of June 2011, out of the total of 576 districts in the country, contingency plans for 209 districts have been completed and the number increased to 350 by mid-2012. The contingency plan consists of two parts, viz., the first part provides information on the resources availability in the district while the second part deals with the strategies for weather related contingencies in respect of agriculture, livestock, poultry and fisheries sector (Venkateswarlu *et al.,* 2011). These plans were made available to State Department of Agriculture for operational use, testing and further refinement. The first part of the plans consists of district profile which includes the agroclimatic zonation, geographical coordinates, names and address of various agencies related to Agriculture, Rainfall distribution, soils and land use classification, gross and net sown area statistics of major crops, sources of irrigation, major field and horticultural crops, livestock and poultry statistics, fisheries, production, proneness of the district to calamities

Pearlmillet based production system

Month/ Fortnight		Crop / Intercropping system (Varieties)	Cultural operations including mid-season corrections	Remarks
June	II	DhainchaPearlmillet	Ploughing and bunding of field boundary Dry seeding, if mid season drought the dhaincha can be kept for seed purpose, avoid green manuring Raising of pearlmillet nursery if irrigation water available.Go for sowing with ridger seeder if rains are 40 to 50mm.	To absorb maximum rain water. To reduce the risk of failure of sowing due to continuous rain. To reduce risk of crusting due to rains after sowing.
July	I	Pearlmillet	Sowing with ridgerseeder on the shoulder of ridges.	To save the crops from excess rain
	II	Pearlmillet PulsesPearlmillet Castor Pearlmillet	Transplanting of 3 weeks old nursery Do sowing of pulses.Sow short duration varieties (HHB67-2 and HHB68)Sowing at 75 to 90cm row spacing. Thinning, Soil mulching by wheel hand hoe	In case of continuous rains To overcome delayed onset Of monsoon. To reduce extra plant population to overcome the drought. To reduce evaporation losses.
August	I	Pearlmillet	Harvest every 3rdrow for fodder, life saving irrigation if possible. Straw mulching after Interculture, Hoeing and interculture.	To manage drought situation. Hoeing and Interculture.
	II	Pulses, clusterbean and castor	Hoeing and Interculture.	To conserve soil moisture.
September	I		Field preparation for rabi crops and moisture conservation through harrowing followed by planking. Advance application of fertilizer for mustard	Better solublization of nutrients.
	II		Blade harrowing for moisture conservation	

and thematic maps of weather resources. As it provides firsthand information about the district resources and its vulnerability to different kind extreme events, it would help the planners to prepare measures keeping in view of the resources for immediate action. As an example, model plan for rainfed district, *viz*., Amaravathi in Maharashtra where drought is the major climate constraint, and Krishna, an irrigated district in Andhra Pradesh where floods and cyclones are common were prepared (Venkateswarlu *et.al*,2011). By the end of August 2012, such comprehensive plans for 309 districts in the country were âccomplished and the same were put on CRIDA website **http://www.crida.in** for immediate use. As the comprehensive plans for each district runs in to roughly 40 pages. CRIDA has brought another publication during 2012 exclusively devoted to crop plans under delayed monsoon conditions viz., (2 to 8 weeks delay)for 100 districts in peninsular India comprising Andhra Pradesh, Karnataka, Tamil Nadu and Maharashtra (Y.G. Prasad et.al. 2012). The same is available in the CRIDA website. As an example the crop plans for districts viz, Mahabubnagar (Andhra Pradesh) and Solapur (Maharashtra) under normal and delayed monsoon rainfall conditions along with the soil maps of the districts (Fig 7.13 & 7.14) which are reserved from the above publications is given below:

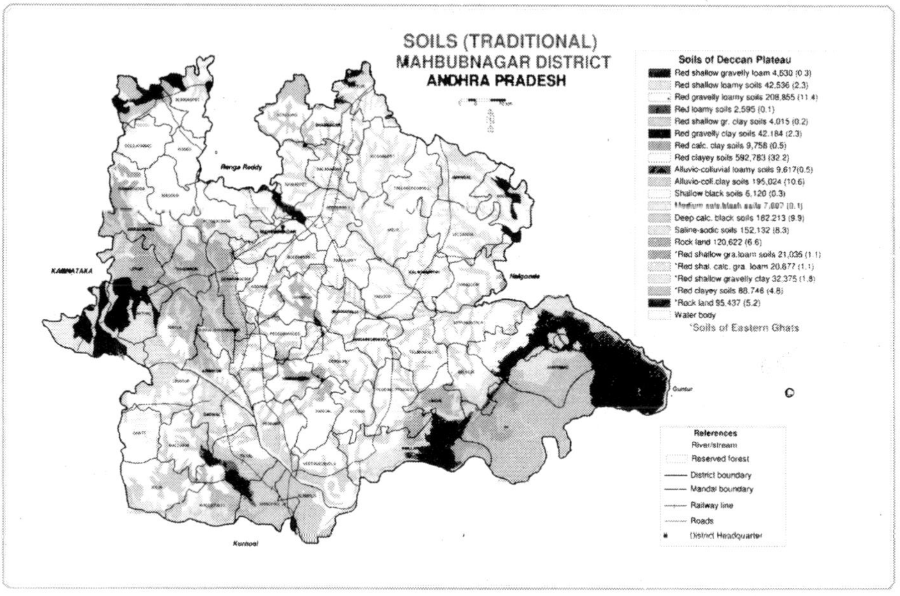

Source: (Y.G. Prasad *et al*., 2012)

Fig. 7.13: Soil map of Mahabubnagar

Mahabubnagar District

Mahabubnagar district falls under Southern Telangana agro climatic zone (AP-7) in Andhra Pradesh. The district receives a mean annual rainfall of 749mm of which 588mm is received during southwest monsoon and 101mm during northeast monsoon. The dominant soil types in the district are black soils and redsoils (Figure 7.13). The district has a cropped area of 566000 ha under rainfed situation (78% of net sown area). The major crops cultivated in the district are maize, castor, rice, sorghum, pigeonpea and cotton.

Normal sowing window for major field crops for Mahabubnagar

Season/ Situation	Maize	Castor	Rice	Sorghum	Pigeonpea
Kharif-Rainfed	3rd week of June to 3rd week of July	2nd week of June to 1st week of July	—	2nd week of June to 4th week of June	3rd week of June to 3rd week of July
Kharif-Irrigated	--	—	1st week of June to 4th week of July	—	—
Rabi-Rainfed	--	--	—	1st week of September to 4th week of September	—
Rabi-Irrigated	3rd week of October to 3rd week of November	—	2nd week of October to 1st week of December	—	—

Contingency crop plan for Mahabubnagar district

Major Farming situation	Normal cropping system	Suggested contingency crops/cropping systems and cultivars			
	Normal onset of monsoon (2nd week of June)	Delay by 2 weeks (4th week of June)	Delay by 4 weeks (2nd week of July)	Delay by 6 weeks (4th week of July)	Delay by 8 weeks (2nd week of August)
Rainfed red Soils	Maize (DHM-103, 105,113) Castor	Maize (DHM-103,105, 113) Castor (Kranthi,Jyothi, GCH-4,6,DCH-519)	Maize (DHM-109, 115,Prakash) Castor +Pigeonpea(1:1) Castor:GCH-4,6,DCH-519 Pigeonpea:Maruthi, Lakshmi,PRG-158	Pigeonpea(PRG-158, Lakshmi,Maruthi) Castor+Pigeonpea (1:1) Castor:Kranthi, Jyothi, Kiran Pigeonpea:Durga, PRG-100	Pearl millet (ICTP-8203, ICMV- 221,HHB-67) Castor+Pigeonpea (1:1) Castor: Kranthi, Jyothi, Kiran Pigeonpea: Durga, PRG-100
	Pigeon pea	Pigeon pea (lRG41, Maruthi,Abhaya)	Pigeonpea (Maruthi,PRG-100)	Pigeonpea	Pigeonpea (Maruthi, Lakshmi, PRG-158)
	Groundnut	Groundnut (Vemana Prasuna, ICGV-91114)	Castor+Pearlmillet/ Finger millet(1:1)	Pigeonpea	Castor+piogeonpea (1:1)
	Sorghum	Sorghum/Sorghum + Pigeonpea Sorghum: PSV-1,Palem-2,CSV-15, CSH-14,18	Pearl millet (ICTP-8203, ICMV-221,HHB-67, RHB-121)	Pearl millet (ICMH-67,RHB-121)	Castor+ Pigeon pea (1:1)/ Horse gram Horsegram: PZM-1,VHG-62,VHG-9
Rainfed black soils	Maize	Maize (DHM-107,111)	Maize (DHM-109,115)	Maize(DHM-109,115, GKMH-1701)	Castor +Pigeon pea(1:1)
	Pigeon pea	Pigeonpea (LRG-30, Abhaya, Lakshmi, Maruthi)	Pigeon pea (LRG-41,MRG-66)	Pigeonpea (PRG-100,Durga)	Sunflower (Modern, DRSF-113)
	Cotton	Cotton (LRA-5166, LK-861)	Cotton (LRA-5166,LK-861)	Fodder sorghum (PC-23,SSG-59-3)	Sunflower (Modern,DRSF-113)
	Sunflower	Sunflower (KSBH-1,53,DRSF-113)	Sunflower (SBSH-44,MOrder, DRSF-108)	Fodder Sorghum	Fodder sorghum (PC-6,23, SSG-59-3)
	Finger millet	Finger millet (Godavari, Ratnagiri)	Fingermillet (Maruthi, Suraj, Champavathi) Champavathi)	Sunflower (Modern, DRSF-113, KBSH-44)	Sunflower

Source: (Y.G. Prasad et.al. 2012)

Solapur District

Solapur district falls under Scarcity agroclimatic zone (MH-6) in Maharashtra. The district receives a mean annual rainfall of 659mm of which 499mm is received during southwest monsoon and 115mm in northeast monsoon. The dominant soil types in the district are medium deep and deep black soils (Figure 7.14).The district has a cropped area of 759000 ha under rainfed situation (82%of net sown area). The major rainfed crops cultivated in the district are sorghum, pigeonpea and irrigated crops are sugarcane, wheat, sorghum, chickpea, sunflower and safflower.

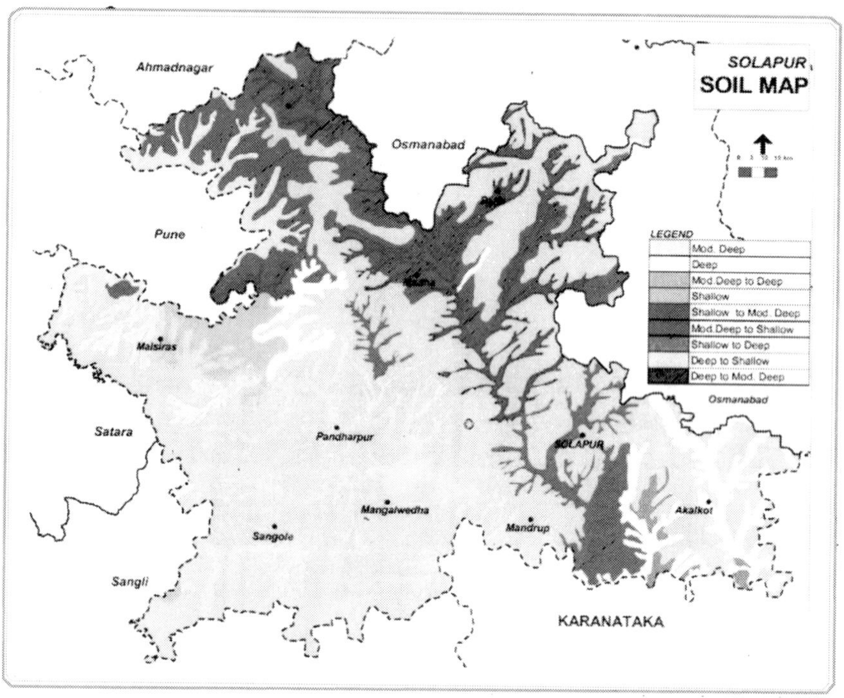

Source: (Y.G. Prasad et al., 2012)

Fig. 7.14: Soil map of Solapur

The normal sowing window for major field crops and contingency crops plans of Solapur district under different Soil conditions are reproduced in the following pages.

Normal sowing window for major field crops in Solapur District

Season/Situation	Sorghum	Pigeonpea	Chickpea	Sunflower	Safflower	Wheat
Kharif- Rainfed	—	2nd week of June to 2nd week of July	—	2nd week of June to 2nd week of August	—	—
Kharif- Irrigated	—	2nd week of July to 2nd week of August	—	2nd week of June to 2nd week of August	—	—
Rabi-Rainfed	2nd week of September to 2nd week of October	—	3rd week of September to 2nd week of October	2nd week of September to 2nd week of October	2nd week of September to 2nd week of October	—
Rabi-Irrigated	4th week of September to 2nd week of October	—	3rd week of Oct to 3rd week of November	1st week of September to 4th week of October	2nd week of September to 4th week of October	1st week of Nov to 1st week of December

Contingency crop plan for Solapur District

Farming situation	Normal Crop Normal onset of monsoon (3rd week of June)	Suggested contingency crops/cropping system and cultivars			
		Delayby2 weeks (1st week of July)	Delayby4 weeks (3rd week of July)	Delayby6 weeks (1st week of August)	Delayby8 weeks (3rd Week of August)
Medium deep to deep black soils	Sunflower	Sunflower+ Pigeonpea (2:1) Pigeonpea: BSMR-736,853 BDN-708 Sunflower: Mordern	Pigeonpea/ Castor Castor(VI-9,DCH-32) Pigeonpea: BSMR-736,853, BDN-708	Pigeonpea+clusterbean (1:2Sunflower+ Pigeonpea(2:1) Foxtailmillet (Arjun) Pigeonpea: BSMR-736, 853,BDN-708 Sunflower: Mordern	Sunflower/)Castor/Castor Ridgegourd (mixedcropping) Foxtailmillet Sunflower: Mordern Castor:VI9 DCH177, DCH32 Foxtailmillet: Arjun
	Pigeonpea	Pearlmillet+Pigeonpea (2:1) Soybean + Pigeonpea (3:1) Pearlmillet: (Shraddha, Saburi,Shanti) Pigeonpea: BSMR-736,853, BDN-708Soybean. JS-335, DS-228, JS9305	Pigeonpea+Clusterbean (1:2) Pigeonpea: BSMR-736,853, BDN-708	Pigeonpea +coriander(1:2) Castor, Foxtailmillet Pigeonpea: BSMR-736,853, BDN-708	Sunflower/ Castor/Castor Ridgegourd (mixed cropping)/ Foxtailmillet Sunflower: Mordern Castor:VI9, DCH177, DCH32
	Greengram	Pigeonpea + Greengram (1:3)Pigeonpea: BSMR-736,853 BDN-708,711 Greengram: TAU-1,T-9	Pigeonpea+Coriander (1:2) Pigeonpea: BSMR-736,853 BDN-708	Pigeonpea+Leafyvegetable (1:2)/Castor/ Foxtailmillet Pigeonpea: BSMR-736, 853 BDN-708 Castor:GCH6, DCH32	Sunflower/Castor/ Castor+Ridgegourd (mixed cropping) Sunflower: Mordern

(*Contd.*)

Farming situation	Normal Crop Normal onset of monsoon (3rd week of June)	Suggested contingency crops/cropping system and cultivars			
		Delayby2 weeks (1st week of July)	Delayby4 weeks (3rd week of July)	Delayby6 weeks (1st week of August)	Delayby8 weeks (3rd Week of August)
				Foxtail milletArjun	Castor: VI9, Aruna, DCH177, DCH32Foxtail millet: Arujun
Fallow- rabi cropping areas	Blackgram	Pigeonpea+Blackgram (1:3) Pigeonpea: BSMR-736,853, BDN-708, 711 Blackgram: TPU-4	Pigeonpea + Leafy vegetable (1:2) Pigeonpea: BSMR-736,853, BDN-708, 711	Pearl millet + Pigeonpea (2:1)Pearl millet: (Shraddha,Saburi, Shanti) Pigeonpea: BSMR-736, 853, BDN-708	Sunflower/ Castor+ Ridge gourd (mixed cropping Foxtail millet (Arjun) Sunflower: Mordern Castor: VI9, GCH-4,5,6, Aruna, DCH177, DCH32 Foxtail millet: Arjun
	Kharif - Fallow-Rabi sorghum / Safflower Sunflower/ Chickpea Kharif - Fallow Strip cropping of Chickpea + Safflower (6:3)/ Strip cropping of Rabi sorghum + Chickpea (6:3)	Kharif - Fallow-Rabi sorghum / Rabi crop Safflower/ Sunflower /ChickpeaStrip cropping of Chickpea + Safflower (6:3), Strip croppingof Rabi sorghum+ Chickpea (6:3) Rabi sorghum: M35-1,Phule Yasodha, Phule Mauli,Phule Vanitha Safflower: N62-8,	Kharif fallow followed by Rabi crop	Kharif fallow followed by rabi crop	Kharif fallow followed by rabi crop

(*Contd.*)

Farming situation	Normal Crop	Suggested contingency crops/cropping system and cultivars			
	Normal onset of monsoon (3rd week of June)	Delayby2 weeks (1st week of July)	Delayby4 weeks (3rd week of July)	Delayby6 weeks (1st week of August)	Delayby8 weeks (3rd Week of August)
	Greengram/ Blackgram/ Cowpea	Bhima Sunflower: LS11Chickpea: Vikas Fallow	Cowpea Cowpea: C-152. Krishkanchan.	-	-
Shallow to Medium deep black soils	Pearl millet	Pearl millet + Horsegram(2:1) Pearl millet + Mothbean (2:1)	Sunflower + Pigeonpea (2:1)	-	-
Shallow black soils	Castor	Castor + Cluster bean (1:2)Castor and ridge gourd (mixed cropping) Castor:GCH-4,5,6, Aruna.	Grasses (Dongari, Madras Anjan)	Grasses (Dongari, Madras Anjan)	Grasses (Dongari, Madras Anjan)
	Horsegram	-	Grasses (Dongari, Madras Anjan)	Grasses (Dongari, Madras Anjan)	Grasses (Dongari, Madras Anjan)
	Mothbean	-	Grasses (Dongari, Madras Anjan)	Grasses (Dongari, Madras Anjan)	Grasses (Dongari, Madras Anjan)

Source: (Y.G. prasad et.al. 2012)

7.10 Drought Proofing Mechanisms

Droughts are not new to the Indian people as many disastrous events like famines occurred in the past that had left out untold miseries in terms of hunger, malnutrition, ill health, migration, loss of human life and as well live stock population. Improved agricultural technologies, increased area under irrigation, availability of high yielding varieties under the aegis of "Green Revolution", better transportation and communication facilities and improved market facilities reduced the vulnerability of a region to droughts considerably in spite of increased food and fodder requirement due to increasing population over the last 60 years. A study taken up under National Agricultural Technology project on "Geo referenced Resources inventory of Rainfed Rice Production System" at CRIDA, Hyderabad during 1999-2002 (Rao *et.al* 2002) revealed that the Paddy yield reduction due to a dry spell duration of above 10 days at reproduction/maturity phase of rice crop in Orissa state was higher during the decade 1970 to 1980 compared to the yield loss during 1990-99 for similar type of dry spell. This was attributed to the increased use of fertilizers and improved irrigation facilities. Thus the technologies developed during last few decades at various locations in the country contributed significantly to increased agricultural production. However fall in total food grain production due to droughts have been noticed in all the recent drought years of 1987, 2002 and 2009 indicating the dominance of nature over the technologies. Production from irrigated regions remains almost stationary and the fluctuation in agricultural production is due to variability in production from rainfed areas. Efforts have been continuing at many locations in the country under All India Coordinated Research Project on Dryland Agricultural of (AICRPDA) ICAR since 1970 to develop suitable drought management technologies for different agro climatic regions in the country to enhance production from rainfed regions. Even after achieving the full irrigation potential, a large area in the country remains under rainfed conditions. Thus the production from this sector assumes importance in the National Agricultural growth. Also crop specific institutes under ICAR and the State Agricultural Universities have been pursuing research in developing drought resistant varieties to minimize the impacts of climate hazards on agricultural production.

Drought proofing mechanism for agricultural crops consists of improving water availability for the crops through watershed development programs, construction of farm ponds for supplemental irrigation, percolation tanks for improving ground water status and enhancing *in-situ* soil moisture conditions by adapting soil and water conservation measures. Also, alternate cropping system, crop choice, agronomic manipulations, proper management of fertilizer and pesticide applications, Integrated Farming System approach for drought risk management, packages, practices for improved crop varieties developed through continuous research efforts by National Agricultural Research System (NARS) consisting of ICAR institutes and SAU and their regional centres helped to improve productivity from rainfed regions. However the acceptance of technologies by the rainfed farmers is mostly limited to switching over to improved crop varieties and adaptation of a few agronomic practices, as the Soil and Water Conservation Technologies falling under Natural Resource Management (NRM) are to be taken at community level. Brief discussion on the above topics is as follows.

7.10.1 Watershed Development for Drought Management

Watershed is the most appropriate ecological area unit for efficient management of land and water resources with required homogeneity in the drought-prone areas. The watershed management can thus play a key role in combating drought and desertification in rainfed areas by improving its water resources. A suitable size of watershed can be selected for making concentrated efforts and for easy implementation of the program/resource. Detailed surveys involving a combination of conventional and participatory approaches, technological options with respect to soil and water conservation, rainwater harvesting, tree plantation, pasture development, livestock management, crop management, and the land management to minimize the risk due to droughts can be taken up (H.P.Singh, 2001). He has further reported that this approach ensures planning on the basis of the total available water resources, conjunctive use of surface and ground water, prioritization for rational use of water and preparation of an integrated action plan for utilization of natural resources for a sustainable productivity and increase the socio-economic capacity of rainfed farmers to face the droughts effectively.

Therefore, the central objective of watershed management is conservation of natural resources of soil and water for increasing productivity in a sustained manner. In this context, runoff management strategy as identified by H.P. Singh (2001) in as follows

- Holding the bulk of rainwater within the field, depending upon the soil water storage capacity: *in situ* water harvesting
- Diversion of excess water for

  Storage in village pond

  Augmenting recharge of deep groundwater

  Recharging shallow aquifers

  Storage in farm ponds

Maintaining adequate supply of rainwater to village pond is the first priority as this is a common property resource (CPR) utilized for various purposes. Recharging of aquifers is the next option as the shallow dug wells are owned by a large number of farmers in rainfed areas. When recharging of aquifers is not possible in the absence of water bearing rocks, the farm pond remains the only option to conserve excess runoff for drought proofing. (H.P. Singh, 2001).

The success of watershed depends upon the availability of amount of runoff for water harvesting through ponds in different agro-climate regions in the country.

Bharat R. Sharma *et al.* (2010) has estimated the water harvesting potentials of selected districts in various States using monthly average rainfall, PET and water holding capacity of soils in the water Balance procedures adopted by FAO across dominant rainfed crops districts. The results are given in the following Table 7.14.

Table 7.10: Water surplus (mm) in different selected districts in India

State	Center	Total Annual Surplus (mm)	State	Center	Total Annual Surplus (mm)
Andhra Pradesh	Adilabad	212	MP	Betul	485
	Chittoor	23	MP	Bhind	494
	East Godavari	53	MP	Bhopal	520
	Karimnagar	57	Chhattisgarh	Bilaspur	632
	Khammam	255	MP	Chhatarpur	410
	Mahabubnagar	58	MP	Chhindwara	439
	Medak	55	MP	Damoh	520
	Nizamabad	171	MP	Datia	360
	Parbhani	134	MP	Dewas	384
	Rangareddy	48	MP	Dhar	263
	Srikakulam	138	MP	Durg	524
	Visakhapatnam	47	MP	Guna	508
	Vizianagaram	264	MP	Gwalior	299
	Warangal	99	MP	Hoshangabad	648
Jharkhand	Bhojpur	405	MP	Indore	353
	Dumka	530	MP	Jabalpur	714
	Gumla	749	MP	Jhabua	173
	Hazaribagh	586	MP	Khandwa	244
	Palamu	562	MP	Khargone	185
	Ranchi	698	MP	Mandla	784
Gujarat	Ahmedabad	243	MP	Mandsaur	328
	Amreli	26	MP	Morena	349
	Baroda	349	MP	Narsinghpur	909
	Bhavnagar	38	MP	Panna	486

(*Contd.*)

State	Center	Total Annual Surplus (mm)	State	Center	Total Annual Surplus (mm)
	Broach	299	Chhattisgarh	Raigarh	918
	Junagadh	148	Chhattisgarh	Raipur	590
	Kaira	312	Chhattisgarh	Raisen	849
	Panch Mahals	249	Chhattisgarh	Rajnandgoan	614
	Rajkot	88	MP	Ratlam	478
	Sabarkantha	300	MP	Rewa	429
	Surat	560	MP	Sagar	674
Haryana	Ambala	175	MP	Satna	416
Karnataka	Belgaum	688	MP	Sehore	458
	Bidar	165	MP	Seoni	735
	Kolar	7	MP	Shahdol	627
	Mandya	5	MP	Shahjapur	444
	Mysore	44	MP	Shivpuri	531
	Shimoga	266	MP	Sidhi	473
Chhattisgarh	Balaghat	772	Chhattisgarh	Surguja	733
Chhattisgarh	Bastar	748	Chhattisgarh	Tikamgarh	449

(*Source*: Bharat R. Sharma *et al.*, 2010)

It can be seen that the highest potential for water harvesting are seen in the Central India comprising Madhya Pradesh and Chhattisgarh, South Gujarat, Jharkhand. Some districts in Andhra Pradesh and Karnataka have low water surplus amount.

7.10.2 Watershed Development in India

Soil and Water conservation programs were taken up in low rainfall regions since independence on extensive areas. The trend has continued under DPAP (Drought Prone Areas Program), DDP (Desert Development Programs) mainly as a resource conservation programs in rainfed areas. During Nineties, soil and water conservation programs were extended to catchments of different dams with an aim to reduce the soil erosion and thus preventing the siltation of reservoirs. A significant improvement in the resource conservation was observed and a realization dawned upon that the rainfed areas could be treated in such a manner that it would help in both reducing the soil erosion from catchments and improve the resource base in catchments. Success of two water shed programs in conserving and developing the resources are worth mentioning here. The success stories of Ralegan Siddhi village in the district Ahmednagar, Maharashtra where a degraded village with many social problem before 1975 changed in to a successful village with community participation by the remarkable social work taken up by Shri Anna Saheb Hazarae. Similarly, promotion of Soil and water conservation measures by Central Soil and Water

Conservation Research and Training Institute (CSWR&TI) Dehradun in the foot hills of Shivalik ranges has rehabilitated the eroded slopes and rainwater harvesting has opened new avenues of income for the people, reduced siltation and increased water availability for rainfed crops. These two are well known examples of watershed development and their success in managing climate risks such as droughts of different intensities in India.

A model watershed program was initiated by ICAR through selection of 47 watersheds across the country with full technical back up by relevant Research Institutes of ICAR. CRIDA (Central Research Institute of Dryland Agriculture), Hyderabad and Central Soil and Water Conservation Research and Training Institute (CSWCRT&I), Dehradun participated in this initiative through the implementation of 30 and 17 watersheds respectively. With the success of these watershed in terms of overall productivity, improvement in resource base, increased ground water levels, the Ministry of Agriculture launched a nationwide watershed development program call NWDPRA (National Watershed Development Program for Rainfed Areas) as separate developmental program within the ministry along with separate allocation of budget. An area of 500 ha was chosen as watershed area.

A committee under the Chairmanship of Dr.Ch. Hanumatha Rao examined in detail the implementation of Watershed Program during early Nineties and suggested to implement the watershed programs even in DPAP and DDP programs. As these are major developmental programs of Rural Development, implementation of watershed programs had moved from Department of Agriculture to Department of Rural Development. Separate guidelines (different from NWDPRA) were formulated by Rural Development Ministry with major focus on socio economic issues. The Programs by Ministry of Rural Development were reviewed at different times subsequently and another set guidelines were brought under the name of Hariyali Guidelines, which withheld the participation of NGOs in the watershed program.

After the constitution of National Rainfed Area Authority (NRAA), by GoI, a need felt to have a common guidelines for watershed programs encompassing the activities of rural development, agriculture and forestry and a new set of guidelines were developed and NGOs participation was accepted to a certain extent within the program implementation. One of the major change was redefining of watershed area from 500 to 4000-5000 ha and constitution of state level nodal agencies for identification, preparation and implementation of watershed programs. Different state governments have started programs of special soil and water conservation programs such as Jaladhara in Madhya Pradesh, APRLP (Andhra Pradesh Rural Livelihood Programs) in Andhra Pradesh, Sujala in Karnataka etc which looked at different facets of watershed development program for improvement. Similarly various NGOs such as Agakhan foundation, MYRADA, BAIF, WASSAN etc and NABARD also worked on watersheds with funding from either Federal/state governments or funding from external sources such as GTZ, Germany, Denmark and SDC etc.

7.10.3 The Success Stories of Watershed as a Drought Proofing Mechanism

The water shed development program was initiated as early as during 5th Five year plan and the real acceleration to the programs was given in 8th five year plan period. Between 1995-96 to 2006-07, the Ministry of Rural Development has sanctioned about 45062 projects out of which 32,411 projects have been completed by 2010-11. (MRD, GOI, 2011) To review the impact of watershed on the rural economy, a workshop was held on "Success stories under watershed management" during February 2011 at New Delhi. A "Compendium of Success Stories under Watershed Development Programs" was brought by the Department of Land Resources, Ministry of Rural Development, GOI (2011). About 55 most prominent success stories in different states viz., Andhra Pradesh, Chhattisgarh, Gujarat, Jharkhand, Karnataka, Kerala, Madhya Pradesh, Nagaland, Orissa, Punjab, Tamil Nadu and at two ICAR Institutes viz, Central Soil and Water Conservation Research and Training Institute (CSWCRTI) and National Research Centre for Agro Forestry (NRCAF) were reported in the above referred publication. The success stories of water sheds and farm ponds have proved their ability in curtailing the drought impacts considerably by minimizing agricultural losses and stabilizing rural economy. As large investments were made by GoI over the years, there is a need to mobilize rural population to properly maintain the watersheds for long term use. A few examples of the success achieved in managing climate risks effectively as reported in the proceedings of the work shop during Feb, 2011 at New Delhi are briefly given below.

Andhra Pradesh

Water shed programs at Guntimadugu village in Kadapa district of Andhra Pradesh were taken under Drought Prone Area Program. In this, series of physical measures to arrest soil erosion, reducing runoff and increasing soil moisture were taken up through soil and water conservation methods. These measures significantly reduced the speed of runoff and increased the ground water recharge. The pre and post watershed development situation in the villages is as follows (Table 7.11).

Table 7.11: Pre and post watershed development in villages of Kadapa district

Item	Unit	Pre-Project Period	Post-Project Period
Status of water table	m	10-12	7 to 9
No. of ground water structures rejuvenated	no.	-	20
Availability of drinking water	Months	8-12	12
Increasing irrigation potential	Acres	1015	1250
Area under single crop	Acres	585	640
Area under double crop	Acres	365	485
Area under multiple crop	Acres	65	125
Area under crop production	Acres	1015	1250
Area under horticulture crop	Acres	25	180
Area under vegetation	Acres	6	15
Area under fodder production/fire wood	Acres	0	6
Milk production	Liters	113	207
No. of SHGs, Co-operative Societies and thrift groups	no.	4	10
People below poverty line	no.	246	208
No. of migrating people	no.	287	93
Women participation in decision making	no.	0	9

(*Source*: Dept. of Land Resources, MRD, GOI, 2011)

It is seen that with improved economic conditions over the years, the migration labour reduced considerably in the watershed area.

Gujarat

Availability of water has always been a major problem to the people of Rajkot district due to the receipt of low rainfall. Therefore, agriculture was practiced at sustenance level and migration of rural population to nearby cities has been a regular feature during summer months. Under the Desert Development Program (DDP), the Gujarat State Watershed Management Agency (GSWMA) was supporting the implementation of watershed program in the Maha River Watershed project spread over an area of 4255 ha covering nine watershed villages in Rajkot district. A series of 79 check dams were constructed on the 35 km long Maha River and its tributary. This has resulted in increase in irrigated area to about 4000 ha in the cropped area. With more fodder availability, the quality of livestock improved and the milk production from a cow increased from 7 lit/day to 10 lit/day. The pre and post watershed project situation at Maha River Watershed project is as follows (Table 7.12).

Table 7.12: The pre and post watershed situation in Gujarat

S.No	Criteria	Pre-watershed scenario (1995-96)	Post-watershed scenario (2006-07)
1.	Cropping intensity	One crop per year	Three crops per year
2.	Crop diversification	Less cash crops grown (only 10% area was used for cotton cultivation)	Area increased under cash crops like cotton, groundnut, castor, cumin and vegetables
3.	Increase in cropping area	Cotton-270ha Groundnut – 82 ha Millet – 285 ha Cumin -0ha Wheat -0 ha	Cotton-2270ha Groundnut – 465ha Millet – 820 ha Cumin -17ha Wheat -164 ha
4.	Increase in production	Cotton-438 ton Groundnut – 123 ton Millet – 314 ton	Cotton-4995 ton Groundnut – 884 ton Millet – 1600 ton
5.	Increase in productivity	Cotton-1.62 ton/ha Groundnut – 1.5 ton/ha Millet – 1.1 ton/ha	Cotton-2.2 ton/ha Groundnut – 1.9 ton/ha Millet – 1.95 ton/ha Cumin – 0.8 ton/ha Wheat -3.12 ton/ha
6.	Increase in irrigated area	Rabi – 400 ha Summer - Nil	Rabi – 2200 ha Summer – 700 ha
7.	Drinking water	Tankers used to provide drinking water in villages	Villages self-sufficient in drinking water
8.	Water table	200-250 feet	50-175 feet

(*Source*: Dept. of Land Resources, MRD, GoI, 2011)

Maharashtra

The Scientists of Central Research Institute for Dryland Agriculture (CRIDA), Hyderabad evaluated the performance of Kadwanchi watershed program spread across 3 villages in an area of 1888 ha in Jalna district of Maharashtra state. This project was operated with the financial assistance from NABARD under Indo-German watershed program between1996-97 to 2001-02. A multi-disciplinary team of Scientists from CRIDA revisited the site after 10 year period (November, 2011) to examine the sustainability aspects. The watershed was rated was highly successful (Osman *et al.*, 2012) on five accounts *viz.*,

1. All structures are intact and safe
2. Dependence mainly on recharge water
3. Change in land use
4. Improvement in infrastructure and
5. Confidence among the rainfed farming community.

The major achievement of the project is in development of water resources that has acted as a shield against drought which resulted in increased quantum of stored water for supplemental irrigation and recharging of ground water. The annual household income before and after in the watershed area is given in Table 7.13.

Table 7.13: Annual household incomes (Rs. In lakhs)

Sources	Before		After		%
	n	Income	n	Income	Change
Crops	38	0.30	38	2.13	610
Vegetables	-	-	14	0.61	-
Fruits	3	2.08	27	6.64	219
Dairy	4	0.14	5	0.25	79
Overall	-	0.40	-	3.21	702

(*Source*: Dept. of Land Resources, MoRD, GOI) n=Sample size

The overall annual household income from different sources after Watershed Development Program (WDP) registered a higher income at 3.21 lakhs than before the project (0.40 lakhs) was initiated, which shows the impact of WDP on the income accrued to be higher at 702 per cent. Because of their improved economy, the villagers could collect about Rs.17.00 lakhs as a donation in a 15 day period to construct a 42 km length road, which they believe that road connectivity is critical for marketing agricultural products.

7.10.4 Watershed Policies and Recommendations

ICRISAT in collaboration with various organizations has brought out a comprehensive assessment of watershed program in India during 2008. Some of the recommendations for strengthening of the watershed program for enhancing the rural livelihoods that includes policies, guidelines, Institutional arrangements, monitoring, technology and gender issue as reported by Wani *et al.* (2008) is given below.

Watershed policies and guidelines

- To enhance the impact of watershed programs, government needs to unify its effort around a new paradigm which shifts the objectives from merely drought-proofing and agricultural production to: sustainably increasing agricultural productivity, reducing poverty, protecting the environment, and building human and natural resource resilience to cope with future challenges, including climate change.

- To help to meet the national goal to conserve, manage and efficiently use scarce water resources, watersheds need to be recognized as the most appropriate framework in which various agencies concerned with surface, ground and drinking water and sanitation can interface to a common purpose.
- The current funding of watershed programs is insufficient to effectively embrace social, environmental, and sustainability objectives.
- We recommend a sum of Rs.20,000/- hectare for integrated watershed development.

Institutional arrangements

- The performance and sustainability of watersheds can be substantially improved by exploring the option of strengthening and supporting small area groups in place of user groups in their planning and execution.

Monitoring and evaluation(M&E)

- Mid-term evaluation, impact assessment after program completion and post-project evaluation after four to five years will enable implementing agencies to make mid-course corrections and governments to adjust policy. M&E information should be put in the public domain.
- There is a need to build capacity within the research establishment to undertake effective technology development for poor people. This requires specific financial allocation, change in mindsets, multi-disciplinary teams with participatory skills and the involvement of poor people from the outset in identifying their particular needs.
- Common property resources can effectively be regenerated as pasture and bio fuel and energy plantations and used to generate income when managed by vulnerable groups. This requires long-term leases, usufruct rights, and financial allocation for development which may need to last beyond project period.

7.10.5 Farm Ponds for Drought Management

Farm ponds constructed or of natural origin in the farm area, are meant either for pre-sowing or protective irrigation during drought period and for livestock consumption. Village ponds are used to meet the needs of drinking water, for both livestock and human population and other needs of such as washing, cleaning and bathing. Water is available in these ponds, for few months to a year depending on rainfall and the catchment's area characteristics. Farm ponds constructed in the fields in rainfed areas could provide one or maximum two supplemental/protective irrigations and a pre sowing irrigation (subject to availability) to post rainy season crops and act as effective measure to manage drought. Scheduling of such limited water supply and setting priorities for water allocation are important. The first task is to cut down conveyance and evaporative losses and to direct water deep into the root zone. The impact of supplemental /lifesaving irrigation of 20 mm at the most critical stages of crop growth during prolonged drought increased the grain yield of majority of crops by 30 to 80% (Venkateswarlu, 1987). Heavy sedimentation, high evaporation, seepage losses, reduced inflows and water pollution limiting their utility. A few recommendations for development of ponds and optimum utilization of stored water as reported by Singh (2001) arc as follows:

1. Reducing sediment input by constructing silt traps at the inlet.
2. Optimizing the depth in relation to surface area.
3. Restricting the entry of animals and human beings in the pond area to minimize the pollution of water. Checking defecation in the catchment area.
4. Cleaning and clearing of bushes in the runoff channels.

On-farm water harvesting through farm ponds on individual holdings is a promise for rainfed areas to provide at least one critical/lifesaving irrigation to Kharif crop to overcome the drought spells. Different state governments (Maharastra, Madhya Pradesh, Gujarat etc.) have initiated special programs such as on farm ponds/small storage structures in order to ensure the sustainability and to improve the livelihoods of people. On-farm water harvesting is a complex process as there is considerable variation in harvestable runoff at different spatial and temporal scales. The factors affecting the harvestable runoff reported by Singh HP (2001) is as follows.

For example

- Black soil would generate more runoff compared to red soil for same rainfall.
- Higher the prevailing moisture more would be the runoff generated.
- Less vegetative cover would generate more runoff.

Thumb rules available for estimation of harvestable runoff indicate that either 10% of rainfall or 30% of available runoff is to be considered for estimation of storage structure.

7.10.6 Success Stories of Farm Ponds

Rained uplands in Eastern India

The on-farm reservoir (OFR) system to mitigate the drought effects on agricultural production system in the rainfed uplands of eastern India was studied. It suggested that the cropping systems should be based on hydrological events considering the water harvesting potential of upland crop fields. The water harvesting potential studies revealed that 85 percent of the supplemental irrigation requirement of the rainfed upland rice during critical growth stages can be met from the surface runoff generated from the rice lands at 50 per cent PE level and the rest can be met from the direct rainfall collected in the OFR (Panigrahi and Panda-2003). The economic analysis by these authors indicated that 12 per cent of the farm area is optimum for the construction of lined OFR that can meet on an average 93 and 33 mm of supplemental irrigation to rice during critical growth stage in wet season and pre sowing irrigation for mustard crop in post monsoon. The BCR, IRR and PBP of the optimum (12%) size of OFR were 1.22, 16.1 % and 13 years respectively (Panigrahi *et.al.*, 2005)

Farm ponds adaptations in drought prone areas of Gujarat (Vittal and Shaita Kishore, 2010)

Farm ponds were constructed in the Surendernagar district in Saurashtra region in Gujarat state. The AKRSP an NGO working in the area adapted the drought coping mechanisms viz., soil and water conservation, water development methods and alternate livelihood development through Self Help Groups (SHG). A total of 416 farm ponds, 105 boribunds, 132 check dams, 18 tanks and 6 water recharging system were constructed, over 46 villages. Significant improvements in livelihood were observed in the study area.

A farmer of the Khintla village in Sayla Lalnka of Surendernagar district constructed a farm pond in his field which has increased the net profit in a year by 4 times due to availability of water to irrigate the land.

Rain water harvesting through farm ponds and shallow dig wells in Bastar region (Patil *et al.*, 2010)

A series of eight farm ponds were constructed in the lowland situation as runoff water can be trapped and stored at different points along the natural drain line covering a command area of 50 ha. Shallow dug wells (15 feet depth) were also constructed to harvest seepage water from the ponds that envisages conservation of maximum rainfall. The total amount of water stored in eight ponds worked out to be 33750 m^3. The farmers were given paddle operated low lift pumps for use in ponds and in dug wells. With this facility, the farmers were able to grow crops in rabi season which was otherwise kept fallow. The farmer could cultivate 1.5 acre of land to grow vegetables in the rabi season after the paddy crop. This could give a return of Rs.65765 from this 1.5 acres of land by utilizing the harvested water through paddle operated low lift pump. About 12.25 per cent increase in paddy yield was realised due to better moisture availability. The net returns obtained by adoption of the technologies were 10.44 times higher than the conventional practices.

Income from Farm pond in Anantapur District

A project to demonstrate the usefulness of farm ponds (both lined and unlined) for increasing the water availability there by increasing the crop production was conducted under an organization BIRD-AP in the chronically drought prone district of Anantapur (Sivarudrappa, 2010). The unlined ponds have served as both percolation ponds for ground water recharge and as an irrigation source. The project was implemented in an area of 361 ha having two tanks, 25 dug wells, 100 bore wells and one-check dam. No farm ponds or percolation tanks existed before the start of the project. A farmer in this area dug two ponds (one with lining and the other without lining) during 2003. Prior to the excavation of farm pond, the ground water table was at 90-150 ft. After the excavation, the ground water level reached to 28 feet due to good recharge from unlined ponds. The farm ponds with lining retained water from June to January and the farmer established 2 acre-mango garden and the water in the lined ponds was exclusively used through drip irrigation system for mango plantation. Using of harvested water from farm ponds to the groundnut crop enhanced the productivity of crop from 4 to 7.9 q/ha. Recharging of exciting defunct well enabled him to cultivate groundnut in rabi season by producing 10 q/ha. With improved ground water conditions, a bore well was drilled in 2006. He was cultivating fruits and vegetables that fetched him additional earnings or Rs.1.71 lakhs during 2007-08 Crop-wise earnings during 2007-08 is as follows:

Table 7.14: Additional income from fruits and vegetables at Anantapur

S.No	Crop	Area (Acre)	Earnings (Rs.)
1.	Banana	1.0	1,00,000
2.	Groundnut	2.0	30,000
3.	Vegetables	1.0	30,000
4.	Paddy	0.5	10,000
5.	Mango	2.0	1,000
6	Sweet orange	1.5	-
		8.0	1,71,000

(*Source*: Sivarudrappa, 2010)

7.11 Rainfed Technologies for Drought Management in Different Agro Climatic Regions

Rainfed regions account for 60 per cent (85m ha) of net cultivated area covering almost all the agro climatic zones in the country. Production from these regions contributes to 44 per cent of country food grain production and supports two thirds of livestock population, influencing the livelihoods of 80 per cent of small and marginal farmers. The gains achieved in irrigated ,system during green revolution is not reflected in the rainfed areas due to climate aberrations and subsistence farming practices followed by resource starved rainfed farmers. Rainfed regions are characterized by diverse climate production system, drought proneness, water scarcity, low soil fertility and the poor economic status of farmers. To achieve the targeted growth rate of Agriculture at 3 to 4 per cent per annum, a need was felt to increase the production from rainfed regions by preparing suitable plans for managing climatic risks. The Research centres of All India Coordinated Research Project for Dryland Agriculture (AICRPDA) spread across the country have evolved location specific interventions since 1970. About 84 technologies relating to crop management, moisture conservation, implements, improved varieties, and water harvesting etc. were compiled by Venkateswarlu *et.al.* (2009).These technologies have contributed greatly in improving productivity from rainfed areas. As some of the technologies such as soil and water conservation measures are community oriented, the governments must initiate and provide adequate funds for such programs to aim at second Green Revolution from rain fed areas. Some of the relevant technologies as reported by Venkateswarlu *et.al.* (2009) suitable for different physiographic regions in the country viz., 1) Southern plateau and hill region (Andhra Pradesh , Karnataka, Tamil Nadu) 2) Central Plateau and hill region (Madhya Pradesh and Rajasthan) 3) Western plateau and hill region (Maharashtra and Madhya Pradesh) 4) Gujarat plains and hills (Gujarat), 5) Eastern plateau and hills region (Jharkhand) 6) Upper Gangetic plains (Uttar Pradesh) 7) Trans Gangetic plains region (Haryana and Punjab) 8) East coast plains and hills region (Tamil Nadu and Orissa) 9) Western Himalayan region (Jammu & Kashmir) are briefly given below.

1) Southern plateau and Hills region (A.P., Karnataka and Tamil Nadu)

a) Water harvesting and supplemental irrigation to rainfed groundnut in Rayalaseema in Andhra Pradesh

Farmers in Kurnool, Anantapur, some parts of Mahabubnagar, Prakasam and Kadapa cultivate groundnut crop solely under rainfed conditions. Crop failure/ low yields in these above regions are due to prolonged dry spells at the critical growth stages.

Protection of groundnut crop during dry spells could be achieved by providing supplemental irrigation from harvested water in farm ponds (Figure 7.15). Irrigation of 1 cm of water through sprinklers at 45 days after sowing (pod formation stage) would increase the groundnut pod yield by 25%.

Source: AICRPDA Main Center & ORP, Anantapur

Fig.7.15: Rainwater harvesting through farm pond and supplemental irrigation to groundnut crop

b) Gravel and sand mulching in sodic soils for moisture conservation in Northern Dry zone of Karnataka

Sodic medium and deep black soils exists to an extent of 2-5 lakh ha in the districts of Bijapur, Bagalkot, Gadag, Koppal, Bellary and part of Dharward, Belgaum, Raichur and Davengere in north Dry zone. Due to low infiltration rate of the soils, rain water is lost as runoff resulting in soil erosion and poor crop yields.

Farmers traditionally apply locally available gravel and sand mixture as mulch for increased yields. Improved technology involves sand application during summer. Before its application, weeds are removed and FYM @ 5t/ha is applied by deep plough. After bringing the soil to fine tilth, nearly 275-300 tractor loads of gravel and sand mixture per hectare is uniformly applied and spread manually to ensure uniform thickness of 10 to 15 cm on the soil surface. Though it is a capital investment technology for land development, it would be effective for 15 to 20 years. Gravel sand mulching helps in conserving rain water *insitu* that enables double cropping of groundnut or green gram in kharif followed by rabi sorghum or sunflower or chickpea. The cost of application for sand mulch can be recovered within 2 years.

c) Improved groundnut varieties for scarce rainfall zone of Andhra Pradesh

Farmers in Kurnool, Anantapur, parts of Prakasam, Mahabubnagar and Kadapa districts covering scarce rainfall zone of Andhra Pradesh planted local bunch of groundnut varieties like TMV-2 that has lost its purity, resulting in sub-optimum yields and economic returns. Improved varieties like K-6, Naryani and IGGV 91114 recommended due to their tolerance to drought, less duration and higher pod yields. The bold seeds of these varieties fetch higher market price. The following Table 7.15 gives the characteristics of the groundnut varieties.

Table 7.15 : Characteristics of Drought tolerant varieties of groundnut

S.No	Variety	Duration (days)	Yield (Kg/ha)	Yield increase (%) over local	Special characters
1	K6	105-110	540	22	Bold seed and drought tolerant
2	Naryani	105-110	699	38	Drought tolerant
3	IGV 91114	105-110	575	27	Drought tolerant
4	Local	130-135	421		

2) Central plateau and hills region (Rajasthan, Madhya Pradesh)

a) Profitable intercropping system for southern Rajasthan

Farmers in Bhilwara, Chittorgarh, Udaipur, Banswara, Dungarpur, Rajsmand and parts of Ajmer grow mixture of two crops as an insurance against drought. Mixed seeding of maize with black gram or green gram is most common in these regions. The proportion of crops is arbitrarily decided by the farmers, which results in low productivity and also the farmers were constrained to take up intercultural and weed control operations.

Intercropping of maize (Navjot/DEM-2/DEAM-3) + black gram (T-9) at 2:2 and groundnut (JL-24/TAG-24)+Sesame (RT-46) at 6:2 are recommended in place of farmers practice for higher returns and their performance over traditional in drought years.

b) Social fencing to develop community pastures in South Rajasthan

The measures taken up by Government to improve the Village Common Property Resources (CPR) through suitable technical interventions like seeding with improved grass species and soil conservation measures have not yielded the desired benefits due to absence of community participation and controlled grazing.

Farmer's participatory approach was taken up to develop community grasslands in some villages in Rajasthan. Participatory Resource Appraisal (PRA) ensured effective participations of villagers. The social fencing supersedes the physical fencing and succeeded in protecting CPRs, which has become a source of regular income to villagers.

3) Western plateau and hills region

a) Percolation tanks for ground water recharge in Malwa region of Madhya Pradesh

In the Vertisol regions of Indore, Dhar, Ujjain, Dewas, Ratlam, Rajgarh, Mandasaur, Jhabua, Sehore and Shajapur districts in Malwa plateau of Madhya Pradesh, heavy runoff is caused due to high intensity rainfall and poor infiltration capacity of these soils. This erodes the soil leading to decline in soil fertility and poor yields.

Storage of runoff is not possible in these regions as the soils in these areas are underlain by basalt rocks. Therefore, construction of percolation tanks allows the downward movement of stored water. Also, provision of diversion bund removes excess water from cultivated fields. Thus both storage or disposal of runoff minimize erosion and increase ground water storage, a protection against drought.

Source: AICRPDA Center, Indore

Fig. 7.16: Percolation tank with diversion bund in the ORP village Baroli near Indore

b) Risk minimizing intercropping system for Vidarbha region of Maharashtra

Traditionally the farmers in the Vidarbha region follow intercropping system of cotton & pigeon pea in 8:2 ratio and sorghum as sole crop. The cropping systems have low yield potential.

In order to stabilize the productivity and for efficient use of moisture from different layers and minimizing the risks due to soil moisture stress conditions, improved intercropping system of cotton+sorghum+pigeon+sorghum in row proportion of 6:1:2:1 with inter row spacing between cotton, sorghum and pigeon pea is 60cm has been recommended for stable yields in the drought prone regions of Vidarbha.

4) Gujarat plains and hills region.

Compartmental bunding for moisture conservation and higher productivity in pearl millet in North Gujarat.

Farmers grow pearl millet in north Gujarat in sandy and sandy loam soils under low and erratic rainfall conditions without any moisture conservation practices resulting in lower yields or total crop failure.

To retain rain water and to decrease soil erosion, compartmental bunds (0.25m height) have been made with the help of bullock operated bund making equipment at the time of sowing. These bunds help in providing more time for rainwater to infiltrate into the soil and conserve soil moisture. These practices gave 41% more grain yield and 18.6% more straw yield compared to conventional method of sowing along the slope.

5) Eastern plateau and hills (Jharkhand)

Improved varieties of rainfed crops for Jharkhand region

Farmers grow long duration Gora rice varieties (100-110 days), susceptible to pests and diseases and yield low. They grow other crops which are also low yielding. The following improved varieties of upland rice, black gram, pigeon pea and mustard produce 40 to 50% higher yields over local checks and are also tolerant to droughts.

Crop	Variety	Duration (days)	Yield (kg/ha)	Increase in yield (%) over local	Special characters
Rice	Birsa Vikas Dhan -108	70-85	1500	40-50	Tolerant to drought, pest and diseases
Blackgram	Birsa urid-1	80	1000	50-80	Tolerant to drought,
Pigeon pea Birsa	Birsa Arhar-1	220	1500	50-80	Tolerant to wilt
Mustard Birsa	Shiwani	105	1200	50-80	High yielding

6) Upper Gangetic plain regions

Inter cropping of pigeon pea + sesame on rainfed uplands of eastern plains of Uttar Pradesh

Farmers generally grow long duration pigeon pea on upland or as mixed crop with short duration pulses and oil seeds under broadcast sowing. This results in low yields due to excess plant population and weed problem.

Therefore planting of sesame between paired rows of pigeon pea (30+90 cm) with recommended crop management practices gave 85% higher productivity over farmers practices.

7)Trans Gangetic plains region (Punjab & Haryana)

Improved varieties of rainfed chickpea for kandi region of Punjab

Farmers in Kandi region of Punjab consisting of Nawanshahr, Hoshiarpur, Gurudaspur and Roopnagar grow local cultivars of chickpea which are prone to dry spells and susceptible to blight.

A superior variety PBG-1 which is resistant to blight with high yielding potential and is tolerant to drought is recommended for their region. The yield advantage of this cultivar is 72% as compared to local cultivars.

8) East Coast plains and Hill regions (Tamil Nadu and Orissa)

Improved varieties of crops for rainfed uplands in Orissa

Farmers in Kanahamal, Boudh, Rayagada, Gajapati and parts of Ganjam districts of Orissa grow local varieties of rice, cow pea and green gram, which are of long duration having low yield potential and are susceptible to moisture stress, pests and diseases.

The following varieties of rice, cow pea and green gram are recommended for upland situations of Orissa.

Crop	Variety	Duration (days)	Yield (kg/ha)	Increase in yield (%) over local	Special characters
Rice	ZHU-11-26	85	2954	94	Short stature, resistant to lodging fertilizer responsive, tolerant to moisture stress, and blast
	Local variety (Saria)	90	1530		
Cowpea	SGL-1	70	525	75	Non-trailing, short duration variety
	Local variety (Kathi jhudunga)	80	300		
Green gram	PDM-54	70	565	101	Early maturing, resistant to yellow mosaic virus, moderately resistant to powdery mildew
	Local variety (Kalamung)	80	280		

9) Western Himalayan Region (Jammu Region)

Agri-horti and Silvi-agriculture systems for Jammu

Farmers in Jammu, Kathua and parts of Udhampur districts of low altitude sub-tropical zone of Jammu generally grow seasonal crops like maize, sorghum, wheat, mustard etc which are subjected to frequent droughts during the season, sometimes leading to total crop failure.

Agri-horti systems (Aonla(NA-7).gauva (Lucknow-4) + gobi sarson in rabi and fodder maize in kharif and silvi agriculture system (Leucaena Leancocphala (K-63) + gobi sarson in rabi season and fodder maize in kharif are recommended for stable income under rainfed conditions in Kandi areas of Jammu.

7.12 Farmer's Awareness on Climate Management

Unpredictable climate and its variability over space and time has been the biggest threat to livelihood sustainability in the recent times, which is an imminent danger to National food security. Added to this, increasing population, illiteracy and poor economic standards in the developing countries exerted tremendous pressure on the natural resources, leading to its over exploitation and misuse. Though technological innovations to manage climate risks are developed for many locations, still they have not yet penetrated into majority of the rural farming communities for various reasons. In some regions, farmers are still dependent on the traditional knowledge that was inherited from their forefathers. In this endeavor, many countries have included climate awareness programs in their National policies on climate change for different strata of people/agencies. The farmers were exposed to the role of weather parameters on agricultural production, how it varied at their locations over the past 100 years and how the climate is likely to behave in the coming years. It is also essential to make them understand the likely impacts of climate extremes such as droughts, heat waves and floods etc., on future food grain production and the measures to be taken to manage these climate deviations. The initiatives taken by GOI in recent past such as awareness of climate should be spread across the country till the farmers are capable of handling climate related risks independently. The efforts made by various countries in this regard especially in developing countries in Asia and Africa are briefly discussed.

7.12.1 India

To increase interaction between the local farming communities with the Agrometeorological Field Units (AMF), Krishi Vigyan Kendras (KVK), Farmers Awareness Program on weather, climate were conducted by India Meteorological Department (IMD) in different agro-climatic regions of the country for the years 2009-10 and 2010-11. A one-day duration workshop was organized by all AMF units by inviting the farmers from nearby villages and exposed them to weather and climate issues that affect the production of their farms. The farmers during the first half of the day were appraised about weather impacts through popular lectures given by eminent Subject Matter Specialists (SMS) like Agrometeorologists, Entomologists, Pathologists, Agronomist etc. and the second half of the day was devoted to the discussions on the existing knowledge of farmers on climate risk management and methods to improve them. During the above period, approximately 7000 farmers attended the program conducted at 61 AMFUs located in SAU/ICAR institutes (IMD, Agrometeorology, 2011). This report by IMD further states that as a part of the program, the farmers were made familiar with the measurement of rainfall by ordinary rain gauge with a view to develop village level rain gauge network in the country. As a beginning, five ordinary rain gauges made of plastic were distributed to the selected group of farmers (progressive) to record and report the rainfall data to the respective AMF unit, which in turn transmitted the data to IMD. Some useful and interesting discussions were held between the experts and the farmers in managing risks in agriculture using advanced technologies are reproduced in Table 7.16. Rearing of livestock is an integral part of agriculture, few AMFU, wherever experts on livestock management are available, discussions on seasonal weather induced diseases were also held.

Some of the farmers queries and experts answers during the program conducted at selected AMF are as follows.

Table. 7.16: Farmers Queries and answers by Experts at selected AMFs

AMFU	Queries	Answer
Faizabad	Farmer was anxious to know how and why the climatic change effect the crop production?	Shrinking of length of winter season, increasing trend of temperature during winter season, decreasing trend of total rainfall and rainfall intensity, increment of untimely and erratic rainfall, westerly wind at the time of flowering of rabi crops (15 Feb.—15 March) is increasing. Consequently, flower shedding and low test weight specially in wheat crops observed hence yield loss in rabi crop is frequently reported.
Jagtiyal	Farmers enquired about rainfall of the last year and crops suitable for the low rainfall conditions.	Different crops suitable for the less rainfall condition were explained. The farmers were suggested to follow the weather forecast particularly during the harvesting stage of the crop.
Kalyani	Is there any possibility of Forecasting cumulative Rainfall of 15 days in advance instead of seven days	It was discussed that so far accuracy is considered, it is not possible to give such accurate forecast.
Palampur	Winter rains and onset of Monsoon should be predicted more accurately and information should be given 10 days advance	The winter rains accuracy for this region is reasonably very good to use in the crop productions. Also the forecast accuracy is good in rabi season for the last fifteen years except for a two year period.
Basar	What is the usefulness of Weather forecast?	Based on the forecast, important agricultural practices can be carried out to minimize damage to standing crop and minimize financial losses.
Basar	What is the importance of Relative humidity in plant growth?	For plant growth, 60% relative humidity is required. For pest and disease incidence a continuous high humidity is necessary.
Aduthurai	How to use the rainfall Forecast in agriculture or how we can apply fertilizer if there is continuous rainfall?	If there is continuous rainfall, we should not apply the fertilizer. If there is no rain for next three days (Inferred from the weather forecast) in the low lying areas, drain the excess stagnating water in the field and apply urea immediately, particularly during rainy days (Rabi).

(*Source*: IMD, Agrometeorology 2011)

The views of some of the progressive farmers have been collected by IMD Agrometeorology (2011) which are given below.

- A progressive farmer attended the program and shared his experience about how he was benefited from the Agromet advisory issued from AAS. He mentioned how he could manage to save potato crop of value of

Rs.3,00,000.00 (rupees three lakh) from the damage due to heavy rainfall(which was at maturity stage in the field) after receiving the forecast of heavy rainfall from AAS, Sonitpur well in advance (AMFU: Sonitpur)

- Farmer explained the concept of choosing crop combinations based on weather parameters and agro ecological situations (AMFU: Kanniwadi)
- A progressive farmer shared his views on various aspects of crop growth in relation to weather (AMFU: Jammu)
- Farmer from Uchahar village, Keshpur, Paschim Midnapur district opined that they followed SRI technique based on weather information for growing rice in their area and they received good yield per Bigha and also try to motivate others to adopt the modern techniques (AMFU: Kharagpur).
- Farmer's revealed that the labour available for agriculture is decreasing at a faster rate due to urbanization and shift to industrial sector. The available agricultural land is also decreasing due to conversion of land for non-agricultural purposes. They fear that rainfall and water available for agriculture is decreasing due to anthropogenic activities and the dry land farmers have to face uncertainties due to aberrant weather conditions (AMFU: Kovilpatti).

Similarly the All India Coordinated Research Project on Agrometeorology (AICRPAM) under its operational Agrometeorological programs surveyed among the farmers representing different agro-ecosystems to assess their understanding and knowledge of the current weather information in managing their crops. Therefore a one day campaign on Farmers Awareness on climate is being conducted as a regular exercise at all the AICRPAM centers from 2008 onwards. The farmer's perception about weather and the respective agricultural operation were obtained from a questionnaire developed at AICRPAM Head Quarters at CRIDA, Hyderabad. The farmers representing 25 AICRPAM centers were grouped as high input use, low input use and horticultural (plantation / hill station) to evaluate the overall response among the groups. The results are given in the Table 7.17 and 7.18. It was noticed that about 69% of farmers from low input use centers are more aware about the importance of rainfall in influencing agricultural production compared with the farmers by high input use region and hills and plantation region. (Rao *et.al.*, 2009 and Rao *et.al.* 2011). Temperature has been identified as the important weather parameters by hill and horticultural farmers and these farmers still follow traditional methods in agriculture. However, farmers from high input region adopt more scientific methods compared with two other categories. More farmers from low input use are aware of weather-based agro-advisories issued by local

Table 7.17: Response of farmers (%) from high input use areas on weather affects on agriculture

S.No.	Question No.	Parameters	Akola	Anand	Anantapur	Bangalore	Hissar	Jabalpur	Kovilpatti	Mohanpur	Parbhani	Solapur	Average
1.	Which is the most important weather parameter affecting agriculture and give the parameters as order of priority?	Rain	93	41	52	66	66	65	62	90	97	80	71
		Temperature	5	14	13	14	19	19	25	10		20	15
		Moisture	2	39	13	12			2				14
		Sunshine		2	10	18			2				8
		No answer		4	12		16	16					12
2.	Do they follow the traditional knowledge in preparing the fields/agricultural operations or do they listen to radio /watch television/ read newspapers in taking the decisions?	Traditional(T)	50	23	29	30	70	60	40	40		75	47
		Scientific (S)	50	36	46	70	30	40	60	50	100	25	51
		Traditional and Scientific		18	21								20
		No answer		23	4								14
3.	Are you aware that the Agrometeo-rological center of this agricultural university is providing advisories based on the forecast and do you really follow up? If not, please mention the reason?	Yes	76	98	80	40	34	45	91	100	100	65	73
		No	24	2	15	60	51	55	6			35	31
		No answer			5		15		3				8

S.No.	Question No.	Parameters	Akola	Anand	Anantapur	Bangalore	Hissar	Jabalpur	Kovilpatti	Mohanpur	Parbhani	Solapur	Average
4.	Do you feel that the pests/diseases are related to weather parameters or the crop growth stages?	Weather (W)	100	75	59	100	92	60	97	95	86	90	85
		Crop growth(CG)		9	30		8	40	3	5	14	10	15
		W +CG		16	11								14
5.	Describe briefly the rainy, hot and cold months in your place? If possible, the values	Right answer	100	86	91	90	98	85	94	100	100	90	94
		No answer	14	9	10	2	15	6					91
6.	Do you feel that climate is changing?	Yes	30	89	77	100	79	72	91	29	100	80	75
		No answer	70	11	23		21	28	9			20	26
	If yes, what are the different parameters that change have been observed?	Parameters		11	56								34
		T&M (rain)											
		Heat &Cold		9	27								18
		Humidity		16	12								14
		Wind		2	5								4
7.	Is this one-day program is useful in learning new things about weather?	Yes	100	93	98	100	84	83	100	100	973	100	96
		No answer		7	2		16	17					9
8.	Are you going to adopt resource conservation practices to cope with climate change?	Yes	90	82	71	100	44	89	94	80	100	90	76
		No answer	10	18	29		56	2	6	20		10	30

(*Source*: Rao, G.G.S.N. *et. al.*, 2009)

Table 7.18: Response of farmers (%) from low input use areas on weather affects on agriculture

S.No.	Question No.	Parameters	Bhubaneswar	Bijapur	Faizabad	Jorhat	Kanpur	Raipur	Jammu	Ranchi	Udaipur	Average
1.	Which is the most important weather parameter affecting agriculture and give the parameters as order of priority?	Rain	8	100	100	95	92	73	85	90	87	89
		Temperature	0			5	8	27	15	7	10	14
		Moisture	1							3	3	3
		Sunshine	5									
		No answer	2									3
2.	Do they follow the traditional knowledge in preparing the yields/ agricultural operations or do they listen to radio/watch television/read newspapers in taking the decisions?	Traditional	8	63	70	93	40	23	46	57	79	61
		Scientific	0	37	30	7	60	77	54	43	21	39
		Traditional and Scientific	2									
		No answer	0									
3.	Are you aware that the Agro-meteorological	Yes	90	25	80	90	60	83	74	92	89	81
		No	10	75	20	10	40	17	26	8	11	19

S.No.	Question No.	Parameters	Bhubaneswar	Bijapur	Faizabad	Jorhat	Kanpur	Raipur	Jammu	Ranchi	Udaipur	Average
	Center of this agricultural university is providing advisories based on the forecast and do you really follow up? If not, please mention the reason?	No answer										
4.	Do you feel that the pests/ diseases are related to weather parameters or the crop growth stages?	Weather	100	28	100	93	100	35	81	94	73	78
		Crop growth		18		7		23	19	4	8	13
		Weather + crop growth		54				42		2	19	29

(*Source*: Rao, VUM *et. al.*, 2011)

Agricultural University compared with other two groups. Farmers from high input use and hill and high rainfall regions are more aware of changing climate and their impacts compared with low input use farmers.

7.12.2. In other countries

Under the United Nations Framework Conventions on Climate Change (UNFCC), it is mandatory that every country prepare their National plans to face the future challenges related climate change. Thus many developing nations have prepared plans and included the Farmers awareness program as a policy to understand the impacts of climate change on agricultural production and to minimize the crop losses. The programs were conducted at Government level as well by Non-Governmental organizations. Some of the experiences of selected countries are briefly described here.

Nepal

Climate change / variability is one of the major challenges that is often encountered by the farmers who are unaware about its impact. In order to face the climate extremes *etc*, few organisations in Nepal have joined together to reduce its impact. Nepal Development Initiative (NEDI) has been playing the role of National Coordinator for the campaign "Family Farming Development Campaign". A one day campaign with an object of raising awareness about climate and its impact on agriculture, adaptation strategies to combat climate change and to increase the farmers participation levels to minimize the impacts during World Environment Day in 2010 was held at Bisundole village of Lalitpur district. After listening the lectures from experts in different disciplines, open discussion with the farmers, the following interesting exchanges transpired are as follows.

Open Discussion and Interaction

Farmer: How to receive prior information on "Weather change"?

Expert: Undoubtedly, Nepali farmers are affected by changed weather patterns in recent years. Weather forecast is normally done for next 24 hours period. This cannot give total idea about what would be the weather in next few days. On the other hand, the Department of Meteorology of Nepal lacks adequate equipments and thus its forecast can't bc fully trusted.

Farmer: What are the ways to minimize climate change impacts on agriculture?

Expert: There are two ways on how to minimize the impacts of climate change in general. They are i) Adaptation and, ii) Mitigation.

Adaptation is the measure taken in order to get tuned with the changed conditions in natural climate. This will help farmers to introduce new technology (seeds, techniques, equipments) in farming so that the dependency on weather is minimized and on the other hand the productivity could be increased.

Climate change is the outcome of human activities. It is therefore, human duty to minimize the hazardous activities so that we would be able to lessen *greenhouse gas emissions*. Thus, *mitigation* helps minimize climate change impacts by reducing greenhouse gas presence in the atmosphere.

Sri Lanka

Dairy a part of agriculture is highly sensitive to weather aberrations. A study was conducted at seven GN Divisions in Anuradhapura district in Sril Lanka to measure the dairy farmers awareness on global warming and climate change issues. (G.De Silva, S. Abesingh, 2011). The data collected and analyzed from 117 farmers revealed that about 67 per cent of the respondents were aware about climate change and its impacts on animals and their products. However, roughly few farmers (23 per cent) had knowledge on global warming and Green House Gases. About 69% of the farmers reported that they understood climate change as changes occurred in weather. All the respondents agreed increased activity of weather parameters viz., temperature, rainfall, flood and flood damage during last few decades. About 33% of the farmers indicated increased frequency of droughts. Majority of the farmers attributed the increased diseases to global warming. Farmers were receiving information on climate change from three sources viz., radio/Television (47%), extension workers (12%) and Newspaper (5%). It was interesting to note that farmers are not aware about dairy farming contributing significantly to global warming by emitting GHG from the dung. After the persuasions, about 84% of the farmers had taken action for disposal of dung, 51% of farmers managed their herd better and 21% changed the food concentrate to their animals.

Pakistan

"Caritas" a humanitarian organisation of Catholic Church in Pakistan organized a three month campaign to combat climate change during 5th March to 05 June 2012, with an objective to spread climate awareness among the people about decreasing agricultural productivity, growing food insecurity and vulnerability of people to natural disasters in Pakistan. Different activities viz., poster competition, awareness seminars, training workshops, tree plantations, livestock management etc were organised. The results of the three months campaign are as follows:

- More than 25000 individuals majority of which were small farmers (both male and female) were given awareness of global warming, depletion of natural resources and best sustainable agriculture practices to enhance food security and soil fertility.
- A total of 3880 plants were planted during tree plantation campaigns in Karachi, Multan, Hyderabad and Rawalpindi with the active involvement of school children, teachers, farmers and other community members.
- Two bio-gas plants were installed in 39/NB (Rawalpindi diocese) in collaboration with Pakistan Domestic bio gas program.
- Ten bio-fermenters were installed in different areas. The farmers using these fermenters have reported reduction in their input cost and less dependency on external chemical fertilizers.

Nigeria

The results of the household survey of local communities on farmers awareness on climate in the Niger delta region of Nigeria has brought out that level of awareness on climate change is still low in the region as nearly 60% of respondents know little or nothing about climate change and its impacts. Further the results that analysed on how the farmers perceived the climate change in their understanding has brought out that about 69% of the respondents reported that they understood climate change as "Change in weather" 11% of the farmers related climate change to "poor yields" while 9% attribute to "ozone layer depletion", some 6% of farmers understood as "heavy rainfall", while 5% observed it as "excessive sunshine". The results further revealed that 56.2% of the respondents perceived it as "increased incidence of drought" and 41.2% of them understood climate change to mean "high wind and heat waves".

Namibia

As per the UNFCC directives for all the participants of the program, Namibia has taken action to develop and implement educational and public awareness program on climate change/variability and its impact. Initially the country has produced Initial communication booklet, poster and fact sheets on climate change in Namibia during 2003. Under the national climate change strategies, promotion of awareness, education, public participation and provision of targeted climate change information has been included as one of the priority areas under climate risk management strategies.

Seychelles

The Government of Seychelles has realized the importance of environment in tourism, industry, a major source of economy. Therefore environmental conservation has always been tied up with National developmental activities always through education sector and, it has been included at the formal education levels. During 1997, the Ministry of Education brought an "Environmental Education Policy" which promoted and supported in understanding valuing, respect and appreciation of the environment that was reflected in all activities. The National curriculum frame work of 2011 emphasized the inclusion of climate change topic at the primary and secondary school level as an option by the interested teachers.

The primary and secondary school science syllabus does not directly address the climate change, but it includes number of topics viz., energy conservation and water conservation at primary level and human impact on the environment, ecosystems, water management, nuclear power, energy sources etc at the secondary school level respectively. On every Environment Day (5 June), the Ministry of Education organizes a National Creative Competition open to all school levels. Similarly during world meteorological day (23 March), a one day workshop on climate change for secondary school students is celebrated. From time to time the Environmental Education Unit in the ministry of Education organizes short training session on climate change issues and their management for in-service teachers.

China

Raising public awareness on climate change is one of the strategic goals of china in addition to develop significant measures in controlling GHG emissions and to promote climate change related science. China planned to strengthening the existing communication system by 2010 with modern information dissemination technologies to promote the education and training to address the climate risk management strategies effectively by exposing the whole society to a friendly social environment.

7.13. Role of ICT in Climate Risk Management through AAS

Information and Communication Technologies (ICT) are the tools that can increase the effectiveness and efficiency in climate risk management of developmental programs that are being operated at various locations in the country. If it is integrated strategically - which includes community Radio, village knowledge centres, mobile phones and interactive media (Internet, e-mails), it can contribute significantly by providing access to relevant information and facilitate knowledge sharing at the community level. Thus ICT help in reducing the vulnerabilities and climate risks faced by the poor and marginalised farming communities from weather extremes. The participants of one day program on learning on ICT and climate change held in Johanesburg in South Africa on 19 December 2008 in the identified following key areas for ICT practitioners in the context climate management *viz*: 1) Sharing knowledge, 2) using ICT in policy advocacy, 3) main streaming of ICT use is to existing program and 4) to facilitate an understanding between people and policy makers (Patrikk P. Kalas 2009). Though ICT can play major role in many fields like, distance learning, e-health, telemedicine, e-government market information and rural development, this part of this book restricts the discussion related to weather based Agro Advisory Services (AAS). In India, addition to dissemination of AAS information by AMF, IMD, ICAR and SAUs, AIR, Doordarshan, and local television network, few NGOs in collaboration with National leaders have been extensively using mobile phones for dissemination of forecast and the weather based advisories to many farmers who have tied up to receive the information. The agencies that have been helping the farmers for better management of climate are as follows.

7.13.1 India

In the era of emerging IT arena in India, many private companies have joined hands with IMD and SAUs to deliver weather based agricultural information to the farmers using multiple modes of information communication systems. As some of the farmers are willing to pay for the information, IMD has tied up with different public and private organizations, which are already working in dissemination of agricultural information to the rural village. The dissemination methodology mainly use Information Technology (IT) which includes Internet and Mobile phone based communication systems. The most advantageous feature of mobile phone is that farmers can communicate with the experts while in the field and request for advice. Currently IMD has partnered with Reuter Market Light (RML), Handygo, Maharashtra State Agriculture Department (Government of Maharashtra), Vritti Solutions, IFFFCO Kisan Sanchar Ltd. (IKSL) and TATA Consultance Services (TCS).

Handygo, a Telecom Value Added Services company is disseminating agromet advisories through IVR (Interactive Voice Response) system to five states (Punjab, Haryana, Maharashtra, Gujarat and West Bengal) in the country. Its rural IVR service No 55678 is designed to serve Farmers by providing information on weather, crops, livestock and fisheries. It gets weather forecast information from IMD for each state and stores in its database. Thereafter, it would be converting these data into voice and then transferring the information in regional languages into the Network where the farmer would be calling and receiving the desired information.

Vritti Solutions Limited, a company working for enabling IT in the rural parts of India have developed interactive software through which the advisories are being sent to the farmers in the Maharashtra State.

IFFCO Kisan Sanchar Limited (IKSL) is providing Value added services to the farmers through the mobile channel. Five Voice messages per day of importance to farmers are sent free of cost to the farmers which cover area of their immediate interest. The uniqueness of the messages is that they are in local language and conveyed by local people. The information covered are recommendation to farmers on best agricultural practices, nutrient related issues, animal husbandry, information on nearest market yards, input on weather & climate and appropriate rural technologies.

State Department of Agriculture, Government of Maharashtra is sending SMS to the farmers in the villages of the state and also through the websitewww.mahaagri.gov.in. In addition the other areas of service, agromet advisories are also being sent through this site.

Tata Consultancy Services (TCS) signed a MOU with IMD, to provide the personalized agro advisory generated under IAAS in local language to the farmers using mKRISHI platform on the compatible handsets of the farmers. It provides an option to give valuable feedback to Agricultural experts for improving their bulletins. Farmers can also send picture of their crops with mobile phone cameras - so that the experts can guide the farmers directly on measure to be taken immediately. The status of State-wise SMS service through different agencies as given by Ramakrishna, (2010) is given in Table 7.19.

Another IT based agro advisories known as e-sagu (e-cultivation) is one of the kind of Information Technology based agro advisory system which is in operation since 2004 at IIIT, Hyderabad. This system is scalable and provides personalized agricultural expert advice to each farmer. (**www.esagu.in** or **www.iiit.ac.in**)

Table 7.19: State wise SMS services of AAS

S. No.	State	Number of SMS			
		RML	Handygo	Vitti Solutions	Government
1.	Maharashtra	119116	500	2300	26000
2.	Haryana	20556	700	-	-
3.	West Bengal	104	700	-	-
4.	Gujarat	16144	400	-	-
5.	Madhya Pradesh	7471	-	-	-
6.	Karnataka	7784	-	-	-
7.	Tamil Nadu	4114	-	-	-
8.	Andhra Pradesh	10410	-	-	-
9.	Rajasthan	22976	-	-	-
10.	Uttar Pradesh	20738	-	-	-
11.	Himachal Pradesh	365	-	-	-
12.	Uttarakhand	113	-	-	-

(Ramakrishna, Y.S. 2010)

The operation of e-Sagu involves a team of agriculture experts who work at the main e-Sagu centre, normally located in a city. The e-Sagu local center comprising few computers and one computer operator is set up for a group of about ten villages. A coordinator, who is an educated and experienced farmer is stationed in the village and is in touch with agricultural experts daily.

Depending on the crop, a coordinator working in the village is assigned with a fixed number of farms to be covered. Likewise many coordinators are employed to cover the entire village. Every day, the coordinator visits a fixed number of farms and takes four to five photographs of each farm. Data containing the photographs and other information is copied on CD and sent to the main system located at IIIT, Hyderabad by a regular courier service. The agricultural scientists at the HQ analyse the information and suggest the next course of action for the farmers through email to the coordinator. The coordinator in turn downloads the advisories and provides the same to the farmers for taking necessary steps.

The expert advice given through e-Sagu has helped the farmers to improve input efficiency by encouraging integrated pest management methods. The impact study shows that the farmers have realized considerable additional monetary benefits equivalent to Rs.3,820 per acre by reducing the fertilizers and pesticide sprays and getting the additional yield. The agriculture scientist can deliver about 150 expert advices per day by reusing the already delivered advices and following a clustered-base approach.

7.13.2 Application of ICT in Providing Advisories in other Countries

Community Radio Practitioners and Community Media have been using ICTs mainly for the internet and mobile phones to manage the climate related risks in local communities around the world. The details of the strategic ICT tools used, the approaches and the locations of the program implementation etc in respect of countries viz., Burkina Faso, Huaral Valley in Peru, Nepal, Ecuador, are given in Table 7.20 (Patrik P.Kalas, 2009).

7.14 Integrated Drought Management Strategies through Farming Systems Approach

Integrated drought management strategy consists of a multi-pronged approach viz., efficient management of resources inclusive of livestock, horticultural and agro forestry systems in agriculture, development of viable technologies, alternate sources of income for rural communities, improving the market facilities and financial institutions, proper coordination among various agencies involved in monitoring and relief measures and dissemination of advisories through better communication facilities etc. Farmers mostly depend for their livelihood on crop production, which is highly variable and often the livelihood of rainfed farmers gets severely affected. Therefore, there is a need to evolve suitable strategies to augment the income throughout the year from farming activity. Many of above mentioned topics have already been covered in earlier sections, hence the topics covering enhancing the economy through integrated approach is discussed in the following paragraphs. Livestock component is an integral part of agriculture. Development of technologies for increased output from these systems reduces the drought impacts significantly by providing stable income. The practices followed by rural population in the Arid zone of India covering western Rajasthan clearly shows the importance of livestock component on their economy unless a famine like condition disrupt the livelihoods of the rural population due to exodus of livestock in search of better pastures. Also fisheries, agro forestry, if integrated with agriculture along with livestock management, which is otherwise known as Integrated Farming System (IFS) has greater potential in enhancing and stabilizing the rural economy. These enterprises not only supplement the income but also help in by increasing the labour employment throughout the year. (Jayanthi *et.al.* 2000, Singh *et al.* 1993 and Singh *et al.,* 1997). IFS is described as a component of farming systems which minimizes climate risks, increase production and profits by improving the utilization of organic wastes and crop residues (Radhamani *et.al.* 2003). The IFS would act as a protective shield against all the ill effects of drought on the rural society and provide them with the necessary economic returns on a continuous basis. This may be integrated with Drought Prone

Table 7.20: Details of the strategic ICT tools used, the approaches and the locations of the program implementation

S.No	Knowledge flash	Approach, locations and program
1.	**South Africa Country: Burkine Faso**	
	Strategic ICT tool used	Interactive community radio
	Innovative approach	Community radios links scientific knowledge at global level to local communities
	Geographic location	Global; Focus on Africa, Burkina Faso
	Key words	Disaster prevention; food security; desertification; human rights-based approach; behaviour change; interactive communication systems
2.	**Country: Peru**	
	Strategic ICT tools used	Telecentres (public access points); information kiosks
	Innovative approach	• Agrarian information system enhanced through telecentres to provide access to information to obtain fair distribution of irrigation water • Alternative energy systems
	Geographic location	Global; focus on Latin America, Peru
	Key words	Agrarian information system; fair distribution of irrigation water; food security; alternative power
3.	**Country : Nepal**	
	Strategic ICT tools used	Traditional and alternative media; internet
	Innovative approach	• Developing the capacity of South Asian media practitioners on the application of ICTs for an online media campaign to advocate prevention, mitigation and adaptation strategies. • Enhancing the partnerships between civil society organizations and media practitioners
	Geographic location	South Asia, Focus on Nepal
	Key words	Capacity development; media
4.	**Country: Ecuador**	
	Strategic ICT tools used	Personal computers for database management; internet portal
	Innovative approach	Capacity development on using ICTs amongst small farmers to document experiences, share information, and build awareness for better natural resource management while providing a platform for lobbying at local and national level
	Geographic location	Latin America, focus on Ecuador
	Key words	Latin America, focus on Ecuador

(*Source*: Patrik P. Kalas-2009)

Area Programs of GoI. Though IFS was mainly initiated in Asian agricultural system, it has also become popular in other countries of the world. This system results in rational use of water with proper distribution among different users. In one of the studies conducted in Tungabhadra project area of Karnataka (Channa basavanna *et.al.* 2009),the productivity, profitability, energy flow, employment generation and water requirement of IFS over conventional rice-rice system recorded 26.3 and 32.3 per cent higher values in respect of productivity and profitability.

The following examples illustrate the economic benefits over the conventional system followed at different locations in India.

In Tamil Nadu, Rice-based farming system model provided a net profit of Rs.11,755/ year form rice –poultry-fish-mushroom in 0.4 ha area over the conventional cropping system (CCS) with rice-rice green manure/pulses that gave a net income of Rs.6,334 only from the same area. Thus the IFS increased the net income and employment from the small farm holding. (Behara *et.al.* 2012). An IFS in northern plain zone in Eastern Uttar Pradesh which is susceptible to drought revealed that rice-pea-okra cropping sequence was most remunerative with highest rice equivalent yield of 17.88 t/ha and had the higher net returns than the conventional rice-wheat system (Singh *et.al* 2006). The IFS situations in pond based integrated farming system in eastern India viz., fish+vegetables+rice gave the highest water productivity, followed by vegetables and irrigated rice (Samra *et.al.* 2003).

The farming systems modules for marginal and small farmers in southern Telangana region of Andhra Pradesh which is characterized by recurring droughts, degraded lands and sub optimum use of inputs was conducted for the years 2005-2008. The small farmers in this area are unable to meet their requirements from crop activity alone as the crop productivity is low due to high rainfall variability. Dairy and horticultural based farming systems modules in Shabad mandal of Ranga Reddy district and a station model in 1.15 ha area was implemented on watershed basis in CRIDA Research Farm. (Annual Report of CRIDA, 2007-08). The results over the year 2005-08 revealed that a farming systems model of (1.15 ha area) with arable crops (0.4725 ha), agroforestry option (0.3496 ha), vegetables (0.1150 ha), grasses (0.1256 ha) and bushes (0.0890 ha) provided a gross income of Rs.16080/- and a net income of Rs.9793 and BC ratio of 2.38.

A comparison of irrigated cotton crop and integrated crop and beef cattle systems was carried out in Texas high plains in USA (Allex *et.al* 2007) from 1999-2005. This region was historically a grazing land ecosystem offering opportunity for pastoral systems and benefits from diversifications. The results demonstrated water saving of about 25 per cent achieved through integration approach. Additional benefits included reduced soil erosion, lower chemical inputs (40 per cent reduction in N-fertilizers), improved soil microbial and enzymatic activities, enhanced e-sequestration and greater rainfall infiltration than in the mono culture system. The prospects of IFS in Bangladesh in view of changing dietary pattern was discussed by Shamim al Mamum *et.al* (2011). The IFS seem to a possible solution to meet the continuous increase in demand for food, stability of income and diverse requirements of food grains, vegetables, milk, eggs, meat etc for improving the nutritional status rural families with limited resources.

Karrou and EL Mourid (2004) from International Centre for Agricultural Research Centre (ICARDA) has brought out significance of use of Scientific approach and involvement of Research Institution for improvement of drought management system in semi-arid and arid agro-pastoral systems in West Asia and North Africa (WANA). The pastoralists' livelihoods in these regions are affected by shortage of rainfall that results in drying up of water resources and declining forage resources for livestock. Use of traditional knowledge enabled the pastoralist to survive and produce under these harsh environments. These strategies summarized by Hazel *et.al.* (2001) are as follows:

1. Mobile or transhuman grazing practices that reduce risks of having insufficient forage in any location.
2. Reciprocal grazing arrangement with more distant communities for access to their resources.
3. Keeping extra animals that can be easily liquidated in a drought, either for food or cash.
4. Investment in digging of wells, cisterns and water harvesting systems.
5. Diversification of crops and storage of surplus grains, straw and forage as a reserve in good rainfall years. Some crops are more tolerant to drought than others.
6. Income diversification into non-agricultural occupations.

The WANA governments launched many programs to provide assistance that solve partially and for a short term the community problems. The crisis management includes supplementary feeds to safeguard livestock, especially, barley, digging of wells, subsidies for irrigation; waiving of debts *etc*. Hazel *et al.* (2001) observed that these policies encourage the practices of unsustainable management techniques in drought prone areas. The study further showed that indigenous and local community strategies for drought management are no longer effective in coping with drought and the Government programs encourage them more dependency on aids and supports. Therefore dependency increased the use of non-adapted drought management practices that worsen the degradation of range lands and pastures. WANA countries have been advised by the author to build their own system based on experiences of some advanced countries on drought risk management plans including integrated Farming system approach.

The UNDP programs in Drought Risk Management (DRM) in Africa and Asia indicated a list of DRM activities that help in developing integrated approach (good practices) for drought management at the local level (Paul Venton, 2012)

Indicative list of DRM activities are as follows:

Immediate safety measures

- Supply food aid and other non-food items to affected communities
- Provide emergency livestock purchases and subsidies to transport animals to market
- Provide supplementary livestock feeding (fodder, forage, hay distribution, water hauling, opening of strategic grazing area, etc.)
- Promote emergency vaccination and de-worming
- Provide seed distribution, stockpiling cereals and low-interest agriculture loans and emergency assistance programs
- Facilitate borehole rehabilitation and water-trucking
- Establish a local coordinating body to ensure emergency response based upon priorities

Short term measures

- Develop water use guidelines based on the types and duration of drought
- Develop emergency water allocation strategies
- Increase communication of climate-related information, with specific advisories.
- Increase local drought monitoring capacity and infrastructure.
- Providing support to farmers for paying premium to drought and crop insurance
- Establish food subsidy programs for drought-affected individuals
- Provide support to most vulnerable groups, such as women and youth.

Mid-term measures

- Expand efforts to promote rainwater harvesting
- Introduce improved soil management techniques that decrease soil erosion and increase water-holding capacity of soil
- Adopt alternative cultivars or crops that are more drought-resistant or heat-tolerant
- Address bottlenecks in seed delivery systems
- Establish system for sharing of experience and capacity development for vulnerable groups in their adaptation measures/responses

Long-term measures

- Investing in business and farm/ranch diversification strategies (e.g., selecting drought tolerant varieties, implementing irrigation where feasible and diversifying away from rain-fed crops to less water-dependent products, such as honey from bee-keeping)
- Address the issues related to deforestation and desertification (land degradation in drylands)
- Review the effectiveness of mid-term measures and strengthening capacities as needed
- Strengthen market access and rural infrastructure
- Reinforce legal, policy and institutional frameworks for drought risk mitigation and dryland development

Source: UNDP (2011)

7.14.1 Drought Adaptation Strategies in Rajasthan and in Maharashtra

Droughts are common to arid parts in the Northwest India comprising western Rajasthan and parts of Gujarat where drought occurs once in every 3 years (Apparao *et al.* 1991). The farmers in the Drought prone regions in other parts of the country include, Punjab, Haryana, West Uttar Pradesh, West Madhya Pradesh, Madhya Maharashtra, Interior Karnataka, Rayalaseema, South Telangana, Parts of Tamil Nadu, Northwest Bihar etc. adjusted their lives and adopted to live with these climatic uncertainties. Venkateswarlu (1987) examined the adaptation and adjustment techniques of farmers to droughts in the arid parts of Rajasthan. He has observed the following farmers strategies to cope up with droughts.

- Reduce consumption of food grains.
- Postpone social engagements such as marriage and others.
- Migrate to better areas with livestock or sell them.
- Diversification of crops and cropping systems that are suitable to climate conditions
- Practicing of inter cropping led to yield stability
- Selling of assets like gold ornaments as a last resort

He has further observed that in addition to regular drought relief measures and on-farm activities such as crop strategies and crop insurance programs, improved crop husbandry with livestock management shall provide the necessary protective shield against droughts.

Similarly some of experiences of farmers in drought prone districts of Andhra Pradesh (Mahabubnagar) and Maharashtra (Sholapur) about adaptation practices to drought collected under an international project on "Vulnerability to Climate Change, Adaptation Strategies and Layers of Resilience (Naveen P. Singh *et.al.*, 2012) revealed that the farmers in the above regions have adapted an integrated approach to improve the economic status of family and simultaneously improve the farm yields. The following are some of the details of the strategies as reported by Naveen P. Singh *et al.* (2012).

- Increased use of ground water due to drying up of common resource properties like, ponds, wells and tanks which is attributed to increased population, cropping intensity and animal grazing pressure that stabilized farm income.
- Crop diversification from coarse cereal to high remunerative crops like cotton, soybean and other commercial crops, fetched more income. Wherever water availability increased due to watershed development or construction of dams, sugarcane is grown. Farmers switched over from longer duration crops to shorter ones to avoid water stress/drought conditions that stabilized production.
- Farmers increased their family income by migration to nearby cities during off season and work as non-farm labour, and petty business etc.
- Farmers found Self Help Groups (SHGS) to be helpful in expanding their livelihood activity and income.

7.14.2 Improvement in Rural Livelihood

To improve the livelihood of the rural poor through efficient management of natural resources a project under National Agricultural Innovation project (NAIP)funded by ICAR entitled "Sustainable rural livelihoods through enhanced farming system productivity and efficient support system in rainfed areas" was operated during the period (2007-08 to 2011-12) at 8 locations *viz.*, RangaReddy, Anantapur, Kadapa, Warangal, Mahabubnagar, Adilabad, Khammam and Nalgonda in Andhra Pradesh which are drought prone regions. The innovativeness of the project was the consortium approach in the project implementation and development of interventions with a holistic approach to address the livelihood security of the community rather than increasing crop productivity alone. Some of the results of the project (Annual Report of NAIP-component –III, 2010) in enhancing the livelihoods and empowerment of information of rural population in the above districts are reporoduced as follows:

1) Farm pond technology at Adilabad

To reap a good harvest of runoff expected in this district, remodeling of the existing farm ponds for longer period water available were carried out under the project. The farmer's apprehension about loss of land with large ponds was countered by encouraging the farmers to cultivate high value crops and vegetables around the pond by erecting trellis that resulted in extra income of Rs.1624/- during 2010. To further strengthen the income generation, farmers were encouraged to release 500 fingerlings per pond which fetched an additional revenue of Rs.2000/- by the end of rainy season.

2) Use of plastic mulch for promoting higher melon production in Anantapur districts

Plastic mulching was promoted in cultivation of melon in this arid district for conservation of moisture. The results obtained in the farmer's field showed watermelon production was 22 t/ac in mulched plot compared to non-mulch plots @ 18 t/ac. In case of muskmelon, it was 14 and 8 t/ac in mulched and non-mulched plots respectively. Based on the benefits obtained by a farmer during the year 2009,this practice was taken by 16 farmers during 2010.

3) Integrated nutrient management in Warangal

Micro nutrients like boron (1kg), Zinc (20kg) and gypsum (100kg) per acre were applied in 50 cotton farmer fields. An average yield increase of 110 kg/ac worth or Rs.5500 was noticed. Azospirillum was introduced as a bio-fertilizer in 50 acres of rice and the yield increment was 150 kg/acre worth Rs.1600/-.

4) Recharge of defunct irrigation wells (open wells) in Warangal

About 45 open irrigation wells in Jaffergudem area had gone dry due to depletion of ground water subjecting the crops to water stress conditions. Out of the above wells, 23 were selected for recharging in the first phase by tapping into the nearby runoff channel and diverting the runoff using a HDPE pipe into the open well through a silt trap. During 2010, about 5650 m^3 of runoff water was diverted into these wells, which are available for irrigation purposes.

5) Knowledge empowerment and agricultural risk management

The state-of-the art on Information and Communication Technologies (ICT) were extensively used for knowledge empowerment of villagers. The village knowledge centre (VKC) with all electronic and communication gadgets established at each cluster facilitates sharing of data, information and knowledge about crop-wise information and others on dairying, back yard poultry, mango cultivation etc placed in Kiosk. The farmers had taken the advantage of VKC by calling IVRS telephone number and access voice modulated information on weather and markets. The farmers can directly get the market information, weather, best management practices for about 13 crops by a call from his/her mobile at Rs.1 / call. Such strategies helped the farmers in drought prone areas to enhance their livelihoods.

7.15 Drought Mitigation Measures

Drought mitigation means actions that can be taken before or at the beginning of drought to help reduce the impacts of drought considerably by adapting permanent measures. These measures are essential for restoring the ecological balance for sustainable production. It involves a wide range of agricultural practices including developing for additional water supplies and conserving soil moisture and livestock management. The soil and water conservation practices minimizes the disruption of the soil structure, composition and natural bio-diversity there by reducing erosion and soil degradation, surface runoff and water pollution. Livestock management consists of reduction in animal population, strategic weaning of calves and segregation of herds and parasites' control. Drought mitigation programs are of not short duration interventions but aim at reducing the impact considerably through long term strategies which need to be included in the country's developmental plans. Droughts can be mitigated by two kinds of measures, either by adopting preventive measures or developing preparedness plans (Anil K.Gupta *et.al.* 2011). Drought preparedness refers to pre-disaster activities to tackle the situation or improve the capabilities of institutions involved in drought management. Preventive and preparedness plan for drought mitigation as suggested by Anil K. Gupta *et al.* (2011) are as follows (Table 7.21).

Table 7.21: Preventive measures and preparedness plan for drought mitigation

Preventive measures	Preparedness plan
• Dams/reservoirs and wetlands to store water • Watershed management • Water rationing • Cattle management • Proper selection of crop for drought-affected areas • Leveling, soil-conservation techniques • Reducing deforestation and fire-wood cutting in the affected areas • Alternative land-use models for water sustainability • Checking of migration and providing alternate employment • Education and training to the people • Participatory community programs	• Improvement in agriculture through modifying cropping patterns and introducing drought-resistant varieties of crops • Management of rangeland with improvement of grazing patterns, introduction of feed and protection of shrubs and trees • Development of water resource system with mproved irrigation, development of improved storage facilities, protection of surface water from evaporation and introduction of drop irrigation system • Animal husbandry activities can help in mitigation with use of improved and scientific methods.

(*Source*:Anil K.Gupta *et.al.*, 2011.)

The role of Krishi Vigyan Kendras (KVK) under Indian Council of Agricultural Research in managing drought during 2009 by the KVKs located in drought stricken districts was brought out by Kokate *et.al* (2010). He reported that these institutions undertook various activities and provided necessary technological inputs like seeds, planting materials, bio fertilizers, bio agents through awareness and skill development programs. In addition to crop specific intervention, major emphasis was given on resource conservation technologies and dissemination of agro advisories through mass media communication system.

7.15.1 Drought Mitigation Plans

Though drought mitigation plans have been in existence for long time, they are not effective due to poor coordination among various departments and insufficient allocation of funds. Therefore, there is a need to revise the strategy for implementing drought mitigation program by adopting the integrated watershed development programs on a mission basis. The following are the recommendations for drought mitigation strategies to be taken up at various levels as reported in Manual for Drought Management (2009) is given below.

1. Setting up a mission/task force on drought mitigation by respective State Government, which is an advisory body with experts from various disciplines for suggesting policies and programs for drought mitigation.
2. Conduct Drought Risk and Vulnerability assessment for each state by analyzing the data for identification of vulnerable regions and to suggest measures for developing new watershed areas, irrigation and agricultural practices and risk management measures.
3. The mission task force should identify programs and measures for drought mitigation plans related to management of irrigation, crop planning, soil and water conservation and agro forestry and suggest new innovations for drought risk management such as viz., weather insurance, trade in water use, wet land banks, inter basin and intra basin, water use planning and urban rain water harvesting.
4. The mission task force should develop Decision Support Systems for drought management by improving the density of network of observational systems and collection of relevant data sets on soil, vegetation and water resources that would help in strengthening of monitoring systems for preparation of management and mitigation strategies.
5. The mission task force should promote education and awareness of mitigation polices and measures by conducting awareness program on water conservation and harvesting, optimal water use and the need for increasing the forest cover etc. for safe guarding the environment.
6. The environment mission task force must encourage community level plans for drought mitigation by involving different community institutions and NGOs actively engaged in drought mitigation activities.

The main programs of drought mitigation strategies are 1) reduce soil erosion and run off, 2) improve the soil moisture and water use, efficiency of which involves various soil and water conservation practices (Manual for Drought Management, 2009) which are briefly reported as follows. Soil water conservation can be carried out in two ways; one is artificial recharging of ground water through watershed development programs and the other by following traditional methods in which use of water collected through ponds for various purposes. A brief description of water conservation methods as given in Manual for Drought Management (2009) are as follows:

7.15.2 Soil water Conservation and Recharging of Ground water Methods

A typical watershed development programs as a drought mitigation measure has several components which depends on the topography, soil characteristics, rainfall intensities and land use. The components are

a) Contour bunding

It is one of the age old soil and water conservation technologies that controls soil erosion, improve soil moisture status and recharge ground water, prevents silting of tanks/reservoirs in the downstream. The practice consists of construction of narrow bunds on contours to collect runoff water behind them, which then is gradually absorbed by the soil. The bunds are usually 30 cm in height and spacing between them depends on the slope and nature of soil. For gentle slopes (2-3%), the bunds are nearly 20 feet apart. It was reported that on average the contour bunds holds 27% higher soil moisture and recorded higher crop yields ranging between 14-180% over the flat surface.

b) Contour trenching

This method consists of excavating shallow/intermittent trenches across the land slope and creation of earthen bund on the downstream side. Plantation of crops is done on the bund and the water collected in the trenches helps in better establishment of plantations on the bund. These are useful where land surface is fairly porous. The size and spacing of trenches should be optimized to intercept about 50% of the peak rainfall in semiarid zone. Periodical maintenance of trenches for de-silting is necessary and may be carried out regularly.

c) Contour cultivation

Contour cultivation (Fig 7.17) with crops and trees is most effective on 2-7 percent slopes. The contour furrows created would form numerous mini barriers wherein water gets collected in the troughs across the flow path of the runoff there by increasing the availability of soil moisture for a longer period.

Fig.7.17: Contour cultivation for soil water conservation

d) Bench terracing

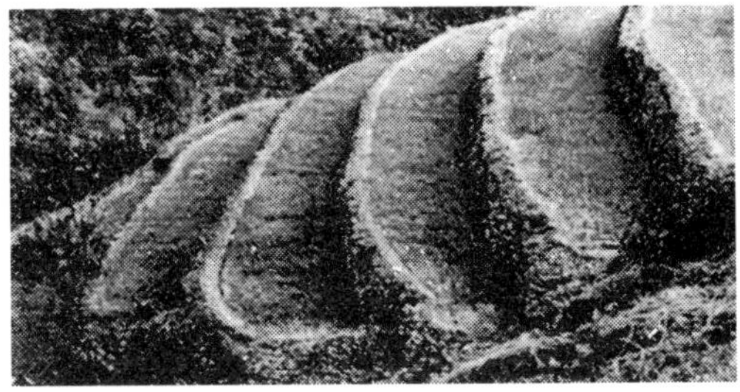

Fig. 7.18: Bench terracing on hill slopes for cultivation

It is practiced on steep hill slopes where the natural forest system (Fig. 7.18) is replaced with agriculture. It involves converting the original ground into well levelled step-like fields that reduces the degree of slope, which helps in reducing the velocity of runoff and increase the Soil moisture.

e) Graded Bunding

These are constructed in high rainfall regions to remove excess water from the field to avoid water stagnation in deep black soils. The channels are wide and shallow.

f) Gully plugging

Fig. 7.19: Bushwood check dam in Degraded Watershed

It is a small conservation structure across gullies and streams in hilly or sloppy areas to reduce the speed of runoff (Fig. 7.19). Plugging measures include vegetative planting and bush wood check dams, boulder bunds, brick masonry and earthen bunds and sand bag plugs.

g) Check dams

Fig. 7.20: Runoff harvesting for irrigation and groundwater recharge

Construction of a masonry embankment across a stream or nalla to hold maximum runoff water to create temporary flooding in upstream with arrangements to drain water at suitable intervals (Fig. 7.20). The stored water allows percolation of water in to deeper soil layers which makes it possible to bring the percolation land under the bed of the nallas for cultivation.

h) Gabion structure

This is common check dam to conserve stream flows with no submergence beyond the stream course. Locally available boulders are stored in a steel wire and put across the stream's width to make a small dam by anchoring it to stream side. The height of such structure is about 0.5m, which is normally used in 10-15 m wide stream.

i) Farm ponds

Farm ponds play significant role in collecting runoff water providing supplemental irrigation during water stress period (Fig. 7.21). In a relatively flat terrain with good soil cover, the pond has an earth section with usually 3:1 side slopes on the waterside and 2:1 side slope on the downstream face.

Fig. 7.21: Water harvesting in farm pond for recycling runoff recharge

j) Percolation tanks

Construction of percolation tanks across second to third order streams helps in ground water recharge. The success of the percolation tanks would be high if they are established in hard rock formation places where the rocks are highly fractured and weathered.

k) Anicuts

An anicut is a small masonry dam constructed across a stream to hold sufficient water and submerge the upstream area during rainy season. Its height is around 22 feet above the ground. These structures are built to recharge ground water and provide irrigation for crops and cater to the drinking water needs of human and animal population.

7.15.3 Traditional Water Harvesting and Conservation Techniques

The use of rain water collection is known to have existed around 4000 years back in Negev desert in Israel which receives less than 15 cm of rainfall annually (Nagarajan, 2003). Under ground water storage volumes up to 300m^3 enough for 10 families and their cattle for year round were found (Pacey and Cullis, 1986) in this region. There were evidences about rain water collection from roof tops and stored in cistern during early Roman times. Elaborate systems of dykes were built and utilized in Thailand some 800 years ago. In India, the

Buddhist monks living in cave temples (Elephanta caves, Udayagir and Chandragiri) had laboriously constructed series of gutters and water cisterns in to the rock faces to avail water supply throughout the year (Nagarajan. 2003). The earlier rulers of India (Chola and Pandya Kings) built rainwater catchment tanks over the region near the hills. Various tanks were constructed across the southern peninsular regions which have been serving both for irrigation and drinking purposes and many are still in use. In Andhra Pradesh alone, there were 58,528 tanks and 43,474 in Karnataka during 1955-56. *Khadins, tankas* and *nadis* in Rajasthan, *Bandharas* and *tals* in Maharashtra, *Bundhis* in Madhya Pradesh and Uttar Pradesh, *Ahars* and *Pynes* in Bihar, *Khuls* in Himachal Pradesh and Jammu & Kashmir, *Puzhals* in Kerala, *Eris* in Tamil Nadu and *Cheruvu* and *Kuntas* in Andhra Pradesh are some of the water harvesting structures that are known to exists since ancient times. The structures used in water harvesting in arid parts of Rajasthan as reviewed by Nagarajan (2003) is as follows:

a) Village pond/tank

Village ponds or tanks are the most common water storage structure in the village where the water is used for different purposes. These ponds have limited catchments that provide requisite quantity of water during the rainy season. Wherever the catchments are small, water from nearby streams is diverted through open channels to fill the ponds. They are known by different names in different regions.

b) Nadis

These are community based water storage structures located near the village which store large volume of water for use of human and livestock (Fig. 7.22). Water from these structures are not used for drinking, used only for bathing, washing etc. The water is usually available for a period of 4 to 6 months. When the water in nadis is exhausted, the villagers use water stored in Tankas. There are about 1436,592 and 1822 nadis in Nagaur, Barmer and Jaisalmer district in Rajasthan and meet about 37% of water needs. (Nagarajan, 2003).

(*Courtesy*: CAZRI, Jodhpur)

Fig. 7.22: Nadi-Runoff harvesting pond in Rajasthan

c) Toba (Tanka or Talai)

It is a dugup pond on a hard plot amidst sand dunes. Water is available for four months only. To minimize evaporation losses, most of these structures are located within the shade tree areas and act as a centre for milk and ghee production.

d) Tankas/Kunds

In arid region of Rajasthan state local population builds unique underground structures of various shapes to collect rainwater for drinking purposes. These structures are called Tankas/Kunds/Kundis in different places. (Fig. 7.23) Since these are closed from all the sides and no light is available inside the structure, hence no bacterial growth takes place and the water remains fresh throughout the year. These are built for both individual households as well for village communities. As these being the only sources of drinking water in remote villages in the extreme arid parts people to those areas protect them and maintain them with utmost care. Just before the onset of rainy season, the catchment area of the Tanka is cleaned to remove all pollutants, human and grazing activity is prohibited. These provide enough water for drinking purposes to the sparse population in areas that receive rainfall around 200 mm/year.

Fig. 7.23: Tankas/Kunds in arid regions of West Rajasthan
(*Courtesy* : CAZRI, Jodhpur)

e) Khadin

It is basically a water harvesting structure on farm land and subsequent use of water saturated land for crop production. (Fig. 7.24) It is a very long (100-300m earthen embankment built across lower hill slopes lying below gravel uplands. The spillways allow excess water to drain off. Apart from the submerged

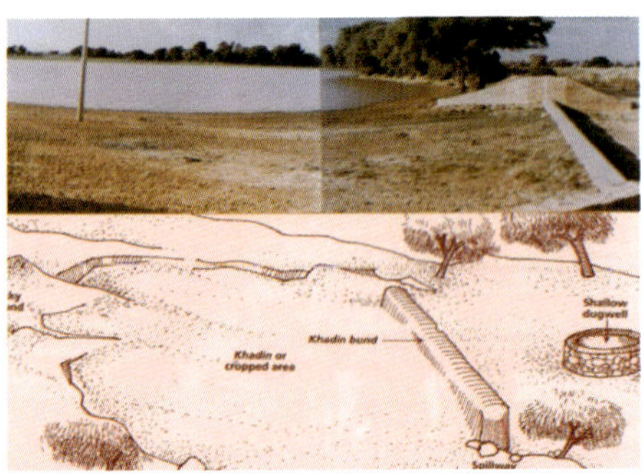

Courtesy: CAZRI, Jodhpur
Fig. 7.24: Khadin in West Rajasthan

area, the Khadin beds are cultivated from top to bottom along the line of reducing moisture. The ponding water induces continuous recharging of ground water.

f) Rain water Harvesting in urban areas

It comprises the collection of rain water from roof tops and stores it in underground tanks for drinking water purpose. This system is very common in the cities like, Jodhpur, Bikaner, Barmer etc., in the Western Rajasthan which helps in preventing the runoff going into drainage there by reducing the load on municipal water supplies. It also helps in reducing urban flooding and recharging water into the aquifer. With severe fall in ground water table in many cities, the local administration made it mandatory for all the new buildings to construct percolation pits to recharge ground water.

The other methods of water saving and conservation especially in raising agricultural crops that had eminated from from the research findings of AICRP on water management are as follows:

- Use of Drip and Sprinkler for irrigating crops.
- Discourage continuous ponding of water in rice and schedule irrigation only after the disappearance of ponding water.
- Reduce the water conveyance losses by proper lining of the canals and field application losses with good land levelling.
- A weed free irrigation channel improves the efficiency of irrigation system and saves the waters.
- Conjunctive use of surface and ground water to increase the agricultural production from saline regions.
- Mulching and weeding improves the soil moisture status under rainfed conditions.

Ultimately, a permanent solution to combat drought in larger part of the country is possible by transferring of water from surplus river basin to the deficit appears to be the only solution that can check and reduce the drought impacts. A National Water Grid for linking resource abundant rivers viz., Brahmaputra, Ganga with other rivers in the deficit region may be initiated on a priority basis to manage drought affectively. Also the Garland Canal proposed long back by experts like eminent engineers like Dr. K.L. Rao, Former Minister of Irrigation and Power of GOI appear to be a long term solution for combating droughts in the coming years though it is an herculean task involving huge investments. A region wise comprehensive action plans may be developed for developing water resources for effectively controlling droughts and stabilize food production.

References

ABARES.2012. Drought in China context, policy and management available at daff.gov.au/abares/publication.

Acharya, N.G Ranga Agricultural University, 2008-09. Andhra Pradesh crop strategies for improving productivity 2008-09. ANGRAU, Hyderabad, 62p.

Allex, V.G. Baker, M.T., Segarra, E and Brown, C.P. 2007. Integrated irrigated crop-livestock systems in Dryland climates. Agronomy.J. 99:346-360.

Anil K. Gupta, Pallavi Tyagi and Vinayk, K.Seghal. 2011. Drought Disaster challenges and mitigation in India: Strategic appraisal. Current Science.100(12), June, 2011:1795-1806.

Annual Report of NAIP Component- III.2010. Central Research Institute for Dryland Agriculture, Hyderabad.

Annual reports, 2005-06, 2006-07 and 2007-08.Central Research Institute for Dryland Agriculture, Hyderabad.

Aparna, R. 2001. The 29 commandments of Jambeshwar. German Research, 1, 8-11.

Apparao, G. 1991. Droughts and southwest monsoon: Training course in monsoon meteorology, 3rd WMO, Asian/African monsoon workshop. Pune, India.

Atef Hamdy. 2004. Drought preparedness and mitigation in the Mediterranean Region (In: Water management for Drought mitigation in the Mediterranean. Eds. A. Hamdy and Trisorio-Liuzzi.G. Option Mediterranean, Ser. B.No.47, CIHE AM-IAMB), 9-47.

Axel Rothange. 2001. Drought management strategies for Namibian Ranchers. AGRICOLA-2001, 99-105.

Behera, U.K., Panigrahi, P. and Sarangi, A. 2012. Multiple water use protocols in Integrated Farming Systems for enhancing productivity water resources management.**26**:2605-2623.

Bharat R. Sharma, Rao, K.V., Vital, K.P.R., Ramakrishna, Y.S. Amara Singh, 2010. Estimating the Pontetial of Rainfed Agriculture in India: Prospects for Water Productivity Improvements. Agricultural Water Management, 97(1): 23-30.

CENSTA(Centre for sustainable development) 2004.The role of Qashqai nomadic communities in reducing variability to recurrent drought and sustainable livelihood development in Iran. Prepared for Rural institutions and Participation Services, FAO. 55p.

Channa Basavanna, A.S. Biradar, D.P. Prabhande, K.N. and Mahabaleswar Hegde. 2009. Development of profitable integrated farming system model for small and medium farmers of Tungabhadra project area of Karnataka. Karnataka. J. Agric. Sci. 22(1):25-27.

De Silva, G and S. Abesinghe. 2011. Global warming and Climate Change: A case study on dairy farmer's awareness In: Proceedings of 16th Int. Forced Envi Symposium conducted by Dept. of Forestry and Environment Sciences, University of Jayawardenepura, Nugegoda, Sri Lanka, 28-29 October 2011.

Department of Land Resources. 2011. Success stories under watershed Development program – a compendium "Work shop on success stories in watershed management 2-3 February 2011, MRD, GOI New Delhi, 168p.

Drought Management Plan Report. 2008. Luxembourg. Official publication of the European Communication. 2007.

Greg Laughlin and Anthony Clark. 2000. Drought Science and Drought policy in Australia: A risk management perspective in Proceedings of Expert Group meeting on Early warning system for drought preparedness and drought management . 5-7 September 2000. Lisbon. Portugal.

Gupta, A.K. and Anjali Singh. 2011. Traditional intellect in Disaster Risk Mitigation: Indian outlook-Rajasthan and Bundelkhand. Ind. Jour. Traditional Knwowledge 10(1):156-166.

Hanumantha Rao, Ch. 1994. Report of the Technical Committee on Drought Prone Areas Programme and Desert Development programme. Ministry of Rural Affairs, GoI, New Delhi.

Hazel, P. Oran, P. and Chaherli, N. 2001.Managing droughts in the low rainfall areas of the Middle East and North Africa, ETPD.Discussion paperN o.78.environment and production Technology Division. International Food Policy Research Institute, Washington, D.C. USA. 78p.

http//www.worldtime.com – 23.August.2012.

India Meteorolgoical Department. 2011. Report on Farmer's awareness program, organized Jointly by IMD, SAU, ICAR, and IIT. Ministry of Earth Sciences, GoI. 12p.

IWMI working paper-85.2004. Drought mitigation in Pakistan. Current status and options for future strategies. 46p

Jayanthi,C., Ranga Swamy, A and China Swamy,C. 2000. Water budgeting for components in low land integrated farming systems. Agricultural Journal.87:411-414.

Kamani, P.R., Munshi, M.A. Makwana, D.K. and Sawali, V.J. 1995. "Badli nu Bhantar Ketlu Saekhu" (Vernacular Gujarati), Krishi Jivan, Spl. Issue on "Varshad Agahi" (May 1995) 26p.

Kokate, K.D. Singh, A.k. Singh, Lakhan Singh. 2010. Drought Mitigation through knowledge Advisory. India. Res. J.Ext.Edu10(1)):1-6.

Karrou M., El Mourid M. 2008. Drought management and planning strategies in semi-arid and arid agro-pastoral systems of West Asia and North Africa: a review. In Lopez-Francos A. (ed). Drought management: Scientific and technological innovations. Zaragoza: CIHEAM-IAMZ., 179-184.

Liverman, D. 1980. Coordination of Response to drought in the Canadian prairies. Proceedings, Association of American Geographers Annual meeting Association of American Geographers, Washington, DC.

Manualfor Drought Management, 2009. Dept. of Agricultural and Cooperation, MOA, GOI, New Delhi.192p.

Manyatsi, A.M. 2011. Applications of Indigenous knowledge systems:inHydrolological Disaster Management in Swaziland. Curr. Res. Jor. Social Sciences 3(4): 353-357.

Mishra, P.K. 2002.Indigenous Technical knowledge on soil and water conservation in Semi-Arid India.NATP. CRIDA, Hyderabad, 151p.

MMP-17 2005. Final Progress Report of the Project. Development of weather based forewarning systems for crop pests and diseases (MM-III-17), CRIDA, Hyderabad.

Mohan Sing Jangira, Bharadwaj S.S. and Sharma, I.M. 2010. Indigenous knowledge and traditions in weather predictions in north-west Himalayas.In proceedings of seminar on Agromteorological services for the farmers (Edited by Vyas Pandey) 10-13 November 2008, Anand Agricultural University, Anand.138-150.

Nagarajan,R, 2003. Droughts Assessment, Monitoring Management and Resources Conservation Capital Publishing Company, New Delhi-312p.

Naveen P.Singh, MCS Bantilan, Byjest,K. and Murthy, M.V.R. 2012. Helping community adapt: Matching climate change perception and policy. Vulnerability to climate change: Adaptation strategies and layers of Resilience, Policy Brief No.18, May 2012. ICRISAT, Patancheru, Hyderabad.

Osman, M., Vijay Anna Borade., Ravi Sankar, K., Soune, S.V., Pandit Warse, Ravi Dupdal, Manoranjan Kumar, Kaushalya Ramachandran, Adake, R.V., Haffis Shaik and Venkateswarlu,.B. 2012. Revisiting Kadwanchi Watershed, Jalna Maharashtra a Success story. Tech. Bulletin 02/2012., CRIDA, Hyderabad, 33p.

Pacey, A. and Cullis, A. 1986.Rain water harvesting, Intermediate Technology publication, London.

Panigrahi and Panda S.N .2003. Optimal sizing of on-farm Reservoir for supplemental irrigation Jr. of Irrig and Drai. Eng. (ASCE), 129(2): 117-128.

Panigrahi, B. Panda, S.N., and Aggarwal .A. 2005. Water balance simulation and economic analysis for optimum size of on farm Reservoir. Water Resource Management 19(3):233-250.

Patil, S.K., D.S.Thakur, D.Khalkho and R.K. Naiak. 2010. Rain water harvesting through farm ponds and shallow dig wells in Bastar region of Chhattisgarh in Proc. Rain water harvesting and research through farm ponds (Eds) KV.Rao, B. Venkateswarlu, K.L. Sharawat, S.P.Wani, P.K. mishra, S.Dixit, K.S. Reddy, Manoranjan Kumar and U.S.Saikia. 21-22 April, 2009. CRIDA, Hyderabadpp.127-134.

Patrik P.Kalas, 2009. Proceedings of planting the knowledge seed adaptation to climate change using ICTs concepts, current knowledge and innovative examples. (Eds.) Patrik P.Kalas, Swis Agency for development and cooperation Alan Finlany open research, Sep. 2009.

Paul Venton.2012. Drought Risk Management : Practioners perspectives from Africa and Asia. UNDP Publication. 45p.

Pisharoty, P.R. 1993. Plant that predicts Monsoon. Honey Bee 4(4):12.

Prasad, Y.G. Venkateswarlu, B. Ravindra Chary, G., Srinivasa Rao, Ch. Rao. K.V. Ramana, D.B.V., Rao., VUM and Reddy Subba.G. and Singh, A.K. 2012.Contingency crop planning for 10 districts in peninsular India, CRIDA, Hyderabad, 302.p.

Qasem Abdon. 2007. Agricultural drought planning and management In. International Conference on Sustainable Development and Management of water in Palestine. 27-30 August 2007. Amman. Jordan.

Radhamani, S., Balasubramanian, Rama Murthy, R.K. and Geethalakshmi, v. 2003. Sustainable integrated farming systems for drylands. A review, Agricultural Reviews.24:204-210.

Ramakrishna Y.S. 2010. Current status of Agrometeorological services in south Asia with special emphasis on Indo-Gangetic Plains. Consultancy Report submitted to IWMI NASC, Complex, NewDelhi.

Ramakrishna, Y.S. G.G.S.N. Rao, and AVMS. Rao., and V.U.M. Rao. 2006. Value added Agromet advisories for Managing Indian and Afghanistan Deserts. Presented at Indian-Afghanistan Symposium on Deserts and Desertification, NASC Complex, New Delhi, 19-21 November, 2006.

Rao, G.G.S.N., Pasupalak, S., Sastry, ASRAS, Upendra Shanker Singh, Padmakar Tripathi, Wadood, A., Abhijit Saha and Nath, KK. 2002. Geo referenced Resource inventory preparation for rainfed rice ecosystem. Final report of NATP funded project RRPS-1, CRIDA, Hyderabad 82p.

Rao. G.G.S.N., Rao., V.U.M., Rao., A.V.M.S, Venkateswarlu, B., Mishra, P.K. and Gogoi, A.K. 2009. Farmers awareness program on climate change in India. Tech. Bulletin No.1/2009. ACIRP on Agrometeorology, CRIDA,Hyderabad. 28p.

Rao., V.U.M., Rao, Bapuji., Khandgonda, I.R.K., Rao, A.V.M.S., Vijaya Kumar., P. Dagar, J.C. and Venkateswarlu, B. 2011. Perception of Indian Farmers on climate change- an assessment and awareness program. Tech. Bulletin., AICRPAM Agrometeorology, CRIDA, Hyderabad, 38p.

Rashid Hassan. 2010. Drought management strategies in South Africa and the potential for economic policy instruments. Policy note, No.02-0810, August 2010. Water Science and Policy Centre, University of California, Riverside.

Ravisankar, K. Pochaiah Mavaty, Murthy, V.R.K. and Ramakrishna, Y.S. 2008.IndigenousRain Forecasting in Andhra Pradesh. Tech. Bulletin, CRIDA, Hyderabad. 67p.

Rossi.G., Bonaccorso B. Nicolosi V., Castiglione L., Cancelliere A. 2010. European perspective on drought preparedness and management : the Italian experience and proposed planning tools. Xerochore Final Conference on Supporting Drought Policies in Europe. Brussels, 23-24 February, 2010. Department of Civil and Environmental Engineering. University of Catania.

Sammy Mutua. 2011. Strengthening Drought early warning at the community and District levels. Analysis of Traditional community warning system in Wajir and Turkana Countries.A report to Oxfam G.B. Kenya. 40p.

Samra, J.S., Sahoo, N. Roy Chowdhary, S, Mohany, r.K. Jeena, S.K. and Verma,H.N. 2003. Sustainable integrated farming system for water logged areas of eastern India. Research Bulletin No.14. WTCER, Bhubaneswar.

Shammim A. Mamum, Fouzia Nusarat and Namota Rani Debi. 2011. Integrated farming system: prospects in Bangladesh. J.Envi. Sci.Nat.Res 4(2):127-136.

Singh H.P. 2001. Developing farming systems and best practices for drought – prone areas in Asia and the Pacific (in) Proceedings of the Asia and Pacific conference on early warming, prevention, preparedness and management of disaster in food and agriculture. 12-15 June 2001. Thailand Annexure XIV.APDC 101/100 p.

Singh, J., Bohra, J.S., Singh, Y. and Singh, J.P. 2006. Development of Farming Systems models for the north-eastern plain zone of Uttar Pradesh. Indian Farming. 56(2):5-11.

Singh, K.P., Singh, S.N., Kumar, H. Kadian, V.S. and Saxena, K.K. 1993. Economic analysis of different farming systems followed on small and marginal land holding in Haryana. Haryana J. Agro. 9:122-125.

Singh, S.N., Saxena, K.K., Singh, K.P., Kumar, H. Kadian, V.S. and 1997.Consistency in income and employment generation in various farming systems. Annals of Argil. Res. 18(3):340-343.

Sinha S.K., K. Kailasanathan and A.K. Vasistha, 1987. Drought management in India: Steps toward eliminating Famines – Planning for drought, (Eds). Donald.A Wilhite, Willian E. Easerling and Deoborah A. Wood., Westview Press, Boulder and London.453-470.

Subba Reddy, G. Ramakrishna, Y.S., Ravindra chary G. and Maruthi Sankar.G.R. 2008. Crop and contingency planning for rain fed regions of India. A compendium by NARS, SAU and Agro-Industries. AICRP on Dryland Agriculture, CRIDA,Hyderabad 174p.

Tamboli , P.M. and Nene, Y.L. 2002. Science in India with special reference to Agricutlrue. Asian Agri-History Foundation. Hyderabad.

Thornthwaite, C.W. 1948. An approach towards a rational classification of climate. Geogr.Rev. 38:85-94.

Thornthwaite, C.W. and Mather, J.R. 1955. The Water Balance: In Climatology, VIII(1), Drexel Institute of Technology, New Jersey, USA 104.

Truby., J.O. & Boulas. 1994. Formulation of mitigation system for long term drought in Colorado. Drought management in a changing water: New direction for water policy IDIC Tec. Report 94-1, University of Nebraska, Lincoln, 239p.

T3wclopclc C. Morcni. 1987. Drought planning and Response: Botswana Experience In planning for droughts (Eds.) Donald A.Wilhite., William E. Easterling and Deborah A. Wood. Westview prem. 445-452.

Varshneya, M.C. 2010. Ancient Knowledge of agro-meteorolgical services, In proceedings of seminar on Agromteorological services for the farmers (Edited by Vyas Pandey) 10-13 November 2008, Anand Agricultural University, Anand.108-116.

Varshneya, M.C. Nanji Kale, Vaidya, V.B; and Vyas Pandey. 2010. Forecasting and Validation of rainfall from Barshi (MS) based on astro-meteorological principle of rainfall conception. In proceedings of seminar on Agromteorological services for the farmers (Edited by Vyas Pandey) 10-13 November 2008, Anand Agricultural University, Anand. 163.-170.

Venkateswarlu, B., Mishra, P.K., Ravindra Chary, G., Maruthi Sankar G.R.and Subba Reddy.G. 2009. Rainfed Farming A compendium of Improved Technologies. AICRP on Dryland Agriculture, CRIDA, Hyderabad. 132p.

Venkateswarlu, B., Singh, A.K. Prasad, Y.G., Ravindra chary, Rao., K.V. Ramana D.B.V, and Rao , VUM. 2011. District level contingency plans for weather aberrations in India. CRIDA, NRM Division, ICAR,Krishi Bhavan, New Delhi, 134p.

Venkateswarlu, J. 1987. Technological socio-political adaptation and adjustment to drought : The Indian Experience. In planning for Drought(Eds).Donald A.Wilhite, William E.Easterling and Deborah A.Wood.West view press.Boulderand London:391-408.

Vittal Kaakaniya and Shaita Kishore.2010.Factors affecting the adaptation of farm ponds in drought prone areas of Gujarat.Sharing experiences of AKRSP.In Proc. Rain water harvesting and research through farm ponds (Eds) KV.Rao, B. Venkateswarlu, K.L. Sharawat, S.P.Wani, P.K. mishra, S.Dixit, K.S. Reddy, Manoranjan Kumar and U.S.Saikia. 21-22 April, 2009. CRIDA, Hyderabad. 111-118.

Wani, SP, Joshi PK, Raju KV, Sreedevi TK, Michael J Wilson, AmitaShah, Diwakar PG, Palanisami K, Marimuthu S, Ramakrishna YS, Meenakshi Sundaram SS and Marcella D'Souza. 2008. Community Watershed as Growth Engine for Development of Dryland Areas – Executive Summary .A Comprehensive Assessment of Watershed Programs in India. Patancheru- 502 324, Andhra Pradesh, India: International Crops Research Institute for the Semi-Arid Tropics. 36 pp.

Wilhite, A. Donald. 1987. The role of Government in planning for drought: Where do we go from here? In. Planning for drought, eds. Donald.A Wilhite, Willian E. Easerling and Deoborah A. Wood., West view press, Boulder and London. 425-444.

Wilhite, A. Donald. 1991. Drought planning and state Government: Current Status Bull. of Amer. Met. Soc. 72: 1531-1536.

Wilhite, A. Donald. 1997. Improving drought management in the west: Report to the western water policy Review Advisory Commission. NDMC, University of Nebraska-Linclon, 45p.

Wilhite, A. Donald. 2003. Drought policy and preparedness: The Australian experience in an International Context. In.Beyond Drought: people, policy and perspective (Eds) Linda Courtenary, Batterill and Melanie Fisher, CSIRO Publishing, Melbourne.

Wilhite, A. Donald., Mark D. Svoboda. 2000. Drought early warning systems in the context of Drought preparedness and mitigation in Early warning system for drought preparedness and drought management (Eds) Donald A.Wilhite., M.V.K. Siva Kumar and Debrooh A. Wood, AGM-2 WMO/TD No.1037. 1-21.

Wilhite, A. Donald., N.J. Rosenberg; and M.H. Glzntz. 1984. Government Response to Drought in the US: Lessons from Mid-1970s, Parts 1-4, Final Report to the Climate Dynamics Program, National Science Foundation Progress Report 84-1 to 84-4. Centre for Agrometeorology and Climatology, University of Nebraska-Linclon.

□□□□

Chapter 8

Conclusions

Review of the state of art of Drought on impact assessment, monitoring and management with special reference to agricultural production has revealed that Drought is still a complex natural disaster and difficult to forewarn in spite of advancements in atmospheric sciences and computational procedures. It is realized only after it is set in and is not known when it is likely to withdraw. A long way one has to travel to predict its occurrence, spread and its continuation to the next season over a region. The existing net work of observational set up in developing countries in Africa, Asia and in South America need to be strengthened further for effective monitoring of significant weather events at regional and micro regional level. With the rapid expansion of communication system over the past two decades, establishment of Automatic Weather Stations (AWS) at as many locations as possible even in remote regions should be taken up on priority basis and funding may be provided by the respective governments or by the International Organizations where ever necessary. Nations with limited resources may also be convinced to strengthen their weather observational net work in view of the increased occurrences of climate extremes and changing climate. Drought vulnerability regions over the globe were delineated based on sparse historical weather data which do not match with real surface conditions. Now, with the availability of weather data from more number of stations both temporally and spatially, drought prone regions over the world may be re-delineated. This would help the planners to allocate funds rationally to the affected regions. Globally every year, some regions or others may be experiencing droughts and more so in drought prone regions. To monitor such events, global cooperation among the countries in data sharing and exchange of information related to adaptation and mitigation strategies and regional cooperation among various Organization/Institutions/ Local NGOs and Line departments is the need of the hour. Already the efforts by international organization like WMO, FAO and few universities in USA, viz., University of Nebraska, University of Princeton in globalizing the drought studies are commendable and worth mentioning in bringing international awareness and movement to manage droughts. Also the advantages of application of ICT for drought monitoring and adaptation strategies were well recognized by different nations and information on droughts is disseminated at regular intervals through websites for different stake holders. The role of ICT especially of mobile technology has undoubtedly proved to be a big boon for the farming community in receiving information about the forecasted weather and the related advisories. Perhaps meeting partially towards the cost of mobile phones and their maintenance by the respective Governments in the form of subsidies would help the farmers in taking timely action to avoid crop damage.

Over the years many researchers have developed numerous drought indices using departures of weather parameters over its long term averages. The concept of integrated drought monitoring adapted by USA is technically and scientifically superior in monitoring droughts efficiently. However, to use this method good amount of weather data along with stream flow, vegetation health, soil moisture status and reservoir levels etc are required. There are many simple methods available which are being widely used by many nations. Out of the indices, the most accepted index is Standardized Precipitation Index (SPI) which requires only long term daily rainfall. Similarly application of Remote Sensing derivatives like NDVI, NDWI etc along with other weather factors need to be strengthened as NDVI alone may not provide the necessary information for declaration of droughts. Though NRSC in India provides the NDVI maps at fortnightly intervals for each state yet, many State Governments still use the old methods in monitoring the drought and its declaration over a region. Monitoring of droughts is done mostly by central agencies in many of the countries through their Meteorological Departments. Whereas, drought declaration is done by the concerned region local Government based on departures of rainfall and other factors related to agricultural production, water levels in reservoirs etc. The coordination among different agencies right from drought monitoring, declaration and relief measures need to be stream lined for efficient drought risk management practices.

There is no uniform code for drought declaration in India as many states follow different systems in declaration of drought. Quite often this procedure results in inordinate delay in declaring relief measures. Therefore, uniform drought code should be developed which is applicable to all climates. Development of Decision Support System (DSS) to manage droughts as done for the State of Andhra Pradesh may be developed for other states too for better management of droughts. Similarly, DSS for different crops grown in under different climatic zones may further be developed to all crops for all regions. Quite often, the social aspects of drought are not considered seriously while preparing plans for droughts. It is essential to incorporate the social components viz., the receptive capacity of farmers to any new technology, social status, family commitments, literacy and etc, to tackle the droughts. At many places, the farmers' wisdom still plays a dominant role in managing the natural calamities. Perhaps, blending of traditional knowledge including the astro meteorological information with the current scientific inputs may give the necessary synergy for managing the droughts.

In view of the threat from global warming and the ongoing climate variability, the projected climate change indicate the increased frequency of extreme climatic events that would affect the food security. Many nations realized this fact and geared up to face the situation arising due to changing climates. Perhaps there is a need to strengthen the existing research infrastructure to integrate the future climates with simulation modeling and to study their impacts on agricultural production and on other biotic elements like pests and diseases. One way to manage the climate risks are through development of efficient weather based agro advisories and their dissemination techniques /methods. The efficiency of existing district level Agro Advisory System should be improved by reliable weather forecast and economically viable advisories. One of the major limitations in agro advisories are the reliability of forecast for a smaller region. The accuracy of it varies around 70 per cent during rainy season. Therefore, there is a need to improve the forecast accuracy up to 90 per cent by utilizing the micro level AWS net work data.

For success of any program, participation of the beneficiary is essential. The initiation of Farmers awareness campaign about climate risk management is one of the bold step taken by GOI to educate the rainfed farmer about dos and don'ts in conserving the natural resources including weather resources. This program need to be expanded further to all the farmers in near future and should continue till the farmers are well versed to cope up with the drought management strategies.

Water harvesting and conservation are only the methods that can effectively minimize the challenges posed by droughts of different intensities. Watershed development and farm pond technologies are relevant mitigation strategies that can reduce the impacts significantly. The success stories of watershed areas and farm ponds in enhancing the economic return have amply proved their capabilities in different rainfall zones as drought proofing mechanism in stabilizing the economic status of rainfed farmers. Similarly experiences gained through the farm pond technologies and the success stories in enhancing the rural livelihoods of the drought prone regions can be expanded further to reap the benefits of these technologies to the small and marginal farmers thus sustaining the production from the water scarcity areas. Methods to increases the efficiency of traditional water harvesting system in low rainfall regions may be given priority as the Indian arid zone is having high population density in the world. Over the years various Research Institutes under ICAR especially of AICRP on Dry land and SAUs, a good number technologies were generated for rainfed regions located in different agroclimatic zone to combat droughts. Somehow they have not penetrated deep in to the rainfed farming community as many of the technologies like Soil Water Conservation methods and Watershed Development etc. under Natural Resource Management are community oriented and individual farmers can't take up such work. Therefore the government must initiate programs in developing the resources to enhance productivity from rainfed areas to achieve the dream of second green revolution from rainfed areas. Perhaps crop diversification and the farming system approach, which can provided stable and sustainable income should be popularized and prioritized and need to be included as a part in Drought prone Area Programs (DPAP) of GOI.

In spite availability of different technologies, the impacts of drought cannot be eradicated totally as large chunk of cultivable area remains rainfed even after reaching fullest irrigation potential. Their impact can be reduced considerably by developing the ground water resources through various soil and water conservation methods and generation of the required power to lift ground water. Inter basin transfer of river water seems to be an ultimate permanent solution for India to reduce the drought impacts considerably. However, it must be taken up with greater precautions as it involve huge expenditure. Development of water resources must be given top priority in all the governmental plans and all the pending projects should be completed at the earliest to safe guard the food and fodder requirement of both human and animal population on a sustainable manner.

□□□□